▶ The social shaping of technology

Second edition

▶ The social shaping of technology

Second edition

Edited by
Donald MacKenzie and **Judy Wajcman**

Open University Press
Buckingham • Philadelphia

Open University Press
Celtic Court
22 Ballmoor
Buckingham
MK18 1XW

email: enquiries@openup.co.uk
world wide web: http://www.openup.co.uk

and
325 Chestnut Street
Philadelphia, PA 19106, USA

First published 1985
Reprinted 1987, 1988, 1990, 1992, 1993, 1994, 1996

First published in this second edition 1999

Copyright © Donald MacKenzie and Judy Wajcman 1999

ISBN   0 335 19914 3 (hb)   0 335 19913 5 (pb)

A catalogue record of this book is available from the British Library

Library of Congress Cataloging-in-Publication Data
The social shaping of technology/edited by Donald MacKenzie and Judy
    Wajcman. – 2nd ed.
        p.   cm.
    Includes bibliographical references and index.
    ISBN 0-335-19914-3 (hc). – ISBN 0-335-19913-5 (pbk)
    1. Technology – Social aspects.   I. MacKenzie. Donald A.
II. Wajcman, Judy.
    T14.5.S6383   1999
    306.4'6 – dc21   98-41239
                                                                                CIP

Typeset by Type Study, Scarborough
Printed in Great Britain by Redwood Books, Trowbridge

▶ Contents

▶ Notes on contributors

Janet Abbate is a lecturer in history, University of Maryland, at College Park.

Michael H. Armacost, formerly US ambassador to Japan, is president of the Brookings Institution.

W. Brian Arthur is the Citibank Professor and a Coopers and Lybrand Fellow at the Santa Fe Institute.

Anne-Jorunn Berg is a research scientist at the Institute of Social Research in Industry, Trondheim, Norway.

Marc Bloch, who died in 1943, was the author of *Feudal Society*.

Harry Braverman, who died in 1976, was the author of *Labor and Monopoly Capital: the Degradation of Work in the Twentieth Century*.

Paul Ceruzzi is curator of aerospace electronics and computing at the National Air and Space Museum, Smithsonian Institute, Washington DC.

Cynthia Cockburn is a professor in the Department of Social Science and Humanities at City University, London.

Ruth Schwartz Cowan is a professor of history at the State University of New York at Stony Brook.

Moyra Doorly is a freelance journalist in London.

Anni Dugdale is a research fellow at the Research School of Social Sciences, Australian National University.

Richard Dyer is professor of film studies at the University of Warwick.

James Fallows is the editor of *U.S. News and World Report*.

James Fleck is professor in business studies and director of Edinburgh University Management School.

Donna Haraway teaches science studies, feminist theory, and women's studies at the University of California, Santa Cruz.

Jeanette Hofmann is a research fellow at the Wissenschaftszentrum, Berlin.

Thomas P. Hughes is Mellon Professor Emeritus in the department of history and sociology of science at the University of Pennsylvania.

Mary Kaldor is Jean Monnet reader in contemporary European studies at the University of Sussex.

Ronald Kline is associate professor of the history of technology at Cornell University.

Eda Kranakis is in the department of history at the University of Ottawa.

Bruno Latour is professor of sociology at the Ecole Nationale Supérieure des Mines, Paris.

David F. Noble is professor of American studies, University of Minnesota.

Nelly Oudshoorn is a professor in the Department of Philosophy of Science and Technology, University Twente and a lecturer in the Department of Science and Technology Dynamics, University of Amsterdam, The Netherlands.

Trevor Pinch is in the Department of Science and Technology Studies, Cornell University.

Shirley Strum is professor of anthropology at the University of California, San Diego.

Lucy Suchman is a principal scientist at the Xerox Palo Alto Research Center, California.

Robert J. Thomas formerly at the Sloan School of Management, MIT.

Rachel N. Weber is an assistant professor in the Urban Planning Program at the University of Illinois, Chicago.

Langdon Winner is professor of political science at the Rensselaer Polytechnic Institute, Troy, New York.

▶ Acknowledgements

13 new?
+4

The editors and publisher wish to make the following acknowledgements for the extracts which constitute the 30 numbered chapters of this book.

Chapter 1. Reprinted from 'Do artifacts have politics?' by Langdon Winner in *Daedalus*, 109: 121–36. © 1980 *Daedalus*, by permission of the publisher.

97

Chapter 2. Reprinted from *Modest_Witness@Second_Millennium* by Donna Haraway, © 1997 Routledge, by permission of the publisher.

Chapter 3. Reprinted from 'The electrification of America: the system-builders' by Thomas P. Hughes in *Technology and Culture*, 20: 125–39. © 1979 The University of Chicago Press, by permission of the publisher.

96

Chapter 4. Reprinted from 'From scientific instrument to everyday appliance: the emergence of personal computers, 1970–77' by Paul Ceruzzi in *History and Technology*, 13: 1–31. © 1996 Harwood Academic Publishers, by permission of the publisher.

97

Chapter 5. Reprinted from *Constructing a Bridge: An Exploration of Engineering Culture, Design, and Research in Nineteenth-Century France and America* by Eda Kranakis. © 1997 MIT Press, by permission of the publisher.

Chapter 6. Reprinted from 'Competing technologies and economic prediction' by W. Brian Arthur in *Options*. © 1984 International Institute for Applied Systems Analysis, by permission of the publisher.

96

Chapter 7. Reprinted from 'Users as agents of technological change: the social construction of the automobile in the rural United States' by Ronald Kline and Trevor Pinch in *Technology and Culture*, 37: 763–95. © 1996 Society for the History of Technology, by permission of the publisher.

87 Chapter 8. Reprinted from 'The meanings of the social: from baboons to humans' by Shirley Strum and Bruno Latour in *Information sur les Sciences Sociales*, 26: 783–802. © 1987 Sage Publications Ltd, by permission of the publisher.

Chapter 9. Reprinted from 'Caught in the wheels' by Cynthia Cockburn in *Marxism Today*, 27: 16–20. © 1983 *Marxism Today*, by permission of the publisher.

97 Chapter 10. Reprinted from *White* by Richard Dyer: 89–94. © 1997 Routledge, by permission of the publisher.

Chapter 11. Reprinted from *Land and Work in Medieval Europe* by Marc Bloch: 156–9. © 1967 Routledge & Kegan Paul, by permission of Routledge & Kegan Paul PLC and the University of California Press.

Chapter 12. Reprinted from *Capital*, by Karl Marx: 435–7, © Lawrence & Wishart, by permission of the publisher.

Chapter 13. Reprinted from *Labor and Monopoly Capital: The Degradation of Work in the Twentieth Century* by Harry Braverman: 192–5. © 1974 Monthly Review Press, by permission of the publisher.

Chapter 14. Reprinted from 'Social choice in machine design: the case of automatically controlled machine tools' by David Noble in A. Zimbalist (ed.) *Case Studies on the Labor Process*: 18–50. © 1979 Monthly Review Press, by permission of the publisher.

Chapter 15. Reprinted from 'The material of male power' by Cynthia Cockburn in *Feminist Review*, 9: 41–58. © 1981 Cynthia Cockburn, by permission of the author.

94 Chapter 16. Reprinted from *What Machines Can't Do: Politics and Technology in the Industrial Enterprise* by Robert J. Thomas. © 1994 University of California Press, by permission of the author.

new? 17 Chapter 18. Reprinted from 'Learning by trying: the implementation of
94 configurational technology' by James Fleck in *Research Policy*, 23: 637–52. © 1994 Elsevier Science, by permission of the publisher.

94 Chapter 19. Reprinted from 'Working relations of technology production and use' by Lucy Suchman in *Computer Supported Cooperative Work*, 2: 21–39. © 1994 Kluwer Academic Publishers, by permission of the publisher.

Chapter 20. Reprinted from 'The "Industrial Revolution" in the home: household technology and social change in the twentieth century' by Ruth Schwartz Cowan in *Technology and Culture*, 17: 1–23. © 1976 The University of Chicago Press, by permission of the publisher.

94 Chapter 21. Reprinted from 'A gendered socio-technical construction: the smart house' by Anne-Jorunn Berg in C. Cockburn and R. Furst-Dilic (eds) *Bringing Technology Home: Gender and Technology in a Changing Europe*: 165–80. © 1994 Open University Press, by permission of the publisher.

Chapter 22. Reprinted from 'A woman's place' by Moyra Doorly, in the *Guardian*. 1983, by permission of the *Guardian*.

new? 23
96

Chapter 24. Reprinted from 'The decline of the one-size-fits-all paradigm' by Nelly Oudshoorn in N. Lykke and R. Braidotti (eds) *Between Monsters, Goddesses and Cyborgs*: 153–72. © 1996 Zed Books, by permission of the publisher.

new? 25

Chapter 26. Reprinted from 'Manufacturing gender in commercial and military cockpit design' by Rachel N. Weber in *Science, Technology and Human Values*, 22: 235–53. © 1997 Sage Publications Inc., by permission of the publisher.

97

Chapter 27. Reprinted from *The National Defense* by James Fallows: 76–93. © 1981 James Fallows. Reprinted by permission of Random House Inc. and the Julian Bach Literary Agency.

Chapter 28. Reprinted from *The Politics of Weapons Innovation: The Thor–Jupiter Controversy* by Michael H. Armacost. © 1969 Columbia University Press, by permission of the publisher.

Chapter 29. Reprinted from 'The weapons succession process' by Mary Kaldor in *World Politics*, 38: 577–95. © 1985–86 The Johns Hopkins University Press, by permission of the publisher.

85/6

new 30

▶ Editors' note

Many friends and colleagues have helped us with ideas, suggestions and references while we have been editing this book. Given a limited amount of space, we have not always been able to take their advice, but we have always been grateful for it. Particular thanks for help are due to Barry Barnes, Merriley Borell, Cynthia Cockburn, Ruth Schwartz Cowan, Scott Davis, Jenny Earle, David Edge, Anne Elder, Moyra Forrest, Martin Fransman, Jim Gillespie, David Greasley, Ken Green, Erie Hanley, Tom Hughes, Lynn Jamieson, Roger Jeffery, Andrew Lyon, Ian McLoughlin, Kath Melia, David Miller, Trevor Pinch, Alan Roberts, Wolfgang Rüdig, Carole Tansley, Howard Wagstaff and Peter Weingart. We are very grateful to Janet Abbate, Rebecca Kennedy and Robin Williams for their help and advice in preparing the second edition. Our special thanks to Claire Atkinson who provided excellent research assistance in preparing this second edition.

The system of referencing we have used is as follows. Each of the reprinted readings – whose sources are given in the acknowledgements on pp. x–xii is self-contained, and the footnotes and references for each will be found immediately after it. Similarly the footnotes to the various editorial introductions we have written come directly after each. But we have gathered together at the end of the book all the references in the editorial material. So a reference in an introduction such as '(Hughes 1983)' can be traced by turning to the bibliography on pp. 443–51.

▶ Preface to the second edition

We live our lives in a world of things that people have made. As human beings, we have both relations to each other (the set of relations we call 'society') and also relations to the things we have made and to our knowledge of those things – in other words relations to 'technology'. Almost everything human beings do – paid work, domestic work, warfare, childcare, healthcare, education, transport, entertainment, even sex and reproduction – involves relations both to other people and to things.

The topic of this book is this intertwining of 'society' and 'technology'. When we produced the first edition, published in 1985, we had a simple polemical purpose. Existing discussion of the intertwining of society and technology was dominated, we felt, by a naive 'technological determinism'. According to this viewpoint, technology was a separate sphere, developing independently of society, following its own autonomous logic, and then having 'effects' on society.

We objected to a simplistic technological determinism both politically and intellectually. Politically, it seemed to us to encourage a passive attitude to an enormously important part of our lives. It discouraged creative engagement with technology, narrowing the apparent range of political possibility to a limited and unattractive set of options: uncritical embracing of technological change, defensive adaptation to it, simple rejection of it. Intellectually, technological determinism seemed to us to reduce the intimate intertwining of society and technology to a simple cause-and-effect sequence. In particular, in its focus on technology's effect on society, it neglected the ways in which the relations between people affect the things that they make. It neglected, in other words, the social shaping of technology. Attention to this, we felt, would make the analysis of technological change more rounded and more accurate. It would, we also hoped, promote a politics of technology that would seek consciously to shape technological change with human betterment and environmental protection in mind.

These remain our convictions, but the context in which we write has changed. By 1985, state-of-the-art research in the history of technology (especially, perhaps, the work of Tom Hughes) was already predicated on the assumption that technology and society were bound together inextricably and that the traffic between the two was emphatically two-way. Since 1985, this perspective has spread to the social sciences. The social shaping of technology, which in the mid-1980s still had something of the excitement of heresy, has now become almost an orthodoxy. Even in the late 1980s, research sponsors, such as the UK Economic and Social Research Council, could scarcely see beyond the question of technology's effects on society. Research on the social shaping of technology was regarded as being of doubtful feasibility and dubious interest. By the late 1990s, however, it had become a legitimate, indeed a fashionable, topic. In the more sophisticated seminar rooms, reference to 'technology' or 'society' now prompts the immediate response that to talk of them as distinct entities is misleading.

In this shift of thinking, social scientists have in one sense merely been catching up with engineers. Successful practising engineers have always known that their work is as much economic, organizational, even political, as it is 'technical'. They know that a design that works technically will still fail if it is too expensive, if it is unattractive to employers or customers, if its 'fit' to the structure of an organization is poor, or if it falls foul of powerful political forces. The most enterprising among engineers even know that these are not simply constraints to be met. Costs can be manipulated, both in projections and in reality; employers can be swayed; markets can be created and shaped; organizations can be restructured; regulatory agencies can be 'captured'; and so on. Long before a social scientist (Law 1987) coined the phrase, successful engineers knew that real-world engineering was 'heterogeneous engineering' – the engineering of social relations as well as of physical things – and that technology was, in the words of Misa (1988) both socially shaped and society-shaping.

Given that the idea of the social shaping of technology is now well established, why produce a new volume devoted to the topic? One answer is that the success of the idea in academic circles (for which we claim no credit: the idea was certainly current before the first edition of this book) has had little resonance in our wider culture. Discussion of technology in the mass media, for example, is still framed in largely technologically determinist terms. The engaged politics we sought to encourage, seeking consciously to shape technology, is little more evident now than in 1985, at least in the Anglo-American world (there has been more progress in continental Europe). The Labour government that came to power in the UK in 1997, for example, seems to see its task as helping the UK ride a predetermined wave of technological advance, rather than as helping fashion that wave with human and environmental welfare directly in mind.

A second reason for continued attention to the social shaping of technology is that the teaching of engineering does not, in our experience, generally reflect the technologist's awareness of the intertwining of the 'social' and the 'technical' in engineering practice. In consequence, engineering graduates can enter their profession with an over-narrow view of the work that awaits them. The economic, social and ethical responsibilities of the engineer – topics covered in most curricula – are not matters external to the technical content of engineering: in design practice, these are issues that interact intimately with apparently 'technical' considerations. We would not suggest that this book on its own is enough to convey the point adequately: it would have to be supplemented with work that delves more deeply into the technical than is possible in a volume designed for a wide readership (for samples of work that takes the analysis into more technical terrain, see MacKenzie 1990 and 1996a).

The third reason for returning to the social shaping of technology is that the success of the idea is actually precarious. We fear a rerun in the social studies of technology of what has happened in the social studies of science. There, in the 1970s and early 1980s, a variety of empirical studies, some contemporary, but mainly historical, offered evidence that the content of scientific knowledge was influenced by the social circumstances of its production (for a useful review of this work, see Shapin 1982). Those who produced this work knew well that the evidence was partial, tentative and patchy, and that the conceptual issues involved were poorly understood, but a wider audience of scholars in the humanities and social sciences grasped eagerly at the conclusion that scientific knowledge was 'a social construct'. The notion became something of a premature orthodoxy, and too little was done to clarify what the ambiguous phrase meant, or to produce empirical work that addressed the subtle and difficult issues that underlie it (to our minds, the best single source is Barnes *et al.* 1996). In consequence, when some natural scientists reacted with hostility to the notion of social construction (in the 'science wars' debate sparked by works such as Gross and Levitt 1994), the field was not as well placed as it might have been.

There is little risk of 'technology wars' paralleling the 'science wars'. That technology is socially shaped, and that successful engineering is heterogeneous engineering are not conclusions that practising engineers are likely to resist too strongly: in our experience, these ideas resonate with an important aspect of their self-image. (It is, of course, also important to engineers to maintain the idea of a sphere of 'the technical' separate from politics, but that boundary is generally drawn around more detailed, especially more quantitative, matters than those considered here: see MacKenzie 1990 for examples.) Rather, what concerns us is that acceptance of the overall notion of the social shaping of technology may shut off empirical enquiry into the specific ways in which this shaping takes place. At too high a level of generality, the notion of social shaping is vacuous, except as a polemical counterpoint to simplistic technological determinism. If the idea of the social shaping of technology has intellectual or political merit, this lies in the details: in the particular ways technology is socially shaped; in the light these throw on the nature both of 'society' and of 'technology'; in the particular outcomes that result; and in the opportunities for action to improve those outcomes.

▶ **THE STRUCTURE OF THE BOOK**

In preparing this second edition of *The Social Shaping of Technology*, we have retained readings from the first edition that seem to us to be of lasting interest, while deleting those that have been superseded by more recent scholarship or that were simply too rooted in the context of the early 1980s. In their place, we have chosen a selection of the best writing of more recent years. Although our introductory chapters have also had to be revised extensively, the overall structure of the book remains the same. After this preface comes the longest of our introductory essays. It begins with the ideas of technological determinism and the social shaping of technology, connecting them to the politics of technology. It then moves to a discussion of the influence on technological change of science, of previous technology, and of economics, arguing that the economic shaping of technology – a pervasively important phenomenon – is a form of social shaping. Some other major pathways via which technology is shaped are then examined, as are theoretical frameworks for the analysis of the relations of technology and society. The introduction to Part One ends with discussion of a particular social relation on which much of this book focuses – gender – and one where there is not yet a large body of material discussing its relation to technological change: ethnicity.

The introductory essay to Part One is followed by a set of readings illustrating its main themes. For the remainder of the book, we organize our discussion into a triad of 'production' (technology in the paid workplace), 'reproduction' (the technology of biological reproduction, and also domestic technology) and 'destruction' (military technology). In each of these areas, a brief introduction is again followed by readings illustrating the shaping of technology in these spheres. Of course, these domains do not exhaust the spheres in which technology is applied: we chose them because they provide a broad spectrum of social relations of different types interwoven with quite different sorts of technology.

One feature of the first edition – the bibliographic essay on other areas of study – has had to be discarded: the literature on technology and society has simply become too large for writing such an essay, with any brevity, to be a feasible proposition. So, therefore, we must simply warn the reader that this book discusses selected technologies from the recent history of the Euro-American world. The broad sweep of technologies outside our chosen domains of production, reproduction and destruction; the more distant past; and the world outside the privileged Euro-American heartlands (see, for example, Appleton 1995 and Schrum *et al.* 1995) – all these are, we regret, neglected here.

Bibliographic references for the readings come at the end of each extract. However, references in this preface and the introductory essays are to be found gathered together in the bibliography at the end of this book.

PART ONE

▶ Introductory essay and general issues

▶ Introductory essay: the social shaping of technology

Technology is a vitally important aspect of the human condition. Technologies feed, clothe, and provide shelter for us; they transport, entertain, and heal us; they provide the bases of wealth and of leisure; they also pollute and kill. For good or ill, they are woven inextricably into the fabric of our lives, from birth to death, at home, in school, in paid work. Rich or poor, employed or non-employed, woman or man, 'black' or 'white', north or south – all of our lives are intertwined with technologies, from simple tools to large technical systems.

When this intertwining is discussed in newspapers or other mass media, the dominant account of it can be summed up as 'technological determinism'. Technologies change, either because of scientific advance or following a logic of their own; and they then have effects on society. The development of computer technology, for example, is often seen as following trajectories that are close to natural laws, the most famous being Moore's law (Moore 1965), describing how the number of components on a state-of-the-art microchip doubles in a fixed, predictable period of time (originally a year; now 18 months). This key technical underpinning of modernity fuels an information and communication technology revolution that, numerous pundits tell, is changing and will change the way we live.

▶ TECHNOLOGICAL DETERMINISM AS A THEORY OF SOCIETY

Technological determinism contains a partial truth. Technology matters. It matters not just to the material condition of our lives and to our biological and physical environment – that much is obvious – but to the way we live together socially. The historian Lynn White, for example, famously attributed the coming about of feudal society – a 'society dominated by an aristocracy of warriors endowed with land' (White 1978: 38) – to the invention, and diffusion to Western Europe, of the stirrup. Prior to the stirrup, fighting on horseback was limited by the risk of falling off. Swipe too vigorously with a sword, or lunge with a spear, and horseborne warriors could find themselves lying ignominiously in the dust. Because the stirrup offered riders a much more secure position on the horse, it 'effectively welded horse and rider into a single fighting unit capable of a violence without precedent' (White 1978: 2). But the 'mounted shock combat' it made possible was an expensive as well as an effective way of doing battle. It required extensive training, armour and war horses. It could be sustained only by a reorganization of society designed specifically to support an élite of mounted warriors able and equipped to fight in this 'new and highly specialized way' (White 1978: 38).

White's account is better read as parable than as real history.[1] Among the Franks, the stirrup may have 'caused' feudalism. But it had no such effect in, say, Anglo-Saxon England prior to the Norman conquest. To explain why the creation of a feudal system was attempted, and to explain why it was possible, inevitably requires reference to a set of social conditions wider than military technology alone: the decline in European trade, which made land the only reliable source of wealth; the possibility (under some circumstances and not others) of seizing land for redistribution to feudal knights; and so on. As a simple cause-and-effect theory of historical change, technological determinism is at best an oversimplification. Changing technology will always be only one factor among many others: political, economic, cultural, and so on. If technology's physical and biological effects are complex and contested matters (and, for example, the literature on perceptions of risk strongly suggests this),[2] it would clearly be foolish to expect its social effects to be any simpler. A 'hard', simple cause-and-effect technological determinism is not a good candidate as a theory of social change.

However, the failure of a 'hard' technological determinism does not rule out a 'soft' determinism (Smith and Marx 1994), and to say that technology's social effects are complex and contingent is not to say that it has *no* social effects. That is our reason for beginning both this collection and its predecessor with the article by Langdon Winner. His is one of the most thoughtful attempts to undermine the notion that technologies are in themselves neutral – that all that matters is the way societies choose to use them. Technologies, he argues, can be inherently political. This is so, he says, in two senses. First, technologies can be designed, consciously or unconsciously, to open certain social options and close others. Thus, Winner claims (though see also Joerges, forthcoming), New York builder Robert Moses designed road systems to facilitate the travel of certain types of people and to hinder that of others. Second, Winner argues that not only can particular design features of technologies be political, but some technologies in their entirety are political. Even if it is mistaken to see technologies as *requiring* particular patterns of social relations to go along with them, some technologies are, in given social circumstances, *more compatible* with some social relations than with others. Hence, argues Winner, basing energy supply around nuclear technology that requires plutonium may enhance pressure for stronger state surveillance to prevent its theft, and thus erode traditional civil liberties. This particular claim may be wrong – natural uranium shows no sign of running out, as it appeared it might when Winner wrote this article, and the relatively modest recycling of spent fuel has to date led to no restrictions on civil liberties – but the general form of the argument demands attention. In adopting a technology, we may be opting for far more – economically, politically, even culturally, as well as technically – than appears at first sight. Because 'hard' technological determinism is an oversimplified theory of technological change, discovering in advance what that 'more' might be is very difficult, and predictions are, in consequence, often off-beam. But the difficulty of the task is not reason for avoiding it.

► TECHNOLOGICAL DETERMINISM AS A THEORY OF TECHNOLOGY

As a theory of society, then, technological determinism is asking a good question, albeit often providing an oversimplified answer. Where we part company with it more decisively is in its aspect as a theory of technology,[3] in its typical assumption that technological change is an independent factor, impacting on society from outside of society, so to speak.

This is a very common way of thinking, but to our minds a mistaken one. Most of the rest of this introductory essay – indeed most of the rest of this book – provides arguments and evidence for its mistakenness, but let us dwell for a moment on why the mistakenness matters. The view that technology just changes, either following science or of its own accord, promotes a passive attitude to technological change. It focuses our minds on how to *adapt* to technological change, not on how to *shape* it. It removes a vital aspect of how we live from the sphere of public discussion, choice, and politics. Precisely because technological determinism is partly right as a theory of society (technology matters not just physically and biologically, but also to our human relations to each other) its deficiency as a theory of technology impoverishes the political life of our societies.

In one of the most influential recent works of social theory, for example, Ulrich Beck (1992) both diagnoses and calls for 'reflexive modernization'. This apparently opaque phrase encodes several linked notions, but the one that is crucial here is the idea that instead of modernization ('progress') being a process that just happens to societies, it should become a process that is actively, and democratically, shaped. Beck's work resonates with the remarkably successful attempt of the German Green Party to bring into the heart of the political process the activities and goals of citizen's initiatives, of investigative journalists, of radical engineers, and of the environmentalist, women's and peace movements. As a vitally important part of 'progress', technological change is a key aspect of what our societies need actively to shape, rather than passively to respond to.

Often efforts to develop a politics of technology are seen as anti-technology, as an attempt to impose upon technology rigid, negative, political controls. The prevalence of that misconception is our reason for including here an extract from the work of Donna Haraway, who has become perhaps the most influential feminist commentator on science and technology. Her dense, playful, poetic, and occasionally oblique prose is sometimes misunderstood as an attack on science and technology, but we see it in a different light. She is sharply critical of those who reject technology in favour of a return to a mythical natural state, and she argues instead for an embracing of the positive potential of science and technology. Of course, there is much in those spheres she would wish to see change, but she eschews an 'ecofeminist' celebration of women's spiritual closeness to an unpolluted nature. Famously, and provocatively, preferring to be a 'cyborg' – a cybernetic organism, such as an animal with a human-made implant – than an ecofeminist 'goddess' (see Haraway 1985), Haraway is, in our reading of her, rephrasing an old theme: the liberatory potential of science and technology. In the passage from her work we have selected, she notes the great power of science and technology to create new meanings and new entities, to make new worlds. While critical of many aspects of the way this happens, such as the wholesale extending of private property (i.e. patenting) to life forms, she warns against any purist rejection of the 'unnatural', hybrid entities produced by biotechnology, admitting at one point (Haraway 1997: 89) her 'frank pleasure' at the introduction into tomatoes of a gene from flounders, which live in cold seas, that enables the tomato to produce a protein that slows freezing. She revels in the very difficulty of predicting what technology's effects will be. The 'lively, unfixed, and unfixing' practices of science and technology produces 'surprises [which] just might be good ones' she comments (Haraway 1997: 280).

▶ DOES SCIENCE SHAPE TECHNOLOGY?

Clearly, any efficacious politics of technology, any systematic attempt to ensure that the surprises are indeed good ones, needs an understanding of technological change. Let us begin to sketch an outline of such an understanding by tackling the most obvious force shaping technology: scientific change. Technology, it is often said, is applied science. Scientists discover facts about reality, and technologists apply these facts to produce useful things. As we have indicated, this view of technological change is a key underpinning of popular forms of technological determinism.

There are several things wrong with the notion of technological change as the application of scientific discovery. First, the notion of 'discovery' – the uncovering of what is already there – is naive. Scientists are, of course, in constant, intimate dialogue with the real, material world, but they are active participants in that dialogue, bringing to it conceptual schema, experimental traditions, intellectual investments, ways of understanding the world, models and metaphors – some drawn from the wider society – and so on (see, for example, Shapin 1982; Barnes *et al.* 1996; Galison 1997).

Furthermore, science and technology have by no means always been closely connected activities. Looking backwards is tricky, because people in previous times did not operate with our notions of 'science' and 'technology' (Mayr 1976), and there is some controversy among historians who have studied the issue (see, for example, Musson and Robinson 1969; Mathias 1972). But it can be concluded that before the latter part of the nineteenth century the contribution of activities we would now think of as science to what we would call technology was often marginal. The water-mill, the plough, the spinning wheel, the spinning jenny, even the steam engine – these crucial inventions were in no real sense the application of pre-existing science (see, for example, Cardwell 1971, 1972). *Rhetoric* about the contribution of science to technology there was in plenty, but the rhetoric often bore little relation to the modest reality of that contribution, and needs to be interpreted differently (Shapin 1972: 335–6).

Where science and technology are connected, as they increasingly have been since the second half of the nineteenth century, it is mistaken to see the connection between them as one in which technology is one-sidedly dependent on science. Technology has arguably contributed as much to science as vice versa – think of the great dependence of science on the computer, without which some modern scientific specialties could scarcely have come into existence.[4] Most importantly, where technology does draw on science, the nature of that relation is not one of technologists passively deducing the 'implications' of a scientific advance. Technology, as the word's etymology reminds us,[5] is knowledge as well as artifacts, and the knowledge deployed by engineers is far from just applied science, as engineer-turned-historian Walter Vincenti (1990) demonstrates. Engineers *use* science. They seek from science resources to help them solve the problems they have, to achieve the goals towards which they are working. These problems and goals are at least as important in explaining what they do as the science that is available for them to use.[6]

▶ THE TECHNOLOGICAL SHAPING OF TECHNOLOGY

If science does not in any simple sense shape technology, what of the notion that technological change follows an autonomous logic – the notion that *technology shapes technology* (see Ellul 1964: 85–94; Winner 1977: 57–73)? To understand the force of this argument, it is necessary to see what is wrong with our common, but wholly mystified, notion of the heroic inventor. According to that notion, great inventions occur when, in a flash of genius, a radically new idea presents itself almost ready-formed in the inventor's mind. This way of thinking is reinforced by popular histories of technology, in which to each device is attached a precise date and a particular man (few indeed are the women in the stereotyped lists) to whom the inspired invention 'belongs'.

One important attack on this inspirational notion of invention was mounted by the group of American writers, most importantly William Ogburn, who from the 1920s onwards set themselves the task of constructing a sociology of technology (Westrum 1991). In a 1922 article, Ogburn and his collaborator Dorothy Thomas argued that far from being the result of unpredictable flashes of inspiration, inventions were *inevitable*. Once the 'necessary constituent cultural elements' are present – most importantly including component technologies – there is a sense in which an invention *must* occur: 'Given the boat and the steam engine, is not the steamboat inevitable?' asked Ogburn and Thomas (1922: 90.) They regarded it as crucial evidence for the inevitability of invention that a great many inventions were in fact made independently by more than one person.

Not the least of the difficulties of this position is that apparent inventions of the same thing turn out on closer inspection to be of importantly different things (Constant 1978). A solidly based critique of the inspirational notion of invention can, however, be constructed directly, drawing on the work of writers such as Ogburn's contemporary Usher (1954), his colleague Gilfillan (1935a, 1935b) and, more recently, historians of technology like Thomas P. Hughes (1971, 1983, 1989; see also pp. 50–63 of this book). Hughes's work is of particular relevance because much of it focuses on classic 'great inventor' figures such as Thomas Edison (credited with the invention of, among other things, the gramophone and the electric lightbulb) and Elmer Sperry (famed for his work on the gyrocompass and the marine and aircraft automatic pilot).

Hughes has no interest in disparaging the achievements of those he writes about – indeed he has the greatest respect for them – but his work demonstrates that invention is not a matter of a sudden flash of inspiration from which a new device emerges 'ready-made'. Largely it is a matter of the minute and painstaking modification of existing technology. It *is* a creative and imaginative process, but that imagination lies above all in seeing ways in which existing devices can be improved, and in extending the scope of techniques successful in one area into new areas.

A vitally important type of technical change altogether escapes our conventional notion of 'invention'. Technical change, in the words of Gilfillan (1935a: 5), is often 'a perpetual *accretion* of little details . . . probably having neither beginning, completion nor definable limits', a process Gilfillan saw at work in the gradual evolution of the ship (1935b). The authors of this process are normally anonymous, certainly not 'heroic inventor' figures, and often skilled craft workers, without formal technical or scientific training; it is probably best seen as a process of collective learning rather than individual innovation. 'Learning by doing' in making things (Arrow 1962) and what Rosenberg (1982: 120–40) calls 'learning by using' – feedback from experience of use into both the design and way of operating things – are both of extreme practical importance. Small changes may add up to eventually considerable changes in design, productivity and effectiveness.

New technology, then, typically emerges not from flashes of disembodied inspiration but from existing technology by a process of gradual change to, and new combinations of, that existing technology. Even what we might with some justification want to call revolutions in technology often turn out to have been long in the making. Constant's important study (1980) of the change in aircraft propulsion from the propeller to the jet shows this clearly. Revolutionary as it was in the context of aircraft propulsion, the turbo jet built upon a long tradition of work on water and gas turbines.

Existing technology is thus, we would argue, an important precondition of new technology. It provides the basis of devices and techniques to be modified, and is a rich set of intellectual resources available for imaginative use in new settings.[7] But is it the *only* force shaping new technology? We would say that it is not, and would argue that this can be seen by examining the two most plausible attempts to claim that existing technology is more than just a precondition of new technology, but is an active shaping force in its development. These attempts focus around the ideas of technological 'paradigm' and technological 'system'.

The idea of 'technological paradigm' (see Constant, 1980; Dosi 1982) is an analogical extension of Thomas Kuhn's idea of the scientific paradigm (1970). In Kuhn's work, 'paradigm' has two main meanings, which are interrelated but distinguishable. In the more basic sense, the paradigm is an exemplar, a particular scientific problem-solution that is accepted as successful and which becomes the basis for future work. Thus Newton's explanation of the refraction of light, in terms of forces acting on the particles he believed light to consist in, formed a paradigm for much subsequent work in optics – researchers sought to produce similar explanations for other optical phenomena (Worrall 1982). The paradigm in this first sense of exemplar plays a crucial part in the paradigm in the second, more famous, wider sense of the 'entire constellation of beliefs, values, techniques, and so on shared by the members of a given [scientific] community' (Kuhn 1970: 175).

The discussion of paradigms in technology has been less profound than it might have been because it (like extensions of Kuhn's ideas to the social sciences) has tended to focus on the second meaning of paradigm, despite Kuhn's explicit statement that the first meaning is 'philosophically . . . deeper' (Kuhn 1970: 175; see also Barnes 1982; Gutting 1984; Laudan 1984). But there is no doubt that the concept of paradigm applied to technological change does point us towards important phenomena. Particular technical achievements have played a crucial role as exemplars, as models for further development (see Sahal 1981a, 1981b). In the field of missile technology, for example, the German V-2 missile played this role in early post-war American and Soviet missile development. Because technological knowledge cannot always be reduced to a set of verbal rules, the presence of a concrete exemplar is a vital resource for thought. The Americans possessed actual German-built V-2s, as well as most of the design team; the Soviets painstakingly constructed, with help from some of the designers, replicas of the original missile (Ordway and Sharpe 1979). To a significant extent the V-2 formed the model from which further ballistic missiles were derived by conscious modification.

If we find technologists operating with a paradigm – taking one technical achievement and modelling future work on that achievement – it becomes tempting to treat this as somehow self-explaining and discuss it in terms of mechanical analogies such as following a technical 'trajectory' (Dosi 1982). But to do this would be to miss perhaps the most fundamental point of Kuhn's (1970) concept of paradigm: the paradigm is not a rule that can be followed mechanically, but a *resource* to be used. There will always be more than one way of using a resource, of developing the paradigm. Indeed groups of technologists in different circumstances often develop the same paradigm differently. American and Soviet missile designers, for example developed significantly different missiles, despite their shared use of the V-2 as a departure point (Holloway 1977, 1982; Berman and Baker 1982). Where this does not happen, where there is congruity in the development and extension of a paradigm, this stands equally in need of explanation.

Just how much can be hidden by considering the further development of a paradigm as simply a 'technological trajectory' following an 'internal logic' emerges from another study by Hughes (1969). Here the 'trajectory' being considered is that of successive processes for synthesizing chemicals by 'hydrogenation' – combination with hydrogen at high temperatures and pressures over catalysts. Hughes examines the trajectory of this work in the German chemical firm I. G. Farben and its predecessors. Beginning with the paradigm instance of the Haber-Bosch process for the synthesis of ammonia, the company moved on to the synthesis of wood alcohol and finally of gasoline (from coal). A 'natural' trajectory, indeed, but one that, Hughes shows, at each stage was conditioned by social factors inside and outside the firm, including, most consequentially, the German state's need for wartime independence from external sources of raw materials. In America, the chemical giant Du Pont adopted synthetic processes for the production of ammonia and wood alcohol (Mueller 1964), but did not, in that very different environment, find the step to the synthesis of gasoline 'natural'. In Germany, moving to gasoline synthesis involved greater and greater links between Farben and the Nazi state, links which eventually led 23 executives of Farben to the dock in the Nuremburg war-crime tribunals.

The idea of technological *system* has been used in the history of technology more widely than that of technological paradigm, and thus the characteristics of explanations framed in its terms are more evident. We will follow its usage by Thomas P. Hughes, who makes it in many ways the central theme of his studies of technology. Typically, and increasingly, technologies come not in the form of separate, isolated devices but as part of a whole, as part of a system. An automatic washing machine, say, can work only if integrated into the systems of electricity supply, water supply and drainage. A missile, to take another example, is itself an ordered system of component parts – warhead, guidance, control, propulsion – and also part of a wider system of launch equipment and command and control networks.

The need for a part to integrate into the whole imposes major constraints on how that part should be designed. Edison, as Hughes shows in the extract from his work in Chapter 3, designed the light-bulb not as an isolated device but as part of a system of electricity generation and distribution, and the needs of the system are clearly to be seen in the design of the bulb.

Further, the integration of technologies into systems gives rise to a particular pattern of innovation that Hughes, using a military metaphor, describes as 'reverse salients' (see, for example, Hughes 1971: 273, 1983: 14; for related observations see Rosenberg 1976: 111–12). A reverse salient is a product of uneven development. It is an area where the growth of technology is seen as lagging, like a military front line which has been pushed forward but where in one particular spot the enemy still holds out. Technologists focus inventive effort, like generals focus their forces, on the elimination of such reverse salients; a successful inventor or engineer defines a reverse salient as a set of 'critical problems' that, when solved, will correct the situation. A typical reverse salient appeared in the development of electricity supply systems. As transmission voltages were increased, power was lost between the lines through electric discharge. Because very high voltages were needed to transmit electricity over large distances, loss between the lines was a reverse salient that threatened the development of the electricity supply system as a whole. Consequently, considerable effort was devoted to solving the critical problems involved (Hughes 1976, 1983).

The focusing of innovation on perceived reverse salients is a phenomenon of great generality. Hughes's judgement is that 'innumerable (probably most) inventions and technological developments result from efforts to correct reverse salients' (1983: 80). While this is thus an important way in which technology (as technological systems) shapes technology, does it imply that *only* technology shapes technology? Hughes's answer is 'no', and the reason for that answer is of considerable importance. A technological system like an electric light and power network is never merely technical; its real-world functioning has technical, economic, organizational, political, and even cultural aspects.[8] Of these aspects, the most obviously important one is economic, and it is to that we turn next.

► THE ECONOMIC SHAPING OF TECHNOLOGY

The very concept of 'reverse salient' makes sense only if a technological system is seen as oriented to a *goal* (Hughes 1983: 80). Otherwise, any metaphors of 'advancing' or of 'backward' parts become meaningless. Language of this kind is dangerous if it is allowed to slip towards vague talk of the 'cultural need' for a technology (Ogburn and Thomas 1922: 92), but the notion of a goal can be given a direct and down-to-earth meaning. Most importantly, talk of a system goal is normally talk about economics, about reducing costs and increasing revenues. Electricity supply systems, for example, have been private or public enterprises, and those who have run them have inevitably been concerned above all about costs, profits and losses. The reverse salient is an 'inefficient or uneconomical component' (Hughes 1983: 80), and for many practical purposes inefficient means uneconomical.

Technological reasoning and economic reasoning are often inseparable. Our extract from Hughes's work demonstrates this in the case of Edison's invention of the light-bulb. Edison was quite consciously the designer of a system. He intended to generate electricity, transmit it to consumers, and to sell them the apparatus they needed to make use of it. To do so successfully he had to keep his costs as low as possible – not merely because he and his financial backers wished for the largest possible profit, but because to survive at all electricity had to compete with the existing gas systems. Crucially, Edison believed he had to supply electric light at a cost at least as low as that at which gas light was supplied. These economic calculations entered directly into his work on the light-bulb. A crucial *system* cost, a reverse salient, was the copper for the wires that conducted electricity. Less copper could be used if these wires had to carry less current. Simple but crucial science was available to Edison as a resource: Ohm's and Joule's laws, from which he inferred that what was needed to keep the current low and the light supplied high was a light-bulb filament with a high electrical resistance, and therefore with a relatively high voltage as compared to current. Having thus determined, economically as much as technologically, its necessary characteristics, finding the correct filament then became a matter of 'hunt and try'.

The precise characteristics of the Edison case are perhaps untypical. Even in his time Edison was unusual in his conscious, individual grasp of the nature of technological systems (therein, perhaps, lay his success), and since his time the inventor-entrepreneur has in many areas been overshadowed by the giant corporation with research and development facilities. Menlo Park, Edison's research and development institution, was only an aspect of the beginning of the great transformation brought about by the large scale, systematic harnessing of science and technology to corporate objectives (Noble 1977). But the essential point remains: typically, technological decisions are also economic decisions.

Paradoxically, then, the *compelling* nature of much technological change is best explained by seeing technology not as outside of society, as some versions of technological determinism would have it, but as inextricably part of society. If technological systems are economic enterprises, and if they are involved directly or indirectly in market competition, then technical change is forced on them. If they are to survive at all, much less to prosper, they cannot forever stand still. Technical change is made inevitable, and its nature and direction profoundly conditioned, by this. And when national economies are linked by a competitive world market, as they have been at least since the mid-nineteenth century, technical change outside a particular country can exert massive pressure for technical change inside it.

The dominant way of thinking about the connection between economics and technology is the 'neoclassical' approach, which is based upon the assumption that firms will choose the technique of production that offers the maximum possible rate of profit. Despite its apparent plausibility, this assumption has been the subject of much criticism within economics. The issues involved are complex (there is a useful review of them in Elster 1983) but they hinge upon whether human decision-making does, or indeed could, conform to the strict requirements of the neoclassical model. For example, how can a firm possibly know when it has found the technique of production that produces maximum profits? Is it not more reasonable to assume that a firm will consider only a very limited range from the set of possible options, and will be happy with a 'satisfactory' (and not necessarily maximum) profit rate? In the new approaches that have developed within economics, inspiration has been found in the work of Joseph Schumpeter (1934, 1939, 1943, 1951), with its emphasis on the aspects of innovation that go beyond, and cannot be explained by, rational calculation.[9]

▶ ECONOMIC SHAPING IS SOCIAL SHAPING

The 'alternative', non-neoclassical economics of technology thus offers a direct bridge to more sociological explanations (MacKenzie 1996a, Ch. 3). Costs and profits matter enormously, but in situations of technical innovation key factors are *future* costs and *future* profits. Since there is an element of uncertainty in these, they cannot be taken as simple, given facts. Estimating costs and profits is part of what Law (1987) calls heterogeneous engineeering: engineering 'social' as well as 'technical' phenomena; constructing an environment in which favoured projects can be seen as viable.[10] Market processes punish those who get this wrong and reward those who get this right, but which outcome will prevail cannot be known with certainty in advance (see, for example, Schon 1982). Nor can it be assumed that market processes will eventually lead to optimal behaviour, as successful strategies are rewarded by the differential growth of firms that pursue them. That standard neoclassical argument may have validity for static environments in which selection has a long time to exercise its effects, but not for situations of technological change. A strategy that succeeds at one point in time may fail shortly thereafter, and the market's 'invisible hand' may simply have insufficient time for the neoclassical economist's optimization to take place.

Furthermore, even if sure calculation of costs and profits – and even optimization – were possible, the economic shaping of technology would still be its social shaping. Economic calculation and economic 'laws' are, after all, specific to particular forms of society, not universal, as Karl Marx famously argued (see, for example, Marx [1867] 1976: 173–6). Even if in all societies people have to try to reckon the costs and benefits of particular design decisions and technical choices, the form taken by that reckoning is importantly variable.

Consider, for example, technical innovation in the former Soviet Union. People there certainly made calculations as to what served their economic interests, and plant managers had greater autonomy to make decisions than is often assumed. But the framework of that calculation was different. Prices were set by central planners of the State Price Committee, rather than being subject to the vagaries of the market as in the West. A price, we might say, was thus a different social relation in the Soviet Union. In its classical form, the system of rewards to Soviet managers hinged upon *quantity* of production in the short run – fulfilling the 'norms' of the plan in the current quarter. The focus on quantity implied that while small technological innovations might be welcomed, larger changes (for example, changes that meant elaborate retooling) were a threat; developing a new product meant courting risks with little promise of commensurate reward if successful. The reforms that Soviet leaders introduced to alleviate this situation often made it worse. Thus economic reforms in 1965 tied the rewards to managers more closely to the profitability of their enterprises. But because the price system was not fundamentally changed, the greatest profits could be earned by concentrating on existing products whose costs of production had fallen well below their (bureaucratically set) prices. Innovation, instead of speeding up, actually slowed (Parrott 1983: 225–6), and the consequences contributed to the eventual dramatic collapse of the Soviet system.

Furthermore, even if we restrict our attention to societies in which prices reflect market competition, we find that economic calculation remains a mechanism of social shaping. Economic calculation presupposes a structure of costs that is used as its basis. But a cost is not an isolated, arbitrary number of pounds or dollars. It can be affected by, and itself affect, the entire way a society is organized. This point emerges most sharply when we consider the *cost of labour*, a vital issue in technical change, because much innovation is sponsored and justified on the grounds that it saves labour costs. To take a classic example, because of the different circumstances of nineteenth-century British and American societies (such as the presence in the USA of a 'frontier' of agricultural land whose ownership by indigenous peoples was largely disregarded), labour cost more in America than in Britain. Hence, argued Habakkuk (1962), there was a much greater stimulus in America than in Britain to search for labour-saving inventions, and thus a different pattern of technological change in the two societies. Habakkuk's claim has in fact proven to be controversial (see Saul 1970 and Uselding 1977 for introductions to the controversy), but the general point remains: the way a society is organized, and its overall circumstances, affect its typical pattern of costs, and thus the nature of technological change within it.

That men are typically paid more than women, for example, is clearly not an arbitrary matter, but one that reflects deep-seated social assumptions and an entrenched division of labour, including unequal domestic and child-rearing responsibilities. The different costs of men's and of women's labour translate into different economic thresholds for machines that have to justify their costs by elimination of men's, or of women's, tasks – a mechanism of the gendered shaping of technology that deserves systematic study (see Cowan 1979).

▶ ## TECHNOLOGY AND THE STATE

Social relations, then, affect technological change through the way that they shape the framework of market calculations. But the market is far from the only social institution that shapes technological change.

From antiquity onwards, states have sponsored and shaped technological projects, often on a vast scale. Lewis Mumford (1964: 3) provided a classic account of this, and it is worth quoting from a short summary of his ideas:

> authoritarian technics . . . begins around the fourth millennium B.C. in a new configuration of technical invention, scientific observation, and centralized political control . . . The new authoritarian technology was not limited by village custom or human sentiment: its herculean feats of mechanical organization rested on ruthless physical coercion, forced labour and slavery which brought into existence [human-powered] machines that were capable of exerting thousands of horsepower.

Seventeenth- and eighteenth-century European states were interested in technical progress as a source of greater national power, population and treasure (Pacey 1976: 174–203). This 'mercantilist' framework carried different implications for the shaping of technology than did straightforwardly capitalist judgements. As Hafter (1979: 55–6) writes, 'while in England there was strong commitment to labor-saving devices, in France the mercantilist notion that work must be found for the largest number of hands prevailed'. As late as 1784, the brocade loom was praised in France because it 'employed twice as many workers' as the plain-cloth loom, it being argued that it was 'the benefit of labor which remains in the towns when the products have left that is the real product of the manufactures' (Hafter 1979: 56).

The single most important way that the state has shaped technology has been through its sponsoring of military technology. War and its preparation have probably been on a par with economic considerations as factors in the history of technology. Like international economic competition, war and the threat of war act coercively to force technological change, with defeat the anticipated punishment for those who are left behind.[11] Military technology is the subject of Part Four of this reader, and we need make only one point here, regarding the extent to which military concerns have shaped 'civilian' technology. Military interest in new technology has often been crucial in overcoming what might otherwise have been insuperable economic barriers to its development and adoption, and military concerns have often shaped the development pattern and design details of new technologies.

Three cases in point are nuclear power, air transport and electronics. The initial work on the technology of nuclear energy was directly military in inspiration, and subsequently the economic drawbacks of nuclear power have often been overridden by state interest in securing fissile material for atomic weapons and in gaining 'autonomous' national energy supplies. These state interests closely shaped reactor design, at least in the early years of nuclear energy (Gowing 1982; Rüdig 1983; Simpson 1983; Hecht 1998). Similarly, the civilian jet airliners of the post-war period were made possible by a generation of work on military jets, and Constant (1980: 166–7) argues that the design of 1930s' British and German civil airliners reflected the ways in which those countries' airlines were 'chosen instruments' of foreign and imperial policy. Much of the development of electronics in this century has been sponsored by the military, especially in the USA. Military need and military support played a crucial role in the development of the digital computer (Goldstine 1972; Dinneen and Frick 1977; Flamm 1988; Edwards 1996). Braun and MacDonald's history (1978) shows the crucial role of military support in the development of semiconductor electronics (and thus in the origins of the microchip). That support was particularly important in the early phase of development when on most commercial criteria solid-state devices were inferior to existing valve technology.

► CASE STUDIES OF THE SHAPING OF TECHNOLOGY

Even in these cases of the shaping of technology by military interests, 'shaping' should not be understood as always being direct and conscious – as the simple imprinting of human will on the material world. What emerged, even in the cases just discussed, was by no means always what sponsors had intended: for example, though the military wanted miniaturization, their originally preferred approach was not the eventually successful integrated circuit. Technologies (especially radically new technologies) typically emerge, or fail to emerge, from processes in which no one set of human actors plays a dominant role, and in which the role of a recalcitrant material world cannot be ignored. The confused, unsuccessful negotiation beautifully described by Latour (1996) is far more typical, even for state-sponsored technologies.

The social shaping of technology is, in almost all the cases we know of, a process in which there is no single dominant shaping force. We have chosen as exemplary of this Paul Ceruzzi's study of the emergence of personal computing (a phrase that includes not just the hardware necessary for personal computing, but also, for example, the software needed to make the hardware useful). He eschews technological determinism, denying that the personal computer or personal computing were simply the outgrowth of changing microchip technology (while accepting that developments in that sphere were crucial). Members of the radical counterculture of the

1960s and 1970s, Ceruzzi points out, wanted to liberate computing from its military and corporate masters: they were pursuing one version of the active politics of technology that we are recommending. Author Ted Nelson, for example, combined technical and social radicalism, for instance in his influential proposal for 'hypertext' (designed to help untrained people find their way through computer-held information organized in more complicated ways than in paper documents, and in one sense a precursor of the enormously successful World Wide Web: see Campbell-Kelly and Aspray 1996).

This kind of countercultural impulse interacted with a largely male hobbyist culture, members of which simply wanted to have computers of their own to play with (part of the development of personal computing was starting to treat computers less seriously). The interaction was, for instance, at the heart of the Californian Homebrew Computer Club, which played an important role in the emergence of personal computing. Steve Wozniak and Steve Jobs, founders of Apple Computer, famously started out making 'blue boxes' that mimicked telephone dial tones, allowing users to make free telephone calls, a laudable goal from a countercultural viewpoint. However, Ceruzzi also shows other strands that came together in personal computing, notably the role of previous developments in time-sharing mainframe computers, such as the BASIC programming language developed for students at Dartmouth College (including humanities students, who were presumed to be less sophisticated technically).

Personal computing was indeed socially shaped, but no one actor determined the shape it was to take, and the outcome was no simple reflection of an existing distribution of power. The mighty IBM Corporation, which dominated the mainframe computer business, notoriously came to personal computing relatively late, and the field's development was eventually seriously to weaken IBM's dominance. Orthodox corporate power has subsequently been re-established in the form of the near monopoly of software supplier Microsoft (the early role of Microsoft's founder, Bill Gates, is discussed by Ceruzzi) and the microprocessor supplier Intel. Nevertheless, the more pessimistic analyses of the development of word processing (Barker and Downing 1980) now seem wide of the mark, in part because some of the aspirations of the counterculture were fulfilled. The computer has indeed come 'to the people' – not all the people, to be sure, but enough to make a difference.

Ceruzzi's study is of the development of an entire field of technology. The other case study we have selected for this introductory section is much narrower in its focus. We have chosen it because it shows social shaping, not just of the overall contours of a technology, but of specific, apparently 'purely technical', features of technological designs, of engineering research, and even of mathematical models of artifacts. Eda Kranakis compares in detail two suspension bridge designs: one by the American, James Finley, inventor of the modern suspension bridge with a flat roadway; the other by Claude-Louis-Marie-Henri Navier, a leading French engineer-scientist. Both Finley and Navier were heterogeneous engineers, but heterogeneous engineers working in very different environments with different goals.

Finley, working in the USA in the early nineteenth century, aimed at a relatively cheap bridge design that could fairly easily be tailored to a specific location by craftworkers with limited mathematical skills. He wanted to make money not primarily by building bridges himself but by getting others to pay to use his patented suspension bridge design. His design crystallized these goals. For example, Finley chose a sag/span ratio (see the figure on p. 89) between 1:6 and 1:7, not because this was in any abstract sense optimal, but because this ratio greatly simplified the calculations that users of his patent had to make.

Navier, in contrast, positively sought sophistication in mathematical modelling. He was a salaried state employee, working in an engineering culture where mathematical competence was deliberately fostered and highly prized, and he was seeking promotion as a mathematical scientist as much as an engineer. Navier's bridge was designed, both in its overall conception and in specific features, to demonstrate the applicability to technology of deductive mathematical reasoning. Kranakis suggests that the particular approach to mathematical modelling taken by Navier was influenced by his career goals, and reminds us that even mathematics is not always a universal language. For example, the French mathematical tradition in which Navier worked differed in its approach to the relevant part of mathematics – the calculus – from the approach taken in Britain. On the Continent an algebraic, symbol-manipulating approach predominated, while many mathematicians in Britain clung to a visual, geometric version of the calculus, a preference that reflected the distinctive cultural and educational role of geometry as the paradigm of absolute knowledge, including theological knowledge (Richards 1979).

Two crucial points about 'the social shaping of technology' can be seen in Kranakis's study. First, she is perfectly well aware that bridges are real physical artifacts, and that their behaviour is in no way reducible to the ensemble of beliefs about them. Bridges built using Finley's patent sometimes collapsed, and Navier's bridge suffered a mishap during construction that opened the project up to eventually fatal criticism. The point is a general one: emphasis on the social shaping of technology is wholly compatible with a thoroughly realist, even a materialist, viewpoint. What is being shaped in the social shaping of artifacts is no mere thought-stuff, but obdurate physical reality. Indeed, the very materiality of machines is crucial to their social role, as Part Two of this reader emphasizes. In producing the first edition of this book, we chose the metaphor of 'shaping', rather than the more popular 'social construction', in part because the latter is too prone to the misconception that there was nothing real and obdurate about what was constructed. (One of the ordinary meanings of 'construction' implies falsehood, as in 'the story he told me was a complete construction'. Although this is emphatically not what is implied when we or others have used the metaphor of 'construction', there is always the risk that this will colour how the metaphor is heard.)

The second point is that 'social shaping' does not necessarily involve reference to wider societal relations such as those of class, gender and ethnicity. These *are* sometimes directly crucial, and we give instances below of this, but often what is more immediately relevant are 'local' considerations, such as engineers' membership of professional communities, the reward structures of those communities, and so on. These are social matters too. The 'social' is not the same as what in old debates about the relationship between science and society used to be called 'external factors'; social processes internal to scientific and technological communities are important too. Often these internal processes are themselves conditioned by wider social and historical matters – for example, the reward structure of nineteenth-century French engineering, with its distinctive emphasis on displays of mathematical competence, emerged out of the clashes of the Revolutionary period (Alder 1997) – but they remain social even when that is not the case.

▶ THE PATH-DEPENDENCE OF TECHNICAL CHANGE

We are aware that case studies of social shaping are unlikely, on their own, to undermine the technologically determinist view of technological change. In the long run, the convinced determinist might say, surely what matters is intrinsic technical efficiency: the intrinsically best technology will ultimately triumph, whatever local contingencies affect particular developments.

There are two answers to be given to this deep-seated determinist assumption. First, of course, is the basic point that the technology that is 'best' from one point of view is not necessarily best from another: what is best for workers may not be best from the point of view of their employers; what men believe to be best may not be best for women, and so on. Throughout this reader, we will see examples of different assessments of what counts as technologically desirable. Second, however, is a subtle and important argument developed in our extract from the work of the economist Brian Arthur, an argument also taken up by the economic historian Paul David.[12]

Arthur's point is a simple one, but broad in its implications. Technologies often manifest increasing returns to adoption. The processes of learning by doing and by using, discussed above, and the frequent focus of inventive effort on removing weak points ('reverse salients') from existing technologies, mean that the very process of adoption tends to improve the performance of those technologies that are adopted. This gives the history, especially the early history, of a technology considerable significance. Early adoptions, achieved for whatever reason, can be built into what may become irreversible superiority over rivals, because success tends to breed success and rejection can turn into neglect and therefore permanent inferiority. The history of technology is a path-dependent history, one in which past events exercise continuing influences. Which of two or more technologies eventually succeed is not determined by their intrinsic characteristics alone, but also by their histories of adoption. The technology that triumphs is not necessarily abstractly best, even if there is consensus about what 'best' means. Path-dependence means that local, short-term contingencies can exercise lasting effects.[13]

The history of personal computing, for example, is full of manifestations of path-dependence. The pervasive qwerty keyboard, so-called because of the letters on the upper left, is in no sense demonstrably optimal. It developed to minimize the frequency with which keys in mechanical typewriters stuck together as a result of adjacent keys being hit in too close succession. That rationale clearly became unnecessary after the development of electronic keyboards and word processing, but proposals for alternate layouts are hopeless: the triumph of qwerty has become in practice irreversible. It would, more generally, be hard to make a case for the intrinsic superiority of the technical system that has come to dominate personal computing: the combination of the IBM personal computer architecture, Microsoft's MS-DOS and Windows operating systems, and the descendants of the Intel 8080 microprocessor. Historical contingency played a clear role in that outcome. For example, in part because of a history of anti-trust litigation against IBM, the corporation was willing to license its architecture and permit others to manufacture clones, while its main rival, Apple, refused to do so; the consequence was an entrenchment of the IBM architecture, and the Intel microprocessors it employs, and the restriction of Apple to niche markets.

The issue of path-dependence needs to be analysed with some care, and some claims for the phenomenon have been criticized by Stan Liebowitz and Stephen Margolis (1990, 1995a, 1995b). If a technology has an actually existing rival that is either demonstrably superior or can quickly and reliably be made so, then lock-in to the inferior variant is, they argue, unlikely to be permanent. There are too many ways in which it can be overcome: for example, manufacturers can offer the 'underdog' technology initially below cost to create a market for it, or governments can subsidize it (this has historically been an important function of military expenditure, for example in helping solid-state electronics overcome its initial disadvantages, as noted above). Arthur is wrong to assert (see p. 111) that the alternatives to qwerty are superior; the evidence for this is at best ambiguous (Liebowitz and Margolis 1990). Whether Apple or IBM personal computers are best is a source of endless dispute, and other putative examples of lock-in to clearly inferior technologies are likewise controversial (see, for example, the discussion of the popular example of VHS and Beta video recorder formats in Liebowitz and Margolis 1995a; for David's reply to the overall critique, see David 1997).

In rightly objecting to neoclassical confidence that the best technology will always triumph, Arthur may have bent the stick too far in the opposite direction in suggesting the likelihood of lock-in to the unequivocally inferior. Arguably, both sides in this debate underestimate the complexity and uncertainty of knowledge of the characteristics of technologies, even the most 'technical' characteristics (MacKenzie 1996b). Apparently easily answered questions about existing technologies, such as what key layout permits fastest typing or how accurate a given missile is (MacKenzie 1990), can turn out to be complex and contested. Yet determining a single characteristic of an actually existing technology is the simplest case: in real historical cases, those involved may have to weigh up the relative import-ance of differing characteristics (the efficiency of the internal combustion engine versus its potential for pollution, for example) and determine the likely effect of development efforts that have not yet taken place.

Complexity and uncertainty, however, increase rather than diminish the importance of path-dependence. If there is an unequivocally superior alternative to what historical processes of technological change have left us with, then, as noted above, there will often be reasons for modest confidence that it will be adopted. If, on the other hand, the characteristics of alternatives are uncertain and contested, then the low-risk course will be the path-dependent one of starting from what history has given us and seeking to improve it.

 ## THEORIZING THE TECHNOLOGY-SOCIETY RELATIONSHIP

A major development in the social studies of technology since the first edition of this book in 1985 is the flowering of theoretical work on the relationship between technology and society. Two theoretical approaches, nascent in the mid-1980s, have particularly close bearing upon the social shaping of technology.

First is the 'social construction of technology' perspective, developed by Wiebe Bijker and Trevor Pinch (Bijker 1995; Bijker *et al.* 1987), and represented here in a succinct extract from the work of Pinch and his colleague Ronald Kline. Its focus is on the very phenomenon that has been underestimated in the debate over path-dependence: the 'interpretative flexibility' of technology. Interpretative flexibility refers to the way in which different groups of people involved with a technology (different 'relevant social groups', in Bijker and Pinch's terminology) can have very different understandings of that technology, including different under-standings of its technical characteristics. Bijker and Pinch's focus is not just on the symbolic meaning of technologies (which in cases like motor cars or aircraft is subject to obvious social variation) but includes also variation in criteria for judging whether a technology 'works'.

The Bijker/Pinch 'social construction of technology' approach draws heavily upon earlier work applying a sociological perspective to scientific knowledge. Those developing the sociology of scientific knowledge, such as Bloor (1976), sought symmetry of explanation. Bloor argued against the then prevalent notion that true scientific knowledge was the result simply of unaided human rationality and causal input from the material world. Instead of invoking social processes only when the credibility of false belief had to be explained, Bloor argued that proper explanation of *all* knowledge, true and false, typically would involve recourse to material input, psycho-logical processes *and* social processes.

There are few more difficult and more contentious topics than what sociology-of-knowledge 'symmetry' should be taken to mean, and certainly not all subsequent authors employed the term in the way Bloor did. For Bijker and Pinch, symmetry means avoiding explaining the success or failure of technologies by whether or not they work. For them, 'machines "work" because they have been accepted by relevant social groups' (Bijker 1995: 270). To our minds, this formulation underplays the extent to which technology always involves interaction between human beings and the material world, but we wholeheartedly agree that historians and sociologists of technology should consider the fact that machines 'work' as something to be explained rather than taken for granted in our explanations. In particular, explanations of success and failure in terms of the *intrinsic* superiority or inferiority of technologies are suspect because of the path-dependence of the history of technology. That one type of machine works better than the alternatives may reflect their histories of adoption and improvement rather than any intrinsic, unalterable features of the technologies involved.

The extract from Kline and Pinch's article ends by citing some of the shortcomings of the approach originally taken by Pinch and Bijker. Of these, two are of particular relevance here. The first is the issue of structural exclusion. In Pinch and Bijker's approach, the social groups relevant from the point of view of a particular technology are typically identified empirically: in historical research, for example, 'we can identify what social groups are relevant with respect to a specific artifact by noting all social groups mentioned in relation to that artifact in historical documents' (Bijker 1995: 46). The trouble, of course, is that the exclusion of some social groups from the processes of technological development may be such that they have no empirically discernible influence on it, and are not, for example, mentioned in documents concerning it: this, for instance, will often be the case with women, ethnic minorities and manual workers.[14] It clearly would be most foolish to assume that gender is irrelevant to the development of a technology just because no women were directly involved and the masculinity of the men involved was never mentioned explicitly in discussion of it; and analogous points hold for class and, especially, ethnicity. The point is a difficult one – we would not claim to have a formula for how to analyse the effects on technological development of structural exclusion – but it needs always to be kept in mind. The influence of 'politics' upon weapons technology is, for example, by no means always the direct one of technologists' compliance with explicit political demands. It can also take the indirect form of the efforts of technologists to keep their technologies as 'black boxes', opaque to scrutiny from the political system. The developers of the US submarine-launched ballistic missile systems, for instance, carefully avoided design options that might lead to political controversy and Congressional involvement, however attractive these options seemed to others (MacKenzie 1990).

The other problem with the original formulation of the Bijker/Pinch approach is one that also manifested itself in the first edition of this book: 'the reciprocal relationship between artifacts and social groups'. The theoretical perspective that has done most to sensitize the field to this issue is what is often called actor-network theory, developed by scholars such as Bruno Latour, Michel Callon, Madeleine Akrich and John Law, and represented here by the extract from the work of Latour and primatologist Shirley Strum. The key point can be conveyed by way of self-criticism. In the first edition of this reader we largely thought of the social shaping of technology in terms of the influence of social relations upon artifacts. The problem with this formulation is its neglect of the valid aspect of technological determinism: the influence of technology upon social relations. To put it in other, more accurate, words, it is mistaken to think of technology and society as separate spheres influencing each other: technology and society are mutually constitutive.

The reason why, from the varied and influential writings of Bruno Latour (see Latour 1987, 1991, 1993, 1996), we have chosen Strum's and his article is that it reveals what the mutual constitution of technology and society means, and why it matters. Their starting point is the developing appreciation in primatology (to which Strum's field observations have contributed centrally) that primate societies – baboon societies in particular – cannot be thought of as having fixed social structures into which individuals simply fit. Primatologists increasingly see baboons as actively, continuously negotiating and renegotiating their relative roles, and see social structure as the outcome of this process rather than as something fixed and given.

Primatologists, in other words, now view baboons very similarly to the way modern sociologists, following the decline of rigid views of social structure, see human actors as creating structure in and through interaction. (The schools of sociology that have emphasized this are known as social interactionism and, especially, ethnomethodology.[15]) Yet there is of course an evident difference between the societies that humans and baboons create: baboon societies are limited in time and space, essentially to the span of face-to-face interaction, while human societies have histories and geographies that go far beyond that span. The difference is made, Strum and Latour argue, by the human use of 'material resources and symbols'. It is the former that is of particular interest here. Material resources – artifacts and technologies, such as walls, prisons, weapons, writing, agriculture – are part of what makes large-scale society feasible. The technological, instead of being a sphere separate from society, is part of what makes society possible – in other words, it is constitutive of society.

To talk of 'social relations' as if they were independent of technology is therefore incorrect, Strum and Latour would argue. Artifacts – things humans have made – are involved in most of the ways human beings relate to each other. Sexual acts (without prophylactics against disease or pregnancy) are one of the few exceptions in which humans interact, baboon-like, with our naked bodies and voices alone, and such exceptions are typically embedded in more material relations. The point is not simply a pedantic issue of choice of words, as a couple of examples of the technological transformation and creation of social relations may make clearer.

Consider first the Marxist accounts of technology discussed in Part Two of this book. In essence, these suggest that production technology 'hardens' earlier relations between workers and capitalists (relations that were closer to pure social relations – in other words, not so strongly mediated by artifacts), so strengthening the subordination of labour to capital. The relation of labour to capital is *not* a social relation, Strum and Latour would point out, but a socio-technical relation; and in that respect it is typical. Second, consider the Internet, whose origins are discussed in Part Four. One does not have to buy into the hype surrounding the Internet to see that it permits the creation of new social groups by facilitating easy communication between geographically widely dispersed people with statistically unusual identities or interests. These newly created, or newly reinforced, groups can in their turn influence technological development.[16]

So we see Strum and Latour's article, despite its apparently esoteric topic, as an ambitious critique of nearly all forms of existing social theory. Because these neglect technology, they implicitly conceive of society as if it were constructed by human beings using their voices and naked bodies alone: most social theory, in other words, is actually baboon theory! This baboon theory cannot, Strum and Latour would point out, answer the fundamental questions of social theory – What is society? How is social order possible? – because satisfactory answers to them, in the case of human society, inevitably involve reference to technology. This aspect of the actor-network position – that its fundamental contribution is to social theory, and not, in the first instance, to the sociology of science and technology, narrowly conceived – is often overlooked in debates about it in the literature of the latter field.

Both society and technology, actor-network theory proposes, are made of the same 'stuff': networks linking human beings and non-human entities ('actors', or, in some versions, 'actants'). In this respect, actor-network theory resembles Hughes's technological systems perspective: a technological system such as an electric light and power network ties inextricably together both material artifacts and human beings – ties together 'technology', on the one hand, and economics, organization, politics and culture on the other.

Actor-network theory, however, differs from Hughes's perspective in its much greater, 'philosophical' ambitions. These again hinge, to a considerable extent, on the treacherous term 'symmetry'. Notoriously (this is the source of much of the controversy surrounding it) actor-network theory calls for symmetry in the analytical treatment of human and non-human actors (see, especially, Callon 1986; for the main critique, see Collins and Yearley 1992). We cannot discuss here the full range of issues this raises (for further discussion see MacKenzie 1996a), but can simply note that one version of the claim is wholly compatible with what we argue here: that the material world is no simple reflection of human will, and that one cannot make sense of the history of technology if the material world is seen as infinitely plastic and tractable. Whether its intractability is interpreted as agency (in the sense of intentionality) is of course another matter, one subject to wide cultural variation; but discussion of this would lead us too far away from the purposes of this volume.

▶ **CONSTRUCTING GENDER; CONSTRUCTING 'COLOUR'**

One author sharply aware of the mutual constitution of society and technology is Cynthia Cockburn, and we reprint here her 1983 article 'Caught in the wheels', which represents a pivotal point in the growing engagement between feminism and technology (for other work from the same period or just before, see Cowan 1979 and McGaw 1982). Cockburn went beyond concerns for 'equal opportunities' – greater representation of women in the traditionally male professions of science and engineering – to ask two further questions: is technology itself shaped by gender, and is gender shaped by technology?

Cockburn's answer to the first of these questions is that 'industrial, commercial, military technologies are masculine in a very historical and material sense'. In part, this gendering arises because artifacts and forms of knowledge associated with women are often simply not regarded as 'technology'. Ruth Schwartz Cowan, for example, noted in 1979 their exclusion from traditional history of technology:

> The indices to the standard histories of technology . . . do not contain a single reference, for example, to such a significant cultural artifact as the baby bottle. Here is a simple implement . . . which has transformed a fundamental human experience for vast numbers of infants and mothers, and been one of the more controversial exports of Western technology to underdeveloped countries – yet it finds no place in our histories of technology.
>
> (Cowan 1979: 52)

We explore the gendering of technology in several of the pieces in this volume and elsewhere (Wajcman 1991a). Here, what is more immediately relevant – and is arguably Cockburn's distinctive contribution to the debate around gender and technology – is her answer to the second question: is gender shaped by technology? Technology, she argues, is 'one of the formative processes of men'. The appropriation of technology by men, and the exclusion of women from many of the domains deemed technical, are processes that leave their mark in the very design of tasks and of machines, as Cockburn discusses in her article on typesetting in Part Two of this book. They are also part of the processes by which, in our society, gender is constituted. Different childhood socialization, different role models, different forms of schooling, gender segregation of occupations, different domestic responsibilities and sometimes plain historical processes of expulsion (as after the First and Second World Wars: see Summerfield 1977 and Enloe 1983, Chapter 7) have all contributed to what Cockburn describes elsewhere as 'the construction of men as strong, manually able and technologically endowed, and women as physically and technically incompetent' (1983: 203).

If gender and technology are mutually constitutive, so are ethnicity and technology, though this is a topic that has been much less thoroughly explored in recent literature. The mutual constitution is most evident in relation to that commonplace marker of ethnicity: skin colour. We end Part One of this book with an extract from the work of Richard Dyer, which can be seen as suggesting two points. First, technology has been shaped by ethnicity, in that conventional valuations of skin colour have been the benchmark in the development of photographic and film technologies: these typically are fine-tuned so that they provide pleasing renditions of 'white' faces, sometimes to the detriment of the reproduction of other skin colourations. Second, technology has helped constitute ethnicity, in that conventional hierarchies of desirability have been reinforced by the reproduction of 'white' faces as 'pleasing flesh tones' rather than (as often happened with 'untuned' photographic technologies) as unpleasantly 'beefy'.

▶ NOTES

1   The classic critique of White is Hilton and Sawyer (1963).

2   See, for example, Douglas and Wildavsky (1982), Luhmann (1993), Adams (1995), Stern and Fineberg (1996). Woolgar (1991: 31–2) misunderstands our discussion of the physical and biological effects of technology in the introduction to the first edition of this book. We do *not* suggest that, in his words, 'some technologies do in fact have self-evident attributes and capacities'; MacKenzie (1990, 1996a, 1996b) argues the opposite, that knowedge of even the most 'technical' attributes of a technology can be analysed sociologically. Our point is that the attributes and effects of *all* technologies are *both* socially negotiated and real (physical, material, biological). An emphasis on the first does not imply indifference to the second. Were we to fall into the latter, we would indeed be guilty of the amoral and apolitical position attributed to students of 'the social construction of technology' by Winner (1993). Both lay and professional perceptions of technological risk, for example, are shaped by social and psychological processes, but to assert this is not to deny (nor to be indifferent to) the possibility of real, material harm.

3   We owe this useful way of formulating this key distinction to Edgerton (1993).

4   For a material, even a technological, history of modern physics, see Galison (1997).

5   'Technology' is derived from the Greek *tekhne*, meaning art, craft, or skill, and *logos*, meaning word or knowledge. The modern usage of 'technology' to include artifacts as well as knowledge of those artifacts is thus etymologically incorrect, but so entrenched that we have chosen not to resist it. While our emphasis in this book is on the social shaping of artifacts, we are of course vitally interested in technological knowledge as well. For an outline framework for the sociological analysis of this, see MacKenzie (1996b).

6   See Barnes and Edge (1982, Part 3), Staudenmaier (1980, 1985), and the interesting studies by Aitken of the origins of the radio (1976) and by Cardwell of the development of the science of heat (1971).

7   For two interesting and wide-ranging discussions of this, see Schon (1963) and Edge (1974–5).

8   For the last of these, see Nye (1990).

9 See, for example, Nelson and Winter (1974), Nelson *et al.* (1976), Nelson and Winter (1982), Coombs *et al.* (1987), Dosi *et al.* (1990) and Stoneman (1995). The neoclassical model has also been used by economic historians to explain choice of technology; see Sandberg (1969), the review of the literature of Uselding (1977) and the critique of Sandberg by Lazonick (1981).

10 See, for example, Gansler (1982); for an interesting and detailed discussion of the legitimatory role of cost estimates even in an 'efficient' project, see Sapolsky (1972: 160–91).

11 It is worth rethinking the example of the stirrup and feudalism with this in mind. Even if White is right in the overall features of his account, any causal effect of the stirrup comes not from technology as such but from military competition. For it was surely military competition that, in White's picture, propagated armed shock combat and the feudal system, as those societies that adopted them triumphed over those that did not.

12 See, for example, David (1992) and Arthur (1994).

13 It is interesting to note the analogy that Arthur drew at the end of his article with problems in weather forecasting. Implicitly, he was referring to theories of 'chaos' in advance of the wider vogue that the notion came to enjoy.

14 See, in addition to the sources cited in the extract, Winner (1993).

15 For an accessible introduction to ethnomethodology, particularly in its relations to more traditional sociology, see Heritage (1984).

16 In autumn 1994, an error was discovered in the implementation of floating-point division in Intel's new Pentium[TM] processor. It was an error that would be triggered only rarely, and 'bugs' in early-release microprocessors are common events: previous generations of Intel chips had had similar errors without provoking much upset. However, the divide bug was seized upon in the Internet newsgroup, **comp.sys.intel:** examples of divisions that would trigger it were circulated; material critical of Intel's originally unalarmed response were circulated widely; bad newspaper and television publicity followed. Intel had eventually to scrap existing stocks of the chip and offer users free replacements, and had to set aside $475 million to cover the costs of doing this. Subsequently, it has been making increasing use of formal, deductive techniques in chip development, techniques which are widely believed to offer the prospect of a reduced risk of bugs. The role of **comp.sys.intel**, it seems to us, was as a 'society' bringing together people with an interest in the detailed behaviour of Intel chips. Without electronic communication it is hard to imagine a sufficient critical mass of people coalescing around such an esoteric matter.

1 ▶ Do artifacts have politics?

Langdon Winner

In controversies about technology and society, there is no idea more pro-
vocative than the notion that technical things have political qualities. At
issue is the claim that the machines, structures, and systems of modern
material culture can be accurately judged not only for their contributions of
efficiency and productivity, not merely for their positive and negative
environmental side effects, but also for the ways in which they can
embody specific forms of power and authority. Since ideas of this kind
have a persistent and troubling presence in discussions about the meaning
of technology, they deserve explicit attention. . . .[1]

It is no surprise to learn that technical systems of various kinds are deeply
interwoven in the conditions of modern politics. The physical arrangements of industrial production, warfare, communications, and the like
have fundamentally changed the exercise of power and the experience of
citizenship. But to go beyond this obvious fact and to argue that certain
technologies *in themselves* have political properties seems, at first glance,
completely mistaken. We all know that people have politics, not things. To
discover either virtues or evils in aggregates of steel, plastic, transistors,
integrated circuits, and chemicals seems just plain wrong, a way of mystify-
ing human artifice and of avoiding the true sources, the human sources of
freedom and oppression, justice and injustice. Blaming the hardware
appears even more foolish than blaming the victims when it comes to
judging conditions of public life.

Hence, the stern advice commonly given those who flirt with the notion
that technical artifacts have political qualities: What matters is not technology itself, but the social or economic system in which it is embedded.
This maxim, which in a number of variations is the central premise of
a theory that can be called the social determination of technology, has
an obvious wisdom. It serves as a needed corrective to those who focus

uncritically on such things as 'the computer and its social impacts' but who fail to look behind technical things to notice the social circumstances of their development, deployment, and use. This view provides an antidote to naive technological determinism – the idea that technology develops as the sole result of an internal dynamic, and then, unmediated by any other influence, molds society to fit its patterns. Those who have not recognized the ways in which technologies are shaped by social and economic forces have not gotten very far.

But the corrective has its own shortcomings; taken literally, it suggests that technical *things* do not matter at all. Once one has done the detective work necessary to reveal the social origins – power holders behind a particular instance of technological change – one will have explained everything of importance. This conclusion offers comfort to social scientists: it validates what they had always suspected, namely that there is nothing distinctive about the study of technology in the first place. Hence, they can return to their standard models of social power – those of interest group politics, bureaucratic politics, Marxist models of class struggle, and the like – and have everything they need. The social determination of technology is, in this view, essentially no different from the social determination of, say, welfare policy or taxation.

There are, however, good reasons technology has of late taken on a special fascination in its own right for historians, philosophers, and political scientists; good reasons the standard models of social science only go so far in accounting for what is most interesting and troublesome about the subject. In another place I have tried to show why so much of modern social and political thought contains recurring statements of what can be called a theory of technological politics, an odd mongrel of notions often crossbred with orthodox liberal, conservative, and socialist philosophies.[2] The theory of technological politics draws attention to the momentum of large-scale sociotechnical systems, to the response of modern societies to certain technological imperatives, and to the all too common signs of the adaptation of human ends to technical means. In so doing it offers a novel framework of interpretation and explanation for some of the more puzzling patterns that have taken shape in and around the growth of modern material culture. One strength of this point of view is that it takes technical artifacts seriously. Rather than insist that we immediately reduce everything to the interplay of social forces, it suggests that we pay attention to the characteristics of technical objects and the meaning of those characteristics. A necessary complement to, rather than a replacement for, theories of the social determination of technology, this perspective identifies certain technologies as political phenomena in their own right. It points us back, to borrow Edmund Husserl's philosophical injunction, *to the things themselves*.

In what follows I shall offer outlines and illustrations of two ways in which artifacts can contain political properties. First are instances in which the invention, design, or arrangement of a specific technical device or system becomes a way of settling an issue in a particular community. Seen in the proper light, examples of this kind are fairly straightforward and easily understood. Second are cases of what can be called inherently political technologies, man-made systems that appear to require, or to be strongly compatible with, particular kinds of political relationships. Arguments about cases of this kind are much more troublesome and closer to the heart of the matter. By 'politics,' I mean arrangements of power and authority in human associations as well as the activities that take place within those arrangements. For my purposes, 'technology' here is understood to mean all of modern practical artifice,[3] but to avoid confusion I prefer to speak of technolog*ies*, smaller or larger pieces or systems of hardware of a specific kind. My intention is not to settle any of the issues here once and for all, but to indicate their general dimensions and significance.

▶ TECHNICAL ARRANGEMENTS AS FORMS OF ORDER

Anyone who has traveled the highways of America and has become used to the normal height of overpasses may well find something a little odd about some of the bridges over the parkways on Long Island, New York. Many of the overpasses are extraordinarily low, having as little as nine feet of clearance at the curb. Even those who happened to notice this structural peculiarity would not be inclined to attach any special meaning to it. In our accustomed way of looking at things like roads and bridges we see the details of form as innocuous, and seldom give them a second thought.

It turns out, however, that the two hundred or so low-hanging overpasses on Long Island were deliberately designed to achieve a particular social effect. Robert Moses, the master builder of roads, parks, bridges, and other public works from the 1920s to the 1970s in New York, had these overpasses built to specifications that would discourage the presence of buses on his parkways. According to evidence provided by Robert A. Caro in his biography of Moses, the reasons reflect Moses's social-class bias and racial prejudice. Automobile-owning whites of 'upper' and 'comfortable middle' classes, as he called them, would be free to use the parkways for recreation and commuting. Poor people and blacks, who normally used public transit, were kept off the roads because the twelve-foot tall buses could not get through the overpasses. One consequence was to limit access of racial minorities and low-income groups to Jones Beach, Moses's widely acclaimed public park. Moses made doubly sure of this result by vetoing a proposed extension of the Long Island Railroad to Jones Beach.[4]

As a story in recent American political history, Robert Moses's life is fascinating. His dealings with mayors, governors, and presidents, and his careful manipulation of legislatures, banks, labor unions, the press, and public opinion are all matters that political scientists could study for years. But the most important and enduring results of his work are his technologies, the vast engineering projects that give New York much of its present form. For generations after Moses has gone and the alliances he forged have fallen apart, his public works, especially the highways and bridges he built to favor the use of the automobile over the development of mass transit, will continue to shape that city. Many of his monumental structures of concrete and steel embody a systematic social inequality, a way of engineering relationships among people that, after a time, becomes just another part of the landscape. As planner Lee Koppleman told Caro about the low bridges on Wantagh Parkway, 'The old son-of-a-gun had made sure that buses would *never* be able to use his goddamned parkways.'[5]

Histories of architecture, city planning, and public works contain many examples of physical arrangements that contain explicit or implicit political purposes. One can point to Baron Haussmann's broad Parisian thoroughfares, engineered at Louis Napoleon's direction to prevent any recurrence of street fighting of the kind that took place during the revolution of 1848. Or one can visit any number of grotesque concrete buildings and huge plazas constructed on American university campuses during the late 1960s and early 1970s to defuse student demonstrations. Studies of industrial machines and instruments also turn up interesting political stories, including some that violate our normal expectations about why technological innovations are made in the first place. If we suppose that new technologies are introduced to achieve increased efficiency, the history of technology shows that we will sometimes be disappointed. Technological change expresses a panoply of human motives, not the least of which is the desire of some to have dominion over others, even though it may require an occasional sacrifice of cost-cutting and some violence to the norm of getting more from less.

One poignant illustration can be found in the history of nineteenth century industrial mechanization. At Cyrus McCormick's reaper manufacturing plant in Chicago in the middle 1880s, pneumatic molding machines, a new and largely untested innovation, were added to the foundry at an estimated cost of $500,000. In the standard economic interpretation of such things, we would expect that this step was taken to modernize the plant and achieve the kind of efficiencies that mechanization brings. But historian Robert Ozanne has shown why the development must be seen in a broader context. At the time, Cyrus McCormick II was engaged in a battle with the National Union of Iron Molders. He saw the addition of the new machines as a way to 'weed out the bad element among the men,' namely, the skilled workers who had organized the union local in Chicago.[6] The new machines, manned by unskilled labor, actually produced inferior castings at a higher cost than the earlier process. After three years of use the machines were, in fact, abandoned, but by that time they had served their purpose – the destruction of the union. Thus, the story of these technical developments at the McCormick factory cannot be understood adequately outside the record of workers' attempts to organize, police repression of the labor movement in Chicago during that period, and the events surrounding the bombing at Haymarket Square. Technological history and American political history were at that moment deeply intertwined.

In cases like those of Moses's low bridges and McCormick's molding machines, one sees the importance of technical arrangements that precede the *use* of the things in question. It is obvious that technologies can be used in ways that enhance the power, authority, and privilege of some over others, for example, the use of television to sell a candidate. To our accustomed way of thinking, technologies are seen as neutral tools that can be used well or poorly, for good, evil, or something in between. But we usually do not stop to inquire whether a given device might have been designed and built in such a way that it produces a set of consequences logically and temporally *prior* to any of its professed uses. Robert Moses's bridges, after all, were used to carry automobiles from one point to another; McCormick's machines were used to make metal castings; both technologies, however, encompassed purposes far beyond their immediate use. If our moral and political language for evaluating technology includes only categories having to do with tools and uses, if it does not include attention to the meaning of the designs and arrangements of our artifacts, then we will be blinded to much that is intellectually and practically crucial.

Because the point is most easily understood in the light of particular intentions embodied in physical form, I have so far offered illustrations that seem almost conspiratorial. But to recognize the political dimensions in the shapes of technology does not require that we look for conscious conspiracies or malicious intentions. The organized movement of handicapped people in the United States during the 1970s pointed out the countless ways in which machines, instruments, and structures of common use – buses, buildings, sidewalks, plumbing fixtures, and so forth – made it impossible for many handicapped persons to move about freely, a condition that systematically excluded them from public life. It is safe to say that designs unsuited for the handicapped arose more from long-standing neglect than from anyone's active intention. But now that the issue has been raised for public attention, it is evident that justice requires a remedy. A whole range of artifacts are now being redesigned and rebuilt to accommodate this minority. . . .

I would offer the following general conclusions. The things we call 'technologies' are ways of building order in our world. Many technical devices and systems important in everyday life contain possibilities for many different ways of ordering human activity. Consciously or not, deliberately or inadvertantly, societies choose structures for technologies that influence how people are going to work, communicate, travel, consume, and so forth over a very long time. In the processes by which structuring decisions are made, different people are differently situated and possess unequal degrees of power as well as unequal levels of awareness. By far the greatest latitude of choice exists the very first time a particular instrument, system, or technique is introduced. Because choices tend to become strongly fixed in material equipment, economic investment, and social habit, the original flexibility vanishes for all practical purposes once the initial commitments are made. In that sense technological innovations are similar to legislative acts or political foundings that establish a framework for public order that will endure over many generations. For that reason, the same careful attention one would give to the rules, roles, and relationships of politics must also be given to such things as the building of highways, the creation of television networks, and the tailoring of seemingly insignificant features on new machines. The issues that divide or unite people in society are settled not only in the institutions and practices of politics proper, but also, and less obviously, in tangible arrangements of steel and concrete, wires and transistors, nuts and bolts.

▶ INHERENTLY POLITICAL TECHNOLOGIES

None of the arguments and examples considered thus far address a stronger, more troubling claim often made in writings about technology and society – the belief that some technologies are by their very nature political in a specific way. According to this view, the adoption of a given technical system unavoidably brings with it conditions for human relationships that have a distinctive political cast – for example, centralized or decentralized, egalitarian or inegalitarian, repressive or liberating. This is ultimately what is at stake in assertions like those of Lewis Mumford that two traditions of technology, one authoritarian, the other democratic, exist side by side in Western history.[7] In all the cases I cited above the technologies are relatively flexible in design and arrangement, and variable in their effects. Although one can recognize a particular result produced in a particular setting, one can also easily imagine how a roughly similar device or system might have been built or situated with very much different political consequences. The idea we must now examine and evaluate is that certain kinds of technology do not allow such flexibility, and that to choose them is to choose a particular form of political life. . . .

Arguments to the effect that technologies are in some sense inherently political have been advanced in a wide variety of contexts, far too many to summarize here. In my reading of such notions, however, there are two basic ways of stating the case. One version claims that the adoption of a given technical system actually *requires* the creation and maintenance of a particular set of social conditions as the operating environment of that system. [This] view is offered by a contemporary writer who holds that 'if you accept nuclear power plants, you also accept a techno-scientific-industrial-military elite. Without these people in charge, you could not have nuclear power.'[8] In this conception, some kinds of technology require their social environments to be structured in a particular way in much the same sense that an automobile requires wheels in order to run. The thing could not exist as an effective operating entity unless certain social as well as material conditions were met. The meaning of 'required' here is that of practical (rather than logical) necessity. Thus, Plato thought it a practical necessity that a ship at sea have one captain and an unquestioningly obedient crew.

A second, somewhat weaker, version of the argument holds that a given kind of technology is strongly *compatible with*, but does not strictly require, social and political relationships of a particular stripe. Many advocates of solar energy now hold that technologies of that variety are more compatible with a democratic, egalitarian society than energy systems based on coal, oil, and nuclear power; at the same time they do not maintain that anything about solar energy requires democracy. Their case is, briefly, that solar energy is decentralizing in both a technical and political sense: technically speaking, it is vastly more reasonable to build solar systems in a disaggregated, widely distributed manner than in large-scale centralized plants; politically speaking, solar energy accommodates the attempts of individuals and local communities to manage their affairs effectively because they are dealing with systems that are more accessible, comprehensible, and controllable than huge centralized sources. In this view, solar energy is desirable not only for its economic and environmental benefits, but also for the salutary institutions it is likely to permit in other areas of public life. . . .[9]

There are, then, several different directions that arguments of this kind can follow. Are the social conditions predicted said to be required by, or strongly compatible with, the workings of a given technical system? Are those conditions internal to that system or external to it (or both)? Although writings that address such questions are often unclear about what is being asserted, arguments in this general category do have an important presence in modern political discourse. They enter into many attempts to explain how changes in social life take place in the wake of technological innovation. More importantly, they are often used to buttress attempts to justify or criticize proposed courses of action involving new technology. By offering distinctly political reasons for or against the adoption of a particular technology, arguments of this kind stand apart from more commonly employed, more easily quantifiable claims about economic costs and benefits, environmental impacts, and possible risks to public health and safety that technical systems may involve. The issue here does not concern how many jobs will be created, how much income generated, how many pollutants added, or how many cancers produced. Rather, the issue has to do with ways in which choices about technology have important consequences for the form and quality of human associations.

If we examine social patterns that comprise the environments of technical systems, we find certain devices and systems almost invariably linked to specific ways of organizing power and authority. The important question is: Does this state of affairs derive from an unavoidable social response to intractable properties in the things themselves, or is it instead a pattern imposed independently by a governing body, ruling class, or some other social or cultural institution to further its own purposes?

Taking the most obvious example, the atom bomb is an inherently political artifact. As long as it exists at all, its lethal properties demand that it be controlled by a centralized, rigidly hierarchical chain of command closed to all influences that might make its workings unpredictable. The internal social system of the bomb must be authoritarian; there is no other way. The state of affairs stands as a practical necessity independent of any larger political system in which the bomb is embedded, independent of the kind of regime or character of its rulers. Indeed, democratic states must try to find ways to ensure that the social structures and mentality that characterize the management of nuclear weapons do not 'spin off' or 'spill over' into the polity as a whole.

The bomb is, of course, a special case. The reasons very rigid relationships of authority are necessary in its immediate presence should be clear to anyone. If, however, we look for other instances in which particular varieties of technology are *widely perceived* to need the maintenance of a special pattern of power and authority, modern technical history contains a wealth of examples.

Alfred D. Chandler in *The Visible Hand*, a monumental study of modern business enterprise, presents impressive documentation to defend the hypothesis that the construction and day-to-day operation of many systems of production, transportation, and communication in the nineteenth and twentieth centuries require the development of a particular social form – a large-scale centralized, hierarchical organization administered by highly skilled managers. Typical of Chandler's reasoning is his analysis of the growth of the railroads.

> Technology made possible fast, all-weather transportation; but safe, regular, reliable movement of goods and passengers, as well as the continuing maintenance and repair of locomotives, rolling stock, and track, roadbed, stations, round-houses, and other equipment, required the creation of a sizable administrative organization. It meant the employment of a set of managers to supervise these functional activities over an extensive geographical area; and the appointment of an administrative command of middle and top executives to monitor, evaluate, and coordinate the work of managers responsible for the day-to-day operations.

Throughout his book Chandler points to ways in which technologies used in the production and distribution of electricity, chemicals, and a wide range of industrial goods 'demanded' or 'required' this form of human association. 'Hence, the operational requirements of railroads demanded the creation of the first administrative hierarchies in American business.'[10]

Were there other conceivable ways of organizing these aggregates of people and apparatus? Chandler shows that a previously dominant social form, the small traditional family firm, simply could not handle the task in most cases. Although he does not speculate further, it is clear that he believes there is, to be realistic, very little latitude in the forms of power and authority appropriate within modern sociotechnical systems. The properties of many modern technologies – oil pipelines and refineries, for example – are such that overwhelmingly impressive economies of scale and speed are possible. If such systems are to work effectively, efficiently, quickly, and safely, certain requirements of internal social organization have to be fulfilled; the material possibilities that modern technologies make available could not be exploited otherwise. Chandler acknowledges that as one compares sociotechnical institutions of different nations, one sees 'ways in which cultural attitudes, values, ideologies, political systems, and social structure affect these imperatives.'[11] But the weight of argument and empirical evidence in *The Visible Hand* suggests that any significant departure from the basic pattern would be, at best, highly unlikely.

It may be that other conceivable arrangements of power and authority, for example, those of decentralized, democratic worker self-management, could prove capable of administering factories, refineries, communications systems, and railroads as well as or better than the organizations Chandler describes. Evidence from automobile assembly teams in Sweden and worker managed plants in Yugoslavia and other countries is often presented to salvage these possibilities. I shall not be able to settle controversies over this matter here, but merely point to what I consider to be their bone of contention. The available evidence tends to show that many large, sophisticated technological systems are in fact highly compatible with centralized, hierarchical managerial control. The interesting question, however, has to do with whether or not this pattern is in any sense a requirement of such systems, a question that is not solely an empirical one. The matter ultimately rests on our judgments about what steps, if any, are practically necessary in the workings of particular kinds of technology and what, if anything, such measures require of the structure of human associations. Was Plato right in saying that a ship at sea needs steering by a decisive hand and that this could only be accomplished by a single captain and an obedient crew? Is Chandler correct in saying that the properties of large-scale systems require centralized, hierarchical managerial control?

To answer such questions, we would have to examine in some detail the moral claims of practical necessity (including those advocated in the doctrines of economics) and weigh them against moral claims of other sorts, for example, the notion that it is good for sailors to participate in the command of a ship or that workers have a right to be involved in making and administering decisions in a factory. It is characteristic of societies based on large, complex technological systems, however, that moral reasons other than those of practical necessity appear increasingly obsolete, 'idealistic,' and irrelevant. Whatever claims one may wish to make on behalf of liberty, justice, or equality can be immediately neutralized when confronted with arguments to the effect: 'Fine, but that's no way to run a railroad' (or steel mill, or airline, or communications system, and so on). Here we encounter an important quality in modern political discourse and in the way people commonly think about what measures are justified in response to the possibilities technologies make available. In many instances, to say that some technologies are inherently political is to say that certain widely accepted reasons of practical necessity – especially the need to maintain crucial technological systems as smoothly working entities – have tended to eclipse other sorts of moral and political reasoning.

One attempt to salvage the autonomy of politics from the bind of practical necessity involves the notion that conditions of human association found in the internal workings of technological systems can easily be kept separate from the polity as a whole. Americans have long rested content in the belief that arrangements of power and authority inside industrial corporations, public utilities, and the like have little bearing on public institutions, practices, and ideas at large. That 'democracy stops at the factory gates' was taken as a fact of life that had nothing to do with the practice of political freedom. But can the internal politics of technology and the politics of the whole community be so easily separated? A recent study of American business leaders, contemporary exemplars of Chandler's 'visible hand of management,' found them remarkably impatient with such democratic scruples as 'one man, one vote.' If democracy doesn't work for the firm, the most critical institution in all of society, American executives ask, how well can it be expected to work for the government of a nation – particularly when that government attempts to interfere with the achievements of the firm? The authors of the report observe that patterns of authority that work effectively in the corporation become for businessmen 'the desirable model against which to compare political and economic relationships in the rest of society.'[12] While such findings are far from conclusive, they do reflect a sentiment increasingly common in the land: what dilemmas like the energy crisis require is not a redistribution of wealth or broader public participation but, rather, stronger, centralized public management – President Carter's proposal for an Energy Mobilization Board and the like.

An especially vivid case in which the operational requirements of a technical system might influence the quality of public life is now at issue in debates about the risks of nuclear power. As the supply of uranium for nuclear reactors runs out, a proposed alternative fuel is the plutonium generated as a by-product in reactor cores. Well-known objections to plutonium recycling focus on its unacceptable economic costs, its risks of environmental contamination, and its dangers in regard to the international proliferation of nuclear weapons. Beyond these concerns, however, stands another less widely appreciated set of hazards – those that involve the sacrifice of civil liberties. The widespread use of plutonium as a fuel increases the chance that this toxic substance might be stolen by terrorists, organized crime, or other persons. This raises the prospect, and not a trivial one, that extraordinary measures would have to be taken to safeguard plutonium from theft and to recover it if ever the substance were stolen. Workers in the nuclear industry as well as ordinary citizens outside could well become subject to background security checks, covert surveillance, wiretapping, informers, and even emergency measures under martial law – all justified by the need to safeguard plutonium.

Russell W. Ayres's study of the legal ramifications of plutonium recycling concludes: 'With the passage of time and the increase in the quantity of plutonium in existence will come pressure to eliminate the traditional checks the courts and legislatures place on the activities of the executive and to develop a powerful central authority better able to enforce strict safeguards.' He avers that 'once a quantity of plutonium had been stolen, the case for literally turning the country upside down to get it back would be overwhelming.' Ayres anticipates and worries about the kinds of thinking that, I have argued, characterize inherently political technologies. It is still true that, in a world in which human beings make and maintain artificial systems, nothing is 'required' in an absolute sense. Nevertheless, once a course of action is underway, once artifacts like nuclear power plants have been built and put in operation, the kinds of reasoning that justify the adaptation of social life to technical requirements pop up as spontaneously as flowers in the spring. In Ayres's words, 'Once recycling begins and the risks of plutonium theft become real rather than hypothetical, the case for governmental infringement of protected rights will seem compelling.'[13] After a certain point, those who cannot accept the hard requirements and imperatives will be dismissed as dreamers and fools.

The two varieties of interpretation I have outlined indicate how artifacts can have political qualities. In the first instance we noticed ways in which specific features in the design or arrangement of a device or system could provide a convenient means of establishing patterns of power and authority in a given setting. Technologies of this kind have a range of flexibility in the dimensions of their material form. It is precisely because they are flexible that their consequences for society must be understood with reference to the social actors able to influence which designs and arrangements are chosen. In the second instance we examined ways in which the intractable properties of certain kinds of technology are strongly, perhaps unavoidably, linked to particular institutionalized patterns of power and authority. Here, the initial choice about whether or not to adopt something is decisive in regard to its consequences. There are no alternative physical designs or arrangements that would make a significant difference; there are, further-more, no genuine possibilities for creative intervention by different social systems – capitalist or socialist – that could change the intractability of the entity or significantly alter the quality of its political effects.

To know which variety of interpretation is applicable in a given case is often what is at stake in disputes, some of them passionate ones, about the meaning of technology for how we live. I have argued a 'both/and' position here, for it seems to me that both kinds of understanding are applicable in different circumstances. Indeed, it can happen that within a particular complex of technology – a system of communication or transportation, for example – some aspects may be flexible in their possibilities for society, while other aspects may be (for better or worse) completely intractable. The two varieties of interpretation I have examined here can overlap and intersect at many points.

These are, of course, issues on which people can disagree. Thus, some proponents of energy from renewable resources now believe they have at last discovered a set of intrinsically democratic, egalitarian, communitarian technologies. In my best estimation, however, the social consequences of building renewable energy systems will surely depend on the specific con-figurations of both hardware and the social institutions created to bring that energy to us. It may be that we will find ways to turn this silk purse into a sow's ear. By comparison, advocates of the further development of nuclear power seem to believe that they are working on a rather flexible technology whose adverse social effects can be fixed by changing the design parameters of reactors and nuclear waste disposal systems. For reasons indicated above, I believe them to be dead wrong in that faith. Yes, we may be able to manage some of the 'risks' to public health and safety that nuclear power brings. But as society adapts to the more dangerous and apparently indelible features of nuclear power, what will be the long-range toll in human freedom?

My belief that we ought to attend more closely to technical objects themselves is not to say that we can ignore the contexts in which those objects are situated. A ship at sea may well require a single captain and obedient crew. But a ship out of service, parked at the dock, needs only a caretaker. To understand which technologies and which contexts are important to us, and why, is an enterprise that must involve both the study of specific technical systems and their history as well as a thorough grasp of the concepts and controversies of political theory. In our times people are often willing to make drastic changes in the way they live to accord with technological innovation at the same time they would resist similar kinds of changes justified on political grounds. If for no other reason than that, it is important for us to achieve a clearer view of these matters than has been our habit so far.

 NOTES

1 I would like to thank Merritt Roe Smith, Leo Marx, James Miller, David Noble, Charles Weiner, Sherry Turkle, Loren Graham, Gail Stuart, Dick Sclove, and Stephen Graubard for their comments and criticisms on earlier drafts of this essay. My thanks also to Doris Morrison of the Agriculture Library of the University of California, Berkeley, for her bibliographical help.

2 Langdon Winner, *Autonomous Technology: Technics-out-of-Control as a Theme in Political Thought* (Cambridge, Mass.: M.l.T. Press, 1977).

3 The meaning of 'technology' I employ in this essay does not encompass some of the broader definitions of that concept found in contemporary literature, for example, the notion of 'technique' in the writings of Jacques Ellul. My purposes here are more limited. For a discussion of the difficulties that arise in attempts to define 'technology,' see Ref. 2, pp. 8–12.

4 Robert A. Caro, *The Power Broker: Robert Moses and the Fall of New York* (New York: Random House, 1974), pp. 318, 481, 514, 546, 951–958.

5 *Ibid.*, p. 952.

6 Robert Ozanne, *A Century of Labor-Management Relations at McCormick and International Harvester* (Madison, Wis.: University of Wisconsin Press, 1967), p. 20.

7 Lewis Mumford, 'Authoritarian and Democratic Technics,' *Technology and Culture*, 5 (1964): 1–8.

8 Jerry Mander, *Four Arguments for the Elimination of Television* (New York: William Morrow, 1978), p. 44.

9 See, for example, Robert Argue, Barbara Emanuel, and Stephen Graham, *The Sun Builders: A People's Guide to Solar, Wind and Wood Energy in Canada* (Toronto: Renewable Energy in Canada, 1978). 'We think decentralization is an implicit component of renewable energy; this implies the decentralization of energy systems, communities and of power. Renewable energy doesn't require mammoth generation sources of disruptive transmission corridors. Our cities and towns, which have been dependent on centralized energy supplies, may be able

to achieve some degree of autonomy, thereby controlling and administering their own energy needs' (p. 16).

10 Alfred D. Chandler, Jr., *The Visible Hand: The Managerial Revolution in American Business* (Cambridge, Mass.: Belknap, Harvard University Press, 1977), p. 244.

11 *Ibid.*, p. 500.

12 Leonard Silk and David Vogel, *Ethics and Profits: The Crisis of Confidence in American Business* (New York: Simon and Schuster, 1976), p. 191.

13 Russel W. Ayres, 'Policing Plutonium: The Civil Liberties Fallout,' *Harvard Civil Rights – Civil Liberties Law Review*, 10 (1975): 374, 413–4, 443.

2 ▶ Modest__Witness @ Second__Millennium

Donna J. Haraway

All language, including mathematics, is figurative, that is, made of tropes [non-literal uses of language, such as metaphor]; constituted by bumps that make us swerve from literal-mindedness. I emphasize figuration to make explicit and inescapable the tropic quality of all material-semiotic processes, especially in technoscience. [A semiotic process is, loosely, one in which meaning is created. Haraway uses the term 'technoscience' in preference to 'science and technology' because she feels the two to have become inseparable.] For example, think of a small set of objects into which lives and worlds are built – chip, gene, seed, fetus, database, bomb, race, brain, ecosystem. This mantralike list is made up of imploded atoms or dense nodes that explode into entire worlds of practice. The chip, seed, or gene is simultaneously literal and figurative. We inhabit and are inhabited by such figures that map universes of knowledge, practice and power. . . . The globalization of the world, of 'planet Earth,' is a semiotic-material production of some forms of life rather than others. Technoscience is the story of such globalization; it is the travelogue of distributed, heterogeneous, linked, sociotechnical circulations that craft the world as a net called the global. The cyborg [cybernetic organism; more generally, 'cyborg' refers to a 'fusion of the organic and the technical'. See below.] life forms that inhabit the recently congealed planet Earth – the 'whole earth' of eco-activists and green commodity catalogs – gestated in a historically specific technoscientific womb. Consider, for example . . . :

1 The apparatuses of twentieth-century military conflicts, embedded in repeated world wars; decades of cold war; nuclear weapons and their institutional matrix in strategic planning, endless scenario production, and simulations in think tanks such as RAND; the immune system-like networking strategies for postcolonial global control inscribed in low-

intensity-conflict doctrines; and post-Cold War, simultaneous-multiple-war-fighting strategies depending on rapid massive deployment, concentrated control of information and communications, and high-intensity, subnuclear precision weapons (Helsel 1993; Gray 1991; Edwards 1996).

2 The apparatuses of hypercapitalist market traffic and flexible accumulation strategies, all relying on stunning speeds and powers of manipulation of scale, especially miniaturization, which characterize the paradigmatic 'high-technology' transnational corporations (Harvey 1989; Virilio and Lotringer 1983; Martin 1992).

3 The apparatuses of production of that technoscientific planetary habitat space called the ecosystem, with its constitutive birth pangs in resource management practices in such institutions as national fisheries in the 1920s and 1930s; in post-World-War II theoretical fascination with all things cybernetic; in the Atomic Energy Commission-mediated research projects in the 1950s for tracing radioisotopes through food chains in the Pacific ocean; in 1970s global modeling practices indebted to the Club of Rome and to international projects such as the United Nations Educational, Scientific, and Cultural Organization's (UNESCO) Man and the Biosphere program; and in the early salvos of widespread 'green war' as a dominant New World Order security concern, with its diplomatic forms played out in 1992 at the Earth Summit in Rio de Janeiro (Escobar 1994; Taylor and Buttel 1992).

4 The apparatuses of production of globalized, extraterrestrial, everyday consciousness in the planetary pandemic of multisite, multimedia, multi-species, multicultural, cyborgian entertainment events such as *Star Trek*, *Blade Runner*, *Terminator*, *Alien*, and their proliferating sequelae in the daily information stream, embedded in transnational, U.S.-dominated, broad-spectrum media conglomerates, such as those forged by the mergers of Time-Warner with CNN and of the Disney universe with Capital Cities, owner of CBS (Gabilondo 1991; Sofia 1992).

The offspring of these technoscientific wombs are cyborgs – imploded germinal entities, densely packed condensations of worlds, shocked into being from the force of the implosion of the natural and the artificial, nature and culture, subject and object, machine and organic body, money and lives, narrative and reality. . . .

So, what kinds of kin are allied in the proprietary forms of life in these days near the end of the Second Christian Millennium? How do we, who inhabit such stories, make psychic and commercial investments in forms of life, where the lines among human, machine, and organic nature are highly permeable and eminently revisable? How useful is my abiding suspicion that 'biology' – the historically specific, congealed embodiments in the world as well as the techno-scientific discourse positing such bodies – is an accumulation strategy? The point is less disreputable if I write that 'biotechnology' – both the discourse and the body constituted as a biotechnics – is an accumulation strategy. But much of what is accumulated is more strange than capital, more kind than alien, more alluring than gold. . . .

In what gets politely called modernity and its afterlife (or half-life),

accelerated production of natural knowledge pervasively structures commerce, industry, healing, community, war, sex, literacy, entertainment,
and worship. The world-building alliances of humans and nonhumans in
technoscience shape subjects and objects, subjectivity and objectivity,
action and passion, inside and outside in ways that enfeeble other modes
of speaking about science and technology. In short, technoscience is about
worldly, materialized, signifying and significant power. . . .

I belong to the 'culture' whose members answer to the 'hey, you!' issuing
from technoscience's authoritative practices and discourses. My people
answer that 'hey, you!' in many ways: We squirm, organize, revel, decry,
preach, teach, deny, equivocate, analyze, resist, collaborate, contribute,
denounce, expand, placate, withhold. The only thing my people cannot
do in response to the meanings and practices that claim us body and soul is
remain neutral. We must cast our lot with some ways of life on this planet,
and not with other ways. We cannot pretend we live on some other planet
where the cyborg was never spat out of the womb-brain of its war-besotted
parents in the middle of the last century of the Second Christian Millennium.

The cyborg is a cybernetic organism, a fusion of the organic and the
technical forged in particular, historical, cultural practices. Cyborgs are not
about the Machine and the Human, as if such Things and Subjects universally existed. Instead, cyborgs are about specific historical machines and
people in interaction that often turns out to be painfully counterintuitive
for the analyst of technoscience. The term cyborg was coined by Manfred
Clynes and Nathan Kline (1960) to refer to the enhanced man who could
survive in extraterrestrial environments. They imagined the cyborgian
man-machine hybrid would be needed in the next great technohumanist
challenge – space flight. A designer of physiological instrumentation and
electronic data-processing systems, Clynes was the chief research scientist
in the Dynamic Simulation Laboratory at Rockland State Hospital in New
York. Director of research at Rockland State, Kline was a clinical psy-
chiatrist. Their article was based on a paper the authors presented at the
Psychophysiological Aspects of Space Flight Symposium sponsored by the
U.S. Air Force School of Aviation Medicine in San Antonio, Texas. Enrap-
tured with cybernetics, Clynes and Kline thought of cyborgs as 'self-regulating man-machine systems' (1960:27). One of their first cyborgs was a
standard white laboratory rat implanted with an osmotic pump designed
to inject chemicals continuously. Exchanging knowing glances with their
primate kin, rodents will reappear in this essay at every turn. Beginning
with the rats who stowed away on the masted ships of Europe's age of
exploration, rodents have gone first into the unexplored regions in the great
travel narratives of Western technoscience.

Consequently, my people are akin to field mice who have entered the
anomaly in evolutionary space – a wormhole – called the laboratory. Like
the science-fictional wormhole in an episode of the television show *Deep
Space Nine*, the laboratory continues to suck us into uncharted regions
of technical, cultural, and political space. Passing through the wormhole of
technoscience, the field mice emerge as the finely tailored laboratory

rodents – model systems, animate tools, research material, self-acting organic-technical hybrids – through whose eyes I write this essay. Those mutated murine eyes give me my ethnographic point of view. Cyborg anthropology attempts to refigure provocatively the border relations among specific humans, other organisms, and machines. The interface between specifically located people, other organisms, and machines turns out to be an excellent field site for ethnographic inquiry into what counts as self-acting and as collective empowerment. I call that field site the culture and practice of technoscience. The optical tube of technoscience transports my startled gaze from its familiar, knowing, human orbs into the less certain eye sockets of an artifactual rodent, a primal cyborg figure for the dramas of technoscience. I want to use the beady little eyes of a laboratory mouse to stare back at my fellow mammals, my hominid kin, as they incubate themselves and their human and nonhuman offspring in a technoscientific culture medium.

The relocated gaze forces me to pay attention to kinship. Who are my kin in this odd world of promising monsters, vampires, surrogates, living tools, and aliens? How are natural kinds identified in the realms of late-twentieth-century technoscience? What kinds of crosses and offspring count as legitimate and illegitimate, to whom and at what cost? Who are my familiars, my siblings, and what kind of livable world are we trying to build? . . .

A transgenic organism contains genes transplanted from one strain or species – or even across taxonomic kingdoms, for example, from fish to tomatoes, fireflies to tobacco, bacteria to humans, or vice versa – to another. Transgenic border-crossing signifies serious challenges to the 'sanctity of life' for many members of Western cultures, which historically have been obsessed with racial purity, categories authorized by nature, and the well-defined self. The distinction between nature and culture in Western so-cieties has been a sacred one; it has been at the heart of the great narratives of salvation history and their genetic transmutation into sagas of secular progress. What seems to be at stake is this culture's stories of the human place in nature, that is, genesis and its endless repetitions. And Western intellectuals, perhaps especially natural scientists and philosophers, have historically been particularly likely to take their cultural stories for universal realities. It is a mistake in this context to forget that anxiety over the pollution of lineages is at the origin of racist discourse in European cultures as well as at the heart of linked gender and sexual anxiety. The discourses of transgression get all mixed up in the body of nature. Transgressive border-crossing pollutes lineages – in a transgenic organism's case, the lineage of nature itself – transforming nature into its binary opposite, culture. The line between the acts, agents, and products of divine creation and human engineering has given way in the sacred-secular border zones of molecular genetics and biotechnology. The revolutionary continuities between natu-ral kinds instaurated by the theory of biological evolution seem flaccid compared to the rigorous couplings across taxonomic kingdoms (not to mention nations and companies) produced daily in the genetic laboratory.

In opposing the production of transgenic organisms, and especially opposing their patenting and other forms of private commercial exploitation, committed activists appeal to notions such as the integrity of natural kinds and the natural *telos* or self-defining purpose of all life forms.[1] From this perspective, to mix and match genes as if organisms were legitimate raw material for redesign is to violate natural integrity at its vital core. Transferring genes between species transgresses natural barriers, compromising species integrity. These same activists and others also emphasize many other arguments for opposition to various biotechnological practices in the New World Order, Inc. The objections include increasing capital concentration and the monopolization of the means of life, reproduction, and labor; appropriation of the commons of biological inheritance as the private preserve of corporations; the global deepening of inequality by region, nation, race, gender, and class; erosion of indigenous peoples' self-determination and sovereignty in regions designated as biodiverse while indigenous lands and bodies become the object of intense gene prospecting and proprietary development; inadequately assessed and potentially dire environmental and health consequences; misplaced priorities for technoscientific investment funds; propagation of distorted and simplistic scientific explanations, such as genetic determinism; intensified cruelty to and domination over animals; depletion of biodiversity; and the undermining of established practices of human and nonhuman life, culture, and production without engaging those most affected in democratic decision-making. I take all of those objections very seriously, and all of them are taken up, if inadequately, in this book, but I do not think simply naming the concerns either decides the direction of effects or describes the cross-cultural polyphony through which scientific practice is constituted worldwide. Effects and practices are multilayered and context-specific, and it is too easy for all parties to fall into dogma where fundamental cultural and material values are both not shared and at stake. What must not be lost from sight in all of this complexity, however, is that power, profit, and bodily rearrangements are at the heart of biotechnology as a global practice. The stakes are immense, just as they are in nuclear culture. . . .

For the moment, however, I want to focus only on the Western theme of purity of type, natural purposes, and transgression of sacred boundaries. The history and current politics of racial and immigration discourses in Europe and the United States ought to set off acute anxiety in the presence of these supposedly high ethical and ontological themes. I cannot help but hear in the biotechnology debates the unintended tones of fear of the alien and suspicion of the mixed. In the appeal to intrinsic natures, I hear a mystification of kind and purity akin to the doctrines of white racial hegemony and U.S. national integrity and purpose that so permeate North American culture and history. I know that this appeal to sustain other organisms' inviolable, intrinsic natures is intended to affirm their difference from humanity and their claim on lives lived on their terms and not 'man's.' The appeal aims to limit turning all the world into a resource for human appropriation. But it is a problematic argument resting on unconvincing biology. History is erased, for other organisms as well as for humans, in the doctrine of types and intrinsic purposes, and a kind of timeless stasis in nature is piously narrated. The ancient, cobbled-together, mixed-up history of living beings, whose long tradition of genetic exchange will be the envy of industry for a long time to come, gets short shrift. More fundamentally, in the midst of a nation where race is everywhere reproduced and enforced, everywhere unspeakable and euphemized, and everywhere deferred and treated obliquely – as in talk of drug wars, urban underclasses, diversity, illegal aliens, wilderness preservation, terrorist viruses, immune defenses against invaders, and crack babies – I cannot hear discussion of disharmonious crosses among organic beings and of implanted alien genes without hearing a racially inflected and xenophobic symphony. Located in the belly of the monster, I find the discourses of natural harmony, the nonalien, and purity unsalvageable for understanding our genealogy in the New World Order, Inc. Like it or not, I was born kin to Pu^{239} [plutonium-239, the human-made element used in atomic bombs] and to transgenic, transspecific, and transported creatures of all kinds; that is the family for which and to whom my people are accountable. It will not help – emotionally, intellectually, morally, or politically – to appeal to the natural and the pure.

Perhaps it is perverse for me to hear the dangers of racism in the *opposition* to genetic engineering and especially transgenics at just the moment when national and international coalitions of indigenous, consumer, feminist, environmental, and development nongovernmental organizations have formed to oppose 'patenting, commercialization and expropriation of human, animal and plant genetic materials.'[2] Although the moral, scientific, and economic issues are far from simple, I oppose patenting of animals, human genes, and much plant genetic material. Genes for profit are not equal to science itself, or to economic health. Genetic sciences and politics are at the heart of critical struggles for equality, democracy, and sustainable life. The global commodification of genetic resources is a political and scientific emergency, and indigenous people are among the key actors in biopolitics, just as they have had to be in nuclear culture. But the tendency by the political 'left' – my area of the political spectrum – to collapse molecular genetics, biotechnology, profit, and exploitation into one undifferentiated mass is at least as much of a mistake as the mirror-image reduction by the 'right' of biological – or informational – complexity to the gene and its avatars [incarnations], including the dollar. . . .

Totipotent stem cells are those cells in an organism that retain the capacity to differentiate into any kind of cell. Stem cells can regenerate the whole array of cell types possible for that life form. Stem cells are the nodes in which the potential of entire worlds is concentrated.

Objects like the fetus, chip/computer, gene, race, ecosystem, brain, database, and bomb are stem cells of the technoscientific body. Each of these curious objects is a recent construct or material-semiotic 'object of knowledge', forged by heterogeneous practices in the furnaces of technoscience. To be a construct does NOT mean to be unreal or made up; quite the opposite. Out of each of these nodes or stem cells, sticky threads lead to every nook and cranny of the world. Which threads to follow is an analytical, imaginative, physical, and political choice. I am committed to showing how each of these stem cells is a knot of knowledge-making practices, industry and commerce, popular culture, social struggles, psychoanalytic formations, bodily histories, human and nonhuman actions, local and global flows, inherited narratives, new stories, syncretic technical/cultural processes, and more.

For example, a seed contains inside its coat the history of practices such as collecting, breeding, marketing, taxonomizing, patenting, biochemically analyzing, advertising, eating, cultivating, harvesting, celebrating, and starving. A seed produced in the biotechnological institutions now spread around the world contains the specifications for labor systems, planting calendars, pest-control procedures, marketing, land holding, and beliefs about hunger and well-being. Similarly, in Joseph Dumit's argument, a database is a technical and utopic object that structures future accessibility. A database 'is an ideal place where all elements are equal in the grid – and everyone can access all of them.'[3] The database is a condensed site for contestations over technoscientific versions of democracy and freedom. Both the genome and the brain are databases – literally – built in the experimental, multidisciplinary, documentary, proprietary, information-management, and other practices of the Human Genome Project and the Human Brain Mapping Project.[4]

I cannot follow here each of my stem cells, much less the much larger set that would be needed for the excessive account of technoscience that I crave. But I try to work out at least some of the knots that constitute genes, databases, chips/computers, seeds, cyborgs, races, and fetuses. My accounts are clearly not exhaustive, nor are they rigorously causal, but they are intended to be more than merely suggestive about the connective tissues, lubricants, codes, and actors in the worlds we must care about. The articulations among the stem cells, and within each of them, are links that matter in what gets affectionately called the 'real world.' How do technoscientific stem cells link up with each other in expected and unexpected ways and differentiate into entire worlds and ways of life? How do the differently situated human and nonhuman actors and actants encounter each other in interactions that materialize worlds in some forms rather than others? My purpose is to argue for a practice of situated knowledges in the worlds of technoscience, worlds whose fibers infiltrate deep and wide throughout the tissues of the planet, including the flesh of our personal bodies.

▶ NOTES

1 Jeremy Rifkin (1984a; 1984b) and his Foundation for Economic Trends and Michael Fox (1983; 1992) have been especially outspoken about purity of type and natural integrity. See also Krimsky (1991: 50–57) and OTA (1989: 98–102, 127–38). Rifkin leads the opposition to Calgene's Flavr Savr tomato and Monsanto's genetically engineered bovine growth hormone under the banner of the Pure Food

Campaign. Pure food is a curious concept to invoke for the tomato, a member of the deadly nightshade family. An American fruit by origin, the tomato was imported into Europe in the sixteenth century but was regarded there as toxic and grown as an ornamental item until the eighteenth century. Well before genetically engineered fruits joined the fray, the tomato has been at the center of struggles over immigration, science, food, and labor in California's agribusiness fields, state research institutions, grocery stores, and kitchens (Hightower 1973). On biotechnology and world agriculture, see Hobbelink (1991), Shiva (1993), and Juma (1989).

2 Press release, June 6, 1995, 'Broad Coalition Challenges Patents on Life,' contact person, Philip Bereano, University of Washington. The press release covered meetings in the Adirondack Mountains to plan oppositional strategies. The group issued a position statement called the 'Blue Mountain Declaration.' Working with indigenous organizations to eliminate funding for the Human Genome Diversity Project emerged at the meeting as a major priority. The coalition's statement did not evoke arguments about purity of natural kinds, but the sanctity of life and opposition to manipulation of the natural world remained important ideological resources. I recognize, and often share, the power and importance of those commitments and languages, but I wish my fellow travelers seemed more nervous and less self-certain in their presence. The historical pedigree, for both 'indigenous' and 'Western' speakers, of those languages, ideologies, and associated actions hardly gives cause for unruffled calm. I think progressive politics have to be rooted in more fraught, unsettled, dirty, hybrid languages and expressions of belief, hope, and action.

3 Joseph Dumit, personal communication, December 4, 1992. Dumit's dissertation (1995) on the development of positron emission tomography (PET) brain imaging focuses on the professional, technical, popular, legal, and industrial interactions that forge new disciplines and discourses. His project examines closely the inter-disciplinary development of computer sciences and the interfacing of such specialties with neurosciences in brain-scanning research.

4 The Human Genome Project haunts many chapters in *Modest__Witness@Second__ Millennium*. On genome databases at the beginning of the 1990s, see 'Genome Issue: Maps and Database,' *Science* 254 (October 11, 1991: 201–7). For the Human Brain Mapping Project, see Roberts (1991) and AAAS (1993). The 1990s is the 'Decade of the Brain,' a designation for transnational technoscience something like the United Nations' Decade of the Woman or Year of the Child. Such labels signal conferences, declarations and high-status locations. Data from molecular neurobiology, systems neuroscience, developmental neurobiology, and genetics, as well as new graphics and data storage capacities of computers, have revolution-ized brain-mapping practices, necessitating major changes in the nature of atlases and research interactions. Nonorganic 'brains' also continued in the 1990s as objects of rapt technoscientific attention in artificial intelligence and robotics research. For example, see Travis (1994). In the last decade of the millennium, the action lies in the 'marriage of computational models and experimentation' (Barinaga 1990: 524–26).

▶ **REFERENCES**

AAAS (American Association of Advancement of Science) (1993) *Mapping the Human Brain*. AAAS Special Symposium, 14–15 February, Boston.

Barinaga, M. (1990) Neuroscience models the brain. *Science*, 247: 524–26.

Clynes, M. E. and Kline, N. S. (1960) Cyborgs and space. *Astronautics*, September 26–7: 75–6.

Dumit, J. (1995) *Mindful Images: PET Scans and Personhood in Biomedical America*. Ph.D. dissertation, History of Consciousness Board, University of California at Santa Cruz.

Edwards, P. (1996) *The Closed World: Computers and the Politics of Discourse in Cold War America*. Cambridge, MA: MIT Press.

Escobar, A. (1994) Welcome to Cyberia: notes on the anthropology of cyberculture. *Current Anthropology*, 35 (3): 211–31.

Fox, M. (1983) Genetic engineering: human and environmental concerns. Unpublished briefing paper.

Fox, M. (1992) *Superpigs and Wondercoin: The Brave New World of Biotechnology and Where It all may Lead*. New York: Lyons and Burford.

Gabilondo, J. (1991) *Cinematic Hyperspace, New Hollywood Cinema and Science Fiction Film: Image Commodification in Late Capitalism*. Ph.D. dissertation, Literature Department, University of California at San Diego.

Gray, C. H. (1991) *Computers as Weapons and Metaphors: The U.S. Military 1940–90 and Postmodern War*. Ph.D. dissertation, History of Consciousness Board, University of California at Santa Cruz.

Harvey, D. (1989) *The Condition of Postmodernity: An Enquiry into the Origins of Cultural Change*. Oxford: Basil Blackwell.

Helsel, S. (1993) *The Comic Reason of Herman Kahn: Conceiving the Limits to Uncertainty in 1960*. Ph.D. dissertation, History of Consciousness Board, University of California at Santa Cruz.

Hightower, J. (1973) *Hard Tomatoes, Hard Times: A Report of the Agribusiness Accountability Project on the Failure of America's Land Grant College Complex*. Cambridge, MA: Schenkman.

Hobbelink, H. (1991) *Biotechnology and the Future of World Agriculture*. London: Zed Books.

Juma, C. (1989) *The Gene Hunters: Biotechnology and the Scramble for Seeds*. Princeton: Princeton University Press.

Krimsky, S. (1991) *Biotechnics and Society: The Rise of Industrial Genetics*. New York: Praeger.

Martin, E. (1992) The end of the body? *American Ethnologist*, 19 (1): 121–40.

OTA (Office of Technology Assessment) (1989) *New Developments in Biotechnology: Patenting Life*. Washington, DC: US Government Printing Office.

Rifkin, J. (1984a) *Algeny: A New Word, a New World*. New York: Penguin.

Rifkin, J. (1984b) Letter to William Garland. *Federal Register*, 49 (184): 37016.

Roberts, L. (1991) A call to action on a human brain project. *Science*, 252 (June 28): 1794.

Shiva, V. (1993) *Monocultures of the Mind: Perspectives on Biodiversity and Biotechnology*. London: Zed Books.

Sofia, Z. (1992) Virtual corporeality: a feminist view. *Australian Feminist Studies*, 15 (Autumn): 11–24.

Taylor, P. and Buttel, F. (1992) How do we know we have global environmental problems? Science and the globalization of environmental discourse. *Geoforum*, 23: 405–26.

Virilio, P. and Lotringer, S. (1983) *Pure War*. New York: Semiotext(e).

3 ▶ Edison and electric light

Thomas P. Hughes

Isaiah Berlin in *The Hedgehog and the Fox* quoted the Greek poet Archilochus, who wrote, 'The fox knows many things, but the hedgehog knows one big thing.' This essay on the 'Electrification of America' is about hedgehogs. Sir Isaiah describes them as those 'who relate everything to a single central vision, one system less or more coherent or articulate.' Foxes, in contrast, pursue many ends, 'often unrelated and even contradictory.' Berlin categorizes Dante, Plato, Lucretius, Pascal, Hegel, Dostoyevsky, Nietzsche, Ibsen, and Proust among the hedgehogs.[1] I want to add Thomas Edison, Samuel Insull, and S. Z. Mitchell.

Edison invented systems, Insull managed systems, and Mitchell financed their expansion. These systems were electric light and power, now usually called utilities. Edison invented the system that took form as the Pearl Street generating station of the New York Edison Illuminating Company, now Consolidated Edison Company; Insull managed electric light and power companies that consolidated into Chicago's Commonwealth Edison Company; and Mitchell provided for the growth of large regional power systems. The three men focused upon one level of the process of technological change, such as invention, management, or finance, but in order to relate everything to a single central vision they had to reach out beyond their special competences: Mitchell managed, Insull financed, and Edison knew management and finance, as well. For this reason, Edison should be called an inventor-entrepreneur, Insull a manager-entrepreneur, and Mitchell a financier-entrepreneur – 'entrepreneur' indicating the organizational, system-building drive of the three men.[2] One hesitates to speak of inventor-hedgehog, manager-hedgehog, or financier-hedgehog.

Edison, Insull, and Mitchell were strong holistic conceptualizers and determined solvers of the problems frustrating the growth of systems. This essay, therefore, is also a history of ideas and a study of problem solving.

Their strong concepts resulted from the need to find organizing principles powerful enough to integrate and give purposeful direction to diverse factors and components. The problems emerged as the system builders strove to fulfill their ultimate visions. Not one of them was satisfied to solve a part of the problem, simply to invent, manage, or finance, for each believed that the invention would not become an innovation, the managerial structure would not evolve, and the financial means would not bring growth unless electric light and power were viewed as a coherent system.

Besides focusing upon systems, directing attention to the men who presided over their growth, this essay identifies stages in the history of electric light and power. Around 1880 when Edison flourished, electric light and power were clearly in the inventive stage, and he is representative of many other leading inventors like Elihu Thomson, William Stanley, and Nikola Tesla; Samuel Insull rose to prominence about one-quarter century later after the technology had been shaped and managing large utilities was an even greater challenge. As a result, the names of utility heads like John Lieb, Alex Dow, and Insull dominate the industry. In the twenties, invention and management remained important, but regional systems financed, organized, and managed by holding companies dominated the scene, and men like Mitchell, Charles Stone and Edward Webster, and, again, Insull were pre-eminent.

[In the extract we have chosen, only the work of Edison is discussed.]

► EDISON: INVENTOR-ENTREPRENEUR

Edison was not a simple tinkerer hunting and trying his way to new inventions. He said that he was no genius of heroic proportions; invention – he explained – was 99 percent perspiration and 1 percent inspiration.[3] His more scrupulous and better informed biographers[4] portray him as more than an inventor; they describe his engineering activities as he developed his inventions and his promotional efforts as he brought them into use. His notebooks give evidence that his concepts were bold and encompassing. Edison's activities covered the broad spectrum from invention to innovation; he approached problem solving systematically, and his inventive method synthesized the technological, economic, and scientific.[5]

In his early days, Edison was content to invent a quadruplex telegraph, a telephone transmitter (the receiver was a necessary afterthought for reasons of competition), or some other component of a technological system. Someone else, not Edison, integrated the components into a commercial system ready for the ultimate consumer. After he moved to Menlo Park to establish his research laboratory in 1876 and when he decided to introduce a system of electric lighting in 1878, his reach was far more extended and sweeping – he was ready to preside over the introduction, onto the market, of a complete system of technology synthesizing components of his own invention. As an inventor-entrepreneur, he coordinated a team of electricians, mechanics, and scientists and cooperated with associates concerned about the financial, political, and business problems affecting the technological system.

After conceiving in general and sweeping terms of a system of incandescent lighting in the fall of 1878, Edison announced his brainchild with a fanfare in the *New York Sun* on October 20, 1878. Always good newspaper copy, he told reporters of plans for underground distribution in mains from centrally located generators in the great cities; predicted that his electric light would be brought into private houses and simply substituted for the gas burners at a lower cost; and confidently asserted that his central station would furnish 'light to all houses within a circle of half a mile.' He spoke not only of his incandescent lamp but of other envisaged components of his system, such as meters, dynamos, and distribution mains. A month earlier he had written privately of his concept: 'have struck a bonanza in Electric Light – indefinite subdivision of light.'[6] He was, in essence, sharing his moment of inspiration with associates and the readers of the *Sun*; he had no generator, no promising incandescent lamp, much less a system of distribution – these were at least a year away. Edison, however, had the concept: 'I have the right principle,' he wrote, 'and am on the right track, but time, hard work and some good luck are necessary too. It has been just so in all of my inventions. The first step is an intuition, and comes with a burst, then difficulties arise – this thing gives out and then that. "Bugs" – as such little faults and difficulties are called – show themselves and months of intense watching, study and labor are requisite before commercial success or failure is certainly reached.'[7] But he had the 'right principle.'

Others also report that Edison had a general concept of his system in the fall of 1878. Francis Jehl, who joined Edison as a laboratory assistant in November and who later published reminiscences of the Menlo Park days, recalled that in October 1878 – twelve months before the construction of a practical incandescent lamp and the announcement of his basic generator design – 'Edison had his plans figured out, as a great general figures out his battle strategy before the first cannon is fired.'[8] The secret, according to Jehl, of his accomplishments 'lay in his early vision, far in advance of realization.'[9]

Edison conceptualized so audaciously and embarked upon the invention of an entire system because he had a laboratory and staff to draw upon. He integrated the men and facilities with his concept just as he did the technical components. At Menlo Park there was a hierarchy of systems. His notebooks show that he assigned to his Menlo Park electricians, mechanics, and scientists problems associated with the various components (various parts of the general problem) of the system. The broad concepts were generally his; the men experimented and calculated within his guidelines. Among those to whom he turned often in the first two years of work on the electric light system were Francis R. Upton, Francis Jehl, Charles Batchelor, and John Kreusi. An analysis of the first 200 of the laboratory notebooks, which begin in November 1878 and cover the years 1879 and 1880, indicates that Francis Upton figured most often in the experimentation and calculations.[10]

Francis Upton did a literature search for Edison in the fall of 1878 in New York City before he joined him at Menlo Park. Just before taking up residence there in December, Upton asked if Edison wanted him to continue the search in Boston, Massachusetts, because the 'Berlin summary of Progress in Physics since 1857 . . . and an index to Poggendorff's Annalen' were there.[11] Edison knew his aspirations to invent a system in a field of technology cultivated by scientists – as well as electricians – could only be fulfilled if he drew upon science; Upton reinforced and supplemented Edison in this regard.[12] Edison's systematic approach ignored disciplinary boundaries; today we would say that he was problem, not discipline, oriented.

Upton had come a long way to Menlo Park. Characterized as a scholar and gentleman by his plainer Menlo Park companions, he had studied at Phillips Academy Andover, Bowdoin College, Princeton University, and under Hermann von Helmholtz at Berlin University. Grosvenor P. Lowrey, Edison's counsel, business, and financial adviser, recommended him to Edison knowing of his need for a physicist and mathematician. Francis Jehl said that whatever Upton did and worked on 'was executed in a purely mathematical manner and any wrangler at Oxford would have been delighted to see him juggle with integral and differential equations. . . .'[13] Upton often concentrated upon the development of a dynamo for the system.

Jehl appears frequently in the notebooks in connection with lamp filament investigations. He also came to Edison in November or December of 1878 on the recommendation of Lowrey. As a boy, he had read 'every scientific paper I could find,' and as a young man, he became a great admirer of Edison's.[14] Lowrey, who was general counsel for Western Union, employed Jehl as an office boy and arranged for him to take an apprentice course in the Western Union repair shops. Jehl also attended Cooper Union evenings studying chemistry, physics, and algebra.

Another member of the electric lighting team was Charles Batchelor. He, too, filled out the Edison system, for he was an ingenious master craftsman, dexterous and sharp-eyed, whose wide-ranging experimental techniques and mechanical aptitude kept him at Edison's right hand. Batchelor was so intimately involved with Edison in all of his work 'that his absence from the laboratory is invariably a signal for Mr. Edison to suspend labor.'[15] John Kreusi, who was in charge of the Menlo Park machine shop, also played a major role. Trained in Switzerland as a fine mechanic, he could adeptly construct Edison's various designs with nothing more than rough sketches and cryptic instructions. He, like Batchelor, had been with Edison in Newark, New Jersey, before the establishment of the Menlo Park laboratory.[16]

Many others at Menlo Park were assigned to work on various components of the evolving electric light system. Dr. Herman Claudius, a former officer in the Austrian telegraph corps, built simulations of the system with batteries for generators, fine wires for the distribution system, and resistors for the load. Jehl reported that Claudius had Kirchhoff's laws of conductor networks at his fingertips.[17] The names of some other Edison pioneers who made it possible for him to invent and develop an entire system include John 'Basic' Lawson, J. F. Ott, Dr. A. 'Doc' Haid, William J. Hammer, Edward H. Johnson, Stockton Griffin, George and William Carman, Martin Force, and Ludwig Boehm.

The availability of these varied talents helps explain the encompassing character of Edison's concept of a system. Furthermore, they were supported by a broad array of expensive machine tools, chemical apparatus, library resources, scientific instruments, and electrical equipment in the Menlo Park laboratory complex.[18] A major reason for the establishment of the Edison Electric Light Company in October 1878 was to acquire funds for additional laboratory equipment and new workers like Upton and Jehl. Obviously, the common characteristics of Edison, his men, and the laboratory were shaped by a systematic, demanding endeavor.

At Menlo Park there was more than a system, there was a community as well. Edison chose Menlo Park because the isolated rural setting insulated the staff from the distractions of an urban environment like Newark. Edison and other married members of the community bought or rented farmhouses in the vicinity; Upton brought his bride to a comfortable house provided with the new Edison light. Others lived at Mrs. Jordan's cozy, nicely appointed boardinghouse located a short walk from the laboratory compound. The meals were undoubtedly country and hearty, and the environment was well ordered. There are scores of anecdotes about the character of life in the laboratory, including accounts of late-hour breaks after especially arduous days. On these occasions the pipe organ at the end of the lab's second floor added to the festive consumption of food and drink. Since the working day sometimes extended nearly twenty-four hours, it can be assumed that Edison was willing to charge the expenses to business.

The system, the community, and the style of invention were essentially Edisonian. Few witnesses or historians challenge the conclusion that the organizing genius was Edison's. Yet there was one man, Grosvenor Lowrey, who during the early years of the electric light project appears to have closely advised Edison on financial and political matters. Edison laid down the guidelines for Batchelor, Kreusi and Upton in the laboratory, but Lowrey often guided Edison when the problems involved Wall Street or New York City politicians. Edison, however, did not step back, immerse himself in technological and scientific problems and leave the 'politics' to Lowrey; the correspondence shows that Edison always had a prominent role in the financial and political scenarios.

Because of his knowledge of the world of legal, business, and financial affairs, Lowrey's strengths complemented Edison's. Born in Massachusetts, Lowrey took up the practice of law in New York City and rose to prominence. He acted as counsel to the U.S. Express Company, Wells Fargo & Company, and the Baltimore & Ohio Railroad. He was also legal adviser to the financial entrepreneur Henry Villard. In 1866 he became general counsel of the Western Union Telegraph Company, a position that brought Edison and him together in connection with telegraph patent litigation. Lowrey was one of those who persuaded Edison to turn to electric lighting.[19] Having observed the sensational publicity given to the introduction of the Jablochkoff arc light in Paris in 1878, Lowrey urged Edison to enter the field and offered to raise the money Edison needed to expand Menlo Park. Not only did he advise Edison, he often encouraged the inventor. Lowrey promised in 1878 that the income from electric lighting patents would be enough to fulfill an Edison dream: 'to set you up forever . . . to enable you . . . to build and formally endow a working laboratory such as the world needs and has never seen.'[20] (At the time the only buildings in the Menlo Park group were the laboratory building, the carpenter shop, and the carbon shed – there were no machine shop, library, or office buildings.) Shortly afterward, Edison gave Lowrey a free hand for this purpose in negotiating the sale of forthcoming electric lighting patents and establishing business associations and enterprises at home and abroad: 'Go ahead. I shall agree to nothing, promise nothing and say nothing to any person leaving the whole matter to you. All I want at present is to be provided with funds to push the light rapidly.'[21]

Lowery had close contacts with the New York financial and political world. His law offices were on the third floor of the Drexel Building – Drexel, Morgan, and Company had the first floor. Working closely with his long-time friend, Egisto P. Fabbri, 'an Italian financial genius'[22] and partner of J. Pierpont Morgan, he obtained the funds for Edison from Drexel, Morgan, and Company. His skill and effectiveness in dealing with politicians and political problems is conveyed by a Menlo Park episode. In December 1879, Lowrey arranged a lobbying extravaganza. The objective was to obtain a franchise allowing the Edison Illuminating Company to lay the distribution system for the first commercial Edison lighting system in New York City. Behind the opposition of some New York City aldermen lay gaslight interests and even lamplighters who might be thrown out of work by the new incandescent light. A special train brought the mayor alderman to Menlo Park. In the dusk they saw the tiny lamps glowing inside and outside the laboratory buildings. After a tour and demonstration by Edison and his staff, someone pointedly complained of being thirsty, which was a signal for the group to be led up to a darkened second floor of the laboratory. Lights suddenly went on to disclose a lavish 'spread' from famous Delmonico's. Lowrey presented Edison and the Edison case after dinner; in due time the franchise was granted.[23] The franchise was as necessary for commercial success as a well-working dynamo.

The organization and early management of the companies formed by Lowrey and Edison in connection with the electric light system have been well told elsewhere.[24] Here it is important to stress that the pristine character of the companies manifested Edison's determination to create a coherent system and his willingness to preside over the broad spectrum of technological change. The first company formed – the Edison Electric Light Company – was essentially a means of funding Edison's inventive activity and obtaining a return of investment by sale or licensing of patents on the system throughout the world. The Edison Electric Illuminating Company of New York was a licensee of the parent Edison Electric Light Company. The EEIC built the first commercial Edison system with its central generating station on Pearl Street in New York City, which was started in September 1882. Because Edison invented and developed all major components for the integrated system – except the boilers and steam engines – he had also to establish the Edison Machine Works to build dynamos, the Edison Electric Tube Company to make the underground conductors, and the Edison Lamp Works to turn out incandescent lamps in quantity. He entered into a partnership with Sigmund Bergmann, a former Edison employee, in a company to produce various accessories.[25] Not only was Edison the pivotal figure in the companies during the early years, he personally supervised the construction of Pearl Street Station. In these companies, Edison was an engineer and a manager, but the focus and the commitment for him remained invention.

Supplemented and complemented by his laboratory staff and by the particular resources of Lowrey, Edison solved problems associated with technological change on various levels and in a systematic integrated way. His systematic approach to problem solving was most clearly demonstrated, however, in the invention of incandescent light technology. Edison could not conceive of technology as distinct from economics – at least, when engaged with the electric light system. After initiating the project he read extensively and deeply about gas-lighting from central stations, especially the economics of it. Also he canvassed the potential lighting market in the Wall Street district in New York where he intended to locate his first central station.[26] His notebooks show that he analyzed the cost of operating the Gramme and the Wallace arc light generators that he had acquired for test purposes.[27] From available literature, he and Upton also determined the cost of operating a Jablochkoff arc-lighting system. Laboratory notes reveal that he was especially concerned about the cost of copper and hoped to reduce it in generator and distribution wiring.[28] As early as December 1878, he estimated that the physical plant needed for one incandescent lamp in his system would require capitalization of $11. At an interest rate of 10 percent on this investment and assuming lamp use of 300 hours a year, the charge per lamp would have to be more than 3.66 mills per hour.[29] Edison was clearly thinking within the context of a capitalistic system.

Perusal of Edison's notebooks should lay to rest the myth that he was a simple inventor tinkering with gadgets. There on page after page are concepts, ingenious experimentation, careful and sustained reasoning, and close economic calculation. Notebook number 120 (probable date, 1880), for example, has thirty pages of calculations (probably Upton's under Edison's instructions) about the costs and income of a central station supplying 10,000 lights. These were probably in anticipation of the Pearl Street system to be built in New York City. By the time the calculations were made Edison and Upton knew enough from experimentation and literature searches to assume that a 1-h.p. steam engine and dynamo could supply eight 16-c.p. incandescent lamps. Therefore they needed about 1200 h.p. for the 10,000 lamp system. To house this power plant, they estimated an iron structure, or building, that would cost $8,500 (Table 3.1). Using a Babcock and Wilcox estimate, they figured $30,180 for boilers and auxiliaries. Kreusi predicted that the steam engines and dynamos would cost

Table 3.1 Edison estimate for ten-thousand-lamp central station

Capital investment:			Depreciation	
Power plant building	$ 8,500	2%	$ 170	
Boilers and auxiliary equipment	30,180	10%	3,018	
Steam engines and dynamos	48,000	3%	1,440	
Auxiliary electrical equipment	2,000	2%	40	
Conductors	57,000	2%	1,140	
Meters	5,000	5%	250	
Total	$150,680		$6,058	

Operating and other expenses:

Labor (daily):

Chief engineer	$ 5.00
Assistant engineer	3.00
Wiper	1.50
Principal fireman	2.25
Assistant fireman	1.75
Chief voltage regulator	2.25
Assistant voltage regulator	1.75
Two laborers	3.00
Total	$20.50

Labor (annual)	$ 7,482

Other:

Executive wages (annual)	$ 4,000
Rent, insurance and taxes	7,000
Depreciation	6,058
Coal (annual)	8,212
($2.80/ton; 3lb/h.p. hour; 5 hours daily; 1,200 h.p.)	
Oil, waste, and water	2,737
Lamps (30,000 at 35c each)	10,500
Total	$45,989
Estimated minimum income from 10,000 installed lamps	$136,875
Expenses	– 45,989
	$ 90,886

Source: Edison Menlo Park Notebook no. 120 (1880).

$50,000. After extensive calculation, they anticipated $57,000 for conductors for the system and $5,000 for meters. It followed that on an annual basis using appropriate rates of depreciation, depreciation charges on the building, boilers, engines, dynamos, meters, and conductors would amount to $6,058. Daily labor charges were taken as: chief engineer ($5.00); engineer ($3.00); wiper ($1.50); principal fireman ($2.25); assistant fireman ($1.75); a chief voltage regulator ($2.25); an assistant voltage regulator

($1.75); and two laborers ($3.00). The total labor for the day would be $20.50, or $7,482 annually (the duty day was twelve hours). For 'executive' help the annual cost would be $4,000; rent, insurance, and taxes were estimated at $7,000. (The rent is not explained.) Finally, coal cost was calculated as $8,212 annually ($2.80 per ton and 3lb/h.p.) and oil, waste, and water taken as one-third of coal, or $2,737. Since the central station would furnish the lamps, the estimate was for 30,000 a year at 35c each, or $10,500 annually. The total cost then was $45,989 annually. To estimate income, Edison and Upton assumed that the 10,000 lights could be sold for five hours daily, or 18,250,000 hours annually. They had learned that 10,000 equivalent (15 c.p.) gaslights used five hours a day took 250,000 ft^3 of gas each day, or 91,250,000 ft^3 annually. Since the gas companies charged customers $1.50 for each 1,000 ft^3, income for the gaslight company from supplying light equivalent to Edison's planned central station would be $136,875. Having decided that the price to the consumers would be the same as gas, the calculations showed for the Edison central station an excess of $90,886 in receipts over expenses. The notation beside the excess was simply 'to pay for patent rights and interest.'[30] The extended analysis continues by calling for the company to capitalize at twice the cost of the plant ($2 \times \$150,680 = \$301,360$) which meant that the receipts would allow a dividend of 30 percent on investment. Apparently the holders of the patent rights and interest (the Edison Electric Light Company) took the excess payable in dividends as the payment on patent rights and interest.

Calculations like these were as much a part of Edison's invention and development of an electric lighting system as his overly publicized and well remembered endeavors to find the lamp filament. As a matter of fact the search for the lamp filament was conditioned by cost analyses like the above. It is known that Edison was determined to discover a high-resistance lamp filament in contrast to the low-resistance one generally tried before him by inventors of incandescent lamps; it is not widely realized that his determination was a logical deduction from cost analysis. To explain this, we must consider the cost analysis once more and also introduce science. In doing so we shall demonstrate that Edison's method of invention and development in the case of the electric light system was a blend of economics, technology (especially experimentation), and science. In his notebooks pages of economic calculation are mixed with pages reporting experimental data, and among these one encounters reasoned explication and hypothesis formulation based on science – the web is seamless. His originality and impact lie as much in this synthesis as in his exploitation of the research facilities at Menlo Park.

Among costs listed above, the $57,000 for conductors was the highest capital item. Of this, $27,000 was for copper conductors, $25,000 for the pipes containing them, and $5,000 for insulation. Early in his project, Edison saw that copper cost, especially that dependent upon the cross sectional area and the length of his conductors, was a major variable in the cost equation. Large and long conductors might have raised the price of electric light above gas. To keep length down, he sought a densely populated consumer area; to keep the cross-sectional area small, he had to reason further using the scientific laws of Ohm and Joule. The notebooks show that Edison and Upton used Joule's Law (heat or energy = current^2 × resistance = voltage × current) to calculate energy expended in the incandescent filaments.[31] They also used an adaptation of it to show energy loss in conductors. Energy loss was taken as proportional to the current^2 × the length of the conductor × a constant dependent upon the quality of the copper used, all divided by the cross-sectional area of the conductor (energy loss proportional to C^2La/S). [32] The formula posed an enigma, for if Edison increased the cross-section of the copper conductors to reduce loss in distribution, then he would increase copper costs which was to be avoided. Obviously, a trade-off – to use the jargon of the engineering profession – was in order. There was, however, another variable, the current, to consider. If current could be reduced, then the cross-sectional area of the conductors need not be so large. But current was needed to light the incandescents, so how was one to reduce it?

To solve the dilemma, Edison reasoned as follows. Wanting to reduce the current in order to lower conductor losses, he realized that he could compensate and maintain the level of energy transfer to the lamps by raising the voltage proportionately ($H = C × V$). Then he brought Ohm's Law into play (resistance = voltage divided by current). It was the eureka moment, for he realized that by increasing the resistance of the incandescent lamp filament he raised the voltage in relationship to the current. (Resistance was the value of the ratio.)[33] Hence his time-consuming search for a high resistance filament – but the notable invention was the logical deduction; the filament was a hunt-and-try affair.

While the essence of Edison's reasoning seems clear from the available evidence, I have yet to find in his notebooks or elsewhere the date when he realized that a high-resistance filament would allow him to achieve the energy consumption desired in the lamp and at the same time keep low the level of energy loss in the conductors and economically small the amount of copper in the conductors. In an essay attributed to Edison and sent to Henry Ford at his request in 1926, Edison stated that in the fall of 1878 he had experimented with carbon filaments but that the major problem with these was their low resistance. He observed that 'in a lighting system the current required to light them in great numbers would necessitate such large copper conductors for mains, etc., that the investment would be prohibitive and absolutely uncommercial. In other words, an apparently remote consideration (the amount of copper used for conductors), was really the commercial crux of the problem.'[34] He provided better evidence about the time of origins of his high-resistance concept in stating that 'about December 1878 I engaged as my mathematician a young man named Francis R. Upton. . . . Our figures proved that an electric lamp must have at least 100 ohms resistance to compete commercially with gas.'[35] Edison then said that he turned from carbon to various metals in order to obtain a filament of high resistance, continuing along these lines until about April 1879 when he had a platinum of great promise because the occluded gases had been driven out of it, thereby increasing its infusibility. Edison, then, established a search for a high-resistance filament between

December 1878 and April 1879. Jehl in his *Reminiscences* maintains that Edison wanted a high-resistance lamp as early as October 1878 and had reached this conclusion by reasoning about his envisaged system of electric lighting. Jehl also states that Edison reasoned to the essentials of his system by applying Joule's and Ohm's laws.

Edison's reasoning can be illustrated with a simple example using approximate, rounded-off values. By 1880 he obtained a carbonized-paper filament with resistance ranging from 130 ohms cold to about 70–80 ohms heated. (He wanted 100 ohms.)[36] Desiring a lamp with candle power equivalent to gas, he found that this filament required – in present day units – the equivalent of about 100 watts. This meant that the product of the voltage across the lamp and the current must equal 100 watts. Since the resistance was 100 ohms, the current had to be 1 amp because by Joule's Law the heat energy was equal to the product of the C^2 and the resistance ($100 = C^2 \times 100$; $C = 1$). It then followed that the voltage must be 100 in order to fulfill the energy need ($H = C \times V$; $V = 100/1 = 100$). Therefore the specifications of the lamps in Edison's system in today's terminology: 100 watts; 100 volts; 1 amp; and 100 ohms.[37]

Space does not permit an analysis of the way in which Edison invented other components of his system. The analysis of the Edison method as revealed in the invention of the high-resistance filament, however, cuts close to the core of his creativity. As he declared, there was an abundance of patient hunt and try, and even his most superficial biographers grasp this. Furthermore, those who want to discount the Edisonian method, as compared to the so-called scientific method of the laboratory scientists who followed him, choose to stress the empirical approach. This superficiality and distortion are regrettable as one more of many instances of the obfuscation of the nature of creativity. What should be stressed are the flashes of insight within a context of ordered desiderata. By ordering priorities, Edison defined the problem and insisted, as many other inventors, engineers, and scientists have, that to define the problem is to take the major step toward its solution. The prime desideratum was an incandescent light economically competitive with gas; the major flash of insight was realizing that Ohm's and Joule's laws defined the relationship between the technical variables in his system and allowed their manipulation to achieve the desired economy. Having accepted this, we can better understand why laboratory assistant Jehl in his memoirs repeatedly refers to Edison's familiarity with Ohm's Law as the explanation for his creative success and better understand why so many interpretations of Edison have been flawed by the simple assumption that he did not use science because he spoke testily of scientists, especially mathematicians.

The invention and development of the incandescent light seem to have been the leading edge of Edison's systematic approach. After the characteristics of the lamp were established, then the problem of generator design was generally defined. The generator, for instance, had to supply 100 volts for the parallel-wired incandescent lights and an amperage equal to the number of lamps times approximately 1 amp. The relationship between generator and lamps was determined by the decision to wire the lamps in parallel which in turn resulted from the need to keep the system voltage at a safe level and to make possible operation of the lamps independent of one another. The Edison system was evolving like a drama with a cast of developing, interacting ideas.

In October 1879, the same month in which he found the first practical filament, Edison announced the generator for his system. Other components followed. In September 1882, the Pearl Street system began to supply light for the Wall Street district. With the opening of the Pearl Street Station of the Edison Electric Illuminating Company, the age of central-station incandescent lighting had begun; the modern age of public electric supply had opened.

▶ **NOTES**

1 Isaiah Berlin, *The Hedgehog and the Fox: An Essay on Tolstoy's View of History* (New York, 1953), p 1.
2 I have discussed my concept of an entrepreneur as one who presides over invention, development, and innovation in *Elmer Sperry, Inventor and Engineer* (Baltimore, 1971), pp. 63–70, 241, and 290–295.
3 Frank L. Dyer and T. C. Martin, *Edison: His Life and Inventions*, 2 vols. (New York, 1930), 2:607.
4 Matthew Josephson, *Edison, a Biography* (New York, 1959); Francis Jehl, *Menlo Park Reminiscences*, 3 vols. (Dearborn, Mich., 1937–41); and Dyer and Martin.
5 There are at least 200 laboratory notebooks for the period 1878–80 when Edison was inventing his electric light system. These, along with many more Edison notebooks, are housed at Edison National Historic Site (National Park Service), West Orange, N.J. For this paper, I have drawn especially upon notebooks for November and December 1878. I am grateful to Arthur Abel, archivist, for guidance in using the archives at West Orange.
6 Telegram from Edison to Theodore Puskas, September 22, 1878. All Edison telegrams and letters cited are in Archives, Edison National Historic Site, West Orange, N.J., unless specified otherwise.
7 Edison to Puskas, November 13, 1878.
8 Jehl, 1:216
9 Ibid., 1:217.
10 'Memorandum of Contents of Notebooks from Edison Laboratory,' Edison Archives West Orange, N.J.
11 Upton to G. Lowrey, December 12, 1878.
12 The periodicals in Edison's library at Menlo Park show him to have been a regular recipient of the leading science and engineering periodicals, U.S. and foreign. Dr. Otto Moses, fluent in German and French, was in charge of the library (Bryon Vanderbilt, *Thomas Edison, Chemist* [Washington, D.C., 1971], p. 40). The library holdings have been reassembled in the restored library and science building at Greenfield Village, Dearborn, Mich. I am grateful to the Edison curator there, Robert G. Koolakian, for showing me these. Also Upton's 1878 abstracts of the scientific and technical periodicals survive at Edison Archives, West Orange, N.J. Edison's library preserved at his West Orange Laboratory also has foreign and U.S. scientific and technical periodicals of the 19th century.

13 Jehl, 2:619.

14 Ibid., 1:15.

15 *New York Herald*, December 21, 1879, quoted in Jehl, 1:393.

16 Jehl, 1:54.

17 Ibid., 2:545.

18 Jehl describes the scientific instruments in his *Reminiscences*, see esp. 1:257-270.

19 Payson Jones, *A Power History of the Consolidated Edison System, 1878–1900* (New York, 1940), p. 27; see also Jones, p. 161, on Lowrey.

20 Lowrey to Edison, October 10, 1878.

21 Edison to Lowrey, October 2, 1878.

22 Lewis Corey, *The House of Morgan*, p. 23, quoted in Jones, p. 162.

23 Jehl, 2:778–85.

24 Harold C. Passer, *The Electrical Manufacturers, 1875–1900* (Cambridge, Mass., 1953).

25 For a chart showing the various Edison companies and their relationships, see Jones, p. 13.

26 Jehl, 1:215, 2:731–32.

27 Menlo Park Notebook no. 6 (December 4, 1878-January 30, 1879), pp. 22–30.

28 See Menlo Park Notebook no. 1 (November 28, 1878–July 24, 1879), section on wire calculations, and Notebook no. 12 (December 20, 1878), pp. 174–75, 232–33. On cost of operating Jablochkoff candle, see Notebook no. 6 (December 4, 1878), p. 57. (If only one date is given, it is for first dated entry in notebook; if two dates, the second is for last dated entry. All notebooks are at Edison Archives, West Orange, N.J.).

29 Notebook no. 6, p. 177; Notebook no. 120 (November–December 1880, approximate date), pp. 71–101.

30 Notebook no. 120, p. 99.

31 Notebook no. 3 (November 21, 1878), p. 107; and Notebook no. 9 (December 15, 1878–March 10, 1879), p. 41, have some of many early entries using Ohm's Law. Notebook no. 6, pp. 11 ff., shows use of Joule's Law. Since the last entry in no. 6 is January 30, 1879 (the first was December 4, 1878, see above, n. 27), Edison was using the law early in the electric light project. Notebook no. 10, p. 13, has the word 'Joule' jotted down where $H = C^2R$ is being used. (This notebook has entries from December 1878 to January 1879.)

32 Notebook no. 12, pp. 174–176.

33 Jehl, 1:362–63, 2:852–54 stresses Edison's reliance upon Ohm's Law. This led me to look to it as an important clue to Edison's use of science and his reason. In a reference often used at Menlo Park – the essay 'Electricity' in the *Encyclopaedia Britannica*, 9th ed. – Ohm's Law is stated as $R = E/C$ (resistance equals electromotive force divided by current), p. 41. In a misleadingly titled article Harold Passer argues cogently that Edison's reasoning was as suggested by Jehl ('Electrical Science and the Early Development of the Electrical Manufacturing Industry in the United States,' *Annals of Science 7* [1951]: 382–92). Passer offers no evidence, however, from the notebooks and other original Edison sources. Passer offers a similar discussion of Edison's reasoning and concepts in *The Electrical Manufacturers, 1875–1900* (Cambridge, Mass., 1953), pp. 82, 84, and 89. Dyer and Martin, 1:244–60, parallels the memorandum attributed to Edison in 1926 (see n. 4 above). Josephson (n. 4 above), pp. 193–204, 211–220, stresses Edison's use of Ohm's Law but does not note the importance of Edison's having used Ohm's Law in conjunction with Joule's in order to conceptualize his system. Josephson also dates Edison's first high-resistance lamp as January 1879 (a platinum filament), but gives no source for the statement (p. 199); nor does he

provide a source for the statement that Edison came to the idea for high resistance in a 'flash of inspiration' after September 8, 1878 (p. 194). A. A. Bright, *The
Electric-Lamp Industry* (New York, 1949), does not offer any additional information on the way in which Edison conceived of his system. Jehl remains the
most helpful published source, despite the book's lack of organization, see
1:214–15; 243–45; 255–56; and 2:820–21; 852–54.

34 T. A. Edison, 'Beginnings of the Incandescent Lamp and Lighting System,' p. 4 (a
typescript in the Edison Archives, West Orange, N. J.). It is dated 1926 and
identified as an item sent to Henry Ford at his request. The item should be used
cautiously because, by 1926, Edison and his patent lawyers had organized history
with priorities in mind.

35 Ibid., p. 5.

36 Notebook no. 52 (July 31, 1879), p. 229. The entry is dated December 15, 1879.

37 In January 1881 Edison conducted an economy test of his electric light system
installed at Menlo Park to demonstrate its practicability. This was a prelude to his
installation of a full-size system at Pearl Street, New York City, in 1882. The test
showed him using two sizes of lamps, 16 and 8 c.p. The 16 c.p. depended upon an
electromotive force of 104.25 volts and a resistance of 114 ohms. C. L. Clarke, 'An
Economy Test of the Edison Electric Light at Menlo Park, 1881,' dated February 7,
1881, and published for the first time in Committee on St. Louis Exposition of
Association of Edison Illuminating Companies, *Edisonia: A Brief History of the
Early Edison Electric Lightning System* (New York, 1904), pp. 166-78. Therefore the
current in the 16 c.p. lamp was .9 amps.

4 ▶ Inventing personal computing

Paul Ceruzzi

'Ready or not, computers are coming to the people.
That's good news, maybe the best since psychedelics.'[1]

Those words introduced a story in the fifth anniversary issue of *Rolling Stone* (December 7, 1972). 'Spacewar: the Fantic Life and Symbolic Death Among the Computer Bums' was written by Stewart Brand, a lanky Californian who had already made a name for himself as the publisher of the *Whole Earth Catalog*. Brand's resumé was unique, even for an acknowledged hero of the counterculture: at Stanford in the 1960s, he had participated in Defense Department-sponsored experiments with hallucinogenic drugs; hence the second sentence of the article. In 1968 Brand had helped Doug Engelbart demonstrate his pioneering research on interactive computing to an audience gathered in a large auditorium for the Fall Joint Computer Conference.[2] He was no stranger to computers or to the novel ways one might employ them as interactive tools.

Brand was right – eventually. Two years after his article appeared, *Popular Electronics* published plans for a computer kit called the 'Altair,' and the floodgates of personal computing opened up. The subsequent flood has washed over nearly every office worker and a sizable fraction of homes in industrialized countries, as personal computers (eventually networked to one another) became as common as telephones.

In a technology whose history seems always to be characterized by a series of 'revolutions,' the invention of the personal computer seems the greatest, and perhaps ultimate, revolutionary event of all. This paper will examine that thesis. In particular, it will examine to what extent 'personal computing' (not necessarily synonymous with the 'personal computer') was simply the result of a natural outcome of advances in semiconductor technology, or whether it was the result of a conscious effort by a group of actors to effect a social transformation of computing.[3]

Because of the speed and magnitude of the phenomenon of personal computing, there have been several studies of its emergence. Most of these are coupled with a study of the equally remarkable emergence of the region known as 'Silicon Valley' as the well-spring of computing activity.[4] Other studies emphasize the technological foundation of these machines in semiconductor electronics – an activity also centered in the Santa Clara Valley of California.[5] Precisely because of the immediacy of the subject, it has been difficult for most accounts to have much historical perspective. What is the 'real' story: the rise of Silicon Valley, the invention of specific machines like the Apple II, the personalities of some of the actors such as Bill Gates, or something else? Ten years is hardly enough to gain historical perspective, but personal computing is common enough, and it has stabilized enough in at least certain respects to venture an attempt.[6] . . .

What triggered Brand's insight was watching a group playing a game called 'Spacewar' on a terminal at the Stanford Artificial Intelligence Laboratory. The software revealed computing as far from the do-not-fold-spindle-or-mutilate punched card environment as one could possibly find. The way this small group of people at Stanford was using a computer was extraordinary for 1972; it went against nearly every tenet of the economics of computing as it then existed. Most mainframe installations of the day required that its users submit problems as decks of punched cards, which a computer center operator would run through the machine in 'batches' along with others' programs. Using a computer interactively, never mind playing a game, would have been scandalous almost anywhere else but the lab that Brand visited.

'Spacewar' was running on a PDP-10. This computer, marketed by the Digital Equipment Corporation of Maynard, Massachusetts, had little in common with the personal computers of the 1970s and 1980s.[7] It was physically large – a full system, including terminals and external storage, easily took up a room of its own. It was expensive – a half million dollars for a typical system. It did not use silicon chips – it used discrete transistors for logic and magnetic cores for memory. Its internal design featured a 36-bit word length – twice that of the standard mini-computers of the 1970s, and more than most workstations of the 1990s. Even DEC's own literature called it a mainframe.[8]

But the PDP-10 was designed from the start to support interactive use. Its time-sharing abilities were not as ambitious as some of the research projects then being publicized, e.g. MIT's Project MAC or Dartmouth College's time sharing system. But it was a commercial system and worked well. Of all the early time sharing systems, the PDP-10 best succeeded in creating the *illusion* that each user was in fact being given the full attention and resources of the computer. That illusion created a mental model of what computing could be – to those fortunate enough to have access to one. . . .

The introduction of personal calculators had several profound effects on the direction of computing technology.

The first was that the calculator created a market where chip suppliers could count on a long production run, and thereby gain economies of scale and a low price. As chip density, and therefore capabilities, increased, chip manufacturers faced a variation of the problem that Henry Ford faced with his Model T: only long production runs of the same product led to low prices, but markets did not stay static long enough. That was especially true of integrated circuits, which by nature became ever more specialized in their function as the levels of integration increased. (The only other exception was in memory chips, one reason why Intel was founded with the goal of focusing on that product.) The calculators offered the first consumer market for logic chips that allowed one to amortize the high costs of setting up production lines for complex integrated circuits. The dramatic drop in prices of calculators between 1971 and 1976 showed just how potent this force was.[9]

The second effect was just as important. Pocket calculators, especially those that were programmable, unleashed the force of personal creativity and energy of masses of individuals. This force had been observed among the 'hacker' culture at MIT and Stanford, as documented above (and as observed with trepidation by at least one MIT professor).[10] Their story is one of the more colorful, among the dry technical narratives of hardware and software design. They and their accomplishments, suitably embellished, have become favorite topics of the popular press. Of course the strange personal habits made a good story, but was it true? The group at MIT, especially the members of the Tech Model Railroad Club, deserve genuine credit. They did some fundamental work in developing time-sharing capabilities for the PDP-6. Developing system software was hard work, not likely to be done well by a salaried employee working normal hours and with a family to go home to in the evening. Time-sharing freed users from the tyranny of submitting decks of cards and waiting for a printout, but it forced advanced users to work late at night, when the time-shared systems were lightly loaded and thus more responsive. The more one wanted to make a computer easy to use and useful to ordinary people the more complex and demanding was the software that one had to write for it.

The assertion that hackers created much of modern interactive computing suffers from the fact that there were so few of them. In sheer numbers there may never have been more than a few hundred people fortunate enough to be allowed to 'hack' (defined here as not doing a specific programming job specified by one's employer) on a computer as powerful as a PDP-6. By 1975, there were over 25,000 HP-65 programmable calculators in use, each one owned by an individual who could do whatever he wished to with it.[11] Who were these people? HP-65 users were not 'strange': nearly all were adult, professional men, including civil and electrical engineers, lawyers, financial people, pilots, and so on. Few were students (or professors), because one had to have $795 cash to get one. Most purchased one because they had a practical need for calculation in their jobs. But this was a *personal* machine – they could take it home with them at night. (Figure 4.1). These users – perhaps 5% or 10% of the owners of HP-65s and Texas Instruments SR-52s, were not known for '[t]heir rumpled clothes, their unwashed and unshaven faces, and their uncombed hair . . .'[12] But their passion for programming made them the intellectual descendants of the Tech Model Railroad Club hackers. And their numbers – only to increase as the prices of calculators dropped – were the first indication that personal computing could be truly a mass phenomenon.

Figure 4.1   Calculator 'widow' (courtesy of Penton Publishing Company)

The companies that made these machines were unprepared for these events. They sold the machines as commodities; they could ill afford a sales force that could walk a customer through the complex learning process needed to get the most out of one. That was of course what IBM salesmen were so well-known for – for multi-million dollar mainframe installations. The calculators were designed to be easy enough to use so that such hand-holding was unnecessary to do simple things. But customers soon wanted to do more, and finding little help from the supplier, they turned to one another. Users' groups, clubs, newsletters, and publications proliferated. This supporting infrastructure was critical to the success of personal computing, and in the following decade it would become an industry all its own.

> 'Heard melodies are sweet, but those
> Unheard are sweeter . . .' – Keats.

The introduction of commodity computing devices was a watershed. For about a decade, ever since time sharing and minicomputers revealed an alternative to batch processing, there arose a small group of prophets and evangelists who raged against the world of punched cards and computer rooms, promising a digital paradise of truly interactive tools at one's fingertips. The most famous was Ted Nelson, whose self-published *Computer Lib* proclaimed (with a raised fist on the cover): 'You can and must understand computers now'.[13] Suddenly, in 1974, some of these dreams were becoming real. Now people had to stop dreaming and deal with the specific abilities – and limits – of actual 'dream machines' (the alternate title to Nelson's book). Some of the dreamers, including Nelson, were unable to make the transition. They dismissed the pocket calculator: it was puny, too cheap, couldn't do graphics, wasn't a 'von-Neumann machine,' etc.[14] For them, the dream machine was better, even if (or because) it was unbuilt.[15] By 1985 there would be millions of *actual* 'dream machines' on the desks and in the homes of ordinary people. Most of these would be based on the Intel 8080 chip architecture: unsuitable for serious computing, used by historical accident. But it was there: a constant source of inspiration and creativity to many who used it, an equal source of frustration for those who knew how much better it could be.

▶ | **THE MICROPROCESSOR**

Calculators showed what advanced integrated circuits could do, but they did not open up a direct avenue to personal, interactive computing. Even though each year saw the introduction of new chips that performed more and more sophisticated calculations, those chips were too specialized, too geared toward mathematics, to form a basis for a general-purpose computer. The architecture of these chips was ad hoc and closely held by each manufacturer. What was needed was a set of integrated circuits – or even a single integrated circuit – that incorporated the basic architecture of a general-purpose computer.[16] Such a chip, called a 'microprocessor', did appear.

In 1964 Gordon Moore, then of Fairchild and soon a cofounder of Intel, noted that from the time of its invention in 1958, the number of circuits that one could place on a single integrated circuit was doubling every year.[17] By simply plotting this progress on a piece of semi-logarithmic graph paper, it was clear that by the mid 1970s the semiconductor companies would be selling chips that would integrate enough logic circuits on a small set of chips to equal those of a 1950s-era mainframe. By the late 1960s Transistor-Transistor Logic (TTL) was well-established, but a new type of semiconductor called 'metal-oxide semiconductor' (MOS), was emerging as a way to place many more logic elements on a single chip.[18] The chip density permitted by MOS brought the concept of a computer-on-a-chip into focus among engineers at Intel, Texas Instruments, and other semiconductor firms. . . .

By early 1974, there were two converging forces at work. From the bottom were engineers at Texas Instruments, Rockwell, and Intel, with their new powerful and inexpensive microprocessors. From the top were those people, some unconnected with the computer industry, who saw a variant of a time-shared PDP-10 as a way to transform computing and make it accessible to ordinary people, for Utopian uses. But rather than these two forces converging, for a while it appeared as though they might pass one another by.

As obvious as it appears in hindsight that the 8080 [Intel microprocessor] would lead to the personal computer, Intel engineers and management did not foresee that path.[19] The steps from the 8080 to the PC were not obvious, just as the Intel 8080 itself was not an inevitable product of improvements in chip density. . . .

 ## MOVING-DOWN: FROM MAINFRAME TO PERSONAL COMPUTER

As low-cost microprocessors were appearing in small systems, developments in larger systems were pushing down from the 'top'. The most important of these was the ascendancy of interactive, easy to learn software. If personal, interactive computing was to gain a foothold, it needed two kinds of software. The first was a way to write applications programs. The second was a standard so that these programs could be stored on floppy disks and used on more than one machine. By the mid-1970s those two requirements were also being met. In both cases they came from a mini-computer and time-shared mainframe culture.

The BASIC programming language was invented at Dartmouth College for its pioneering time sharing system. It first ran there in 1964, on a large GE-235 computer. Its developers were John Kemeney and Thomas Kurtz, of the Dartmouth faculty, who acknowledged that Dartmouth students did much of the actual coding of the BASIC compiler.[20] In a speech to the American Museum of Natural History in 1971, and published as the influential book *Man and the Computer*, Kemeney (by then President of Dartmouth) argued that 'next to the original development of general-purpose high-speed computers the most important event was the coming of man-machine interaction.'[21] At Dartmouth he set up a system that operated like the University library: open and free access to all, with costs absorbed by the University as a necessary educational expense. Students of humanities as well as the sciences were encouraged to use it – a truly revolutionary concept at the time. BASIC, an interactive, easy to learn language, was crucial to Kemeney's scheme. As Kemeney reported, it succeeded beyond his expectations.

Kemeney's argument for free access was compelling. Students used the GE mainframe whenever they wished, for whatever use they wished:

> The computation Center is run in a manner analogous to Dartmouth's million-volume open-stack library. Just as any student may go in and browse in the library or check out any book he wishes without asking for permission or explaining why he wants that particular book, he may use the computation center without asking permission or explaining why he is running a particular program. . . .

> At Dartmouth we do not consider . . . recreational uses frivolous . . . [F]or many inexperienced users the opportunity of playing games against a computer is a major factor in removing psychological blocks that frighten the average human being away from free use of machines.[22]

Two aspects of this system are worth keeping in mind. The first was that it was intended from the start to be run on a large, time-shared mainframe computer, accessed through Teletype terminals. The second was that the model of free access like the campus library did not prevail, although it worked well for a small and wealthy private college like Dartmouth. Elsewhere, e.g. large state universities, time sharing systems appeared. But they required some form of accounting and payment, even if the payment often was in 'funny money' – funds not paid by the student or professor but included as an overhead on a large NSF or Defense Department research grant.[23]

It would turn out that it was precisely this notion of free use, with no accounting either for the number of computer cycles one uses, or for serious vs. frivolous uses of the computer, that later on would define the 'personal' in personal computing. But not by time-sharing systems. Although limited, the Dartmouth example was significant. It was a compelling example of free, interactive usage that was not tied to the political or social agendas of the counterculture. Most of all, it produced the BASIC programming language, which we shall see was a key element in the spread of personal computing a few years later (although we shall also see that significant modifications to BASIC were also required).

While this was going on, IBM was enjoying a surge of revenues from installations of its System/360 and System/370 mainframes. The stunning success of this line of large systems was however jeopardized by IBM's troubles in supplying a suitable operating system for it.[24] They eventually delivered a workable version of 'OS/360,' but the company also embarked on developing other systems for smaller installations. In 1966 IBM released 'DOS' (Disk Operating System), which was far more successful.[25]

More significant for this story was the pioneering work done, once again, by the Digital Equipment Corporation. Beginning with DECtape, Digital had pioneered in developing software that allowed one to go quickly to data written in the middle as well as the ends of the spool of tape, in spite of tape's inherently linear ordering of data. The inspiration for this work was the experience many of DEC's employees had with the experimental TX-0 computer at MIT, one of the first systems that had a conversational, inter-active feel to it. As disk storage became available, these employees carried over that feel to DEC's disk operating systems as well. Though known as hardware innovators, DEC engineers were the ones who wrote system software that would have as much of an influence.[26] The PDP-10's operating systems had powerful DECtape as well as disk storage abilities. In the late 1960s DEC produced OS/8 for the PDP-8, which had the 'feel' of the PDP-10 but ran on a machine with very limited memory. OS-8 opened up every-one's eyes at DEC; it showed that small computers could have just as sophisticated capabilities as mainframes, without the bloat that character-ized mainframe system software. Some models of the PDP-11 had an operating system called RT-11 (offered in 1974) that was similar to OS/8, and which further refined the concept of managing data on disks as well as in core.[27] These were the roots of personal computer operating systems.

▶ **THE PERSONAL COMPUTER, 1974**

At this point, around 1974, one can observe the two technological trajec-tories crossing each other: more and more interactive, conversational systems from minicomputer and mainframe companies, more and more powerful chips, especially microprocessors, from semiconductor com-panies. Left to the companies pushing these trajectories, they would not have converged.

Here is where the electronics hobbyists, cousins of the pocket calculator aficionados, come in. This community had a long history of technical innovation – it was radio amateurs, for example, who opened up the high-frequency radio spectrum for long-distance radio communications after the First World War. After World War II the hobby expanded beyond amateur radio to include high-fidelity music reproduction, automatic controls, and simple robotics as well. A cornucopia of surplus radio equipment from the U.S. Army Signal Corps found its way into individuals' hands at low prices, further fueling the phenomenon (a city block in lower Manhattan known as 'Radio Row,' where the World Trade Center now stands, was the most famous source of surplus electronic gear).[28] The shift from vacuum tubes to solid-state integrated circuits made it much harder for an individual to build a circuit on a 'breadboard' at home, but that was compensated by the fact that inexpensive TTL chips now contained whole circuits themselves.[29] The hobby was evolving rapidly from analog to digital applications, but it was healthy. This group supplied the key component needed to make the transition from the microprocessor to the personal computer; an infrastruc-ture of support that neither the minicomputer companies nor the chip makers could provide.

Part of this infrastructure was a variety of electronics magazines. Some were aimed at particular segments, e.g. *QST* for radio amateurs. Two of them, *Popular Electronics* and *Radio Electronics*, were of general interest and sold at newsstands; they covered high-fidelity audio, shortwave radio, television, and assorted gadgets for the home and car. Each issue typically had at least one construction project. As part of the projects, the magazine would have a small electronics company supply a printed circuit board, already etched and drilled, and also some of the more specialized components that readers might have difficulty finding locally.

Beginning in 1971 a few hints of what was to come appeared in these magazines. A machine called the 'Kenbak-1' was advertised in the September 1971 issue of *Scientific American*. The ad stated that it was suitable for 'private individuals' but was really intended for schools. It did not use a microprocessor, and its capabilities were too limited to have much effect. In March 1974 the 'Scelbi-8H,' was advertised in a tiny ad in the back of the March 1974 issue of *QST*. The Scelbi was based on the Intel 8008 – probably the first computer marketed to the public to use a microprocessor. The ad indicated that 'Kit prices for the new Scelbi-8H mini-computer start as low as $440!'[30] The company went on to be an important player in the early personal computer phenomenon.[31] In July 1974, *Radio Electronics* announced a kit also based on the Intel 8008. The story appeared on the cover under the headline 'Build the Mark-8: Your Personal Minicomputer.'[32] The project was much more ambitious than what typically appeared in *R-E*; the article gave only a simple description of the Mark-8 but gave readers the option of sending away for a $5.00 booklet for complete plans. The Mark-8's appearance in *R-E* was a strong factor in the decision by its rival *Popular Electronics* to introduce the Altair kit six months later.[33] These kits were just a few of many similar projects described in advertisements and articles that appeared in hobbyist magazines. They reflected a conscious effort by the community to bring digital electronics, with all its promise but also complexity, to amateurs who up to then had primarily worked with analog radio or audio equipment.

1974 was the *annus mirabilis* of personal computing. It began with the introduction of the HP-65 programmable calculator in January. That summer Intel announced its 8080 processor chip. In late December subscribers to *Popular Electronics* received their January 1975 issue in the mail. The cover showed a prototype of the 'Altair' minicomputer, with an article inside describing how the magazine's readers could obtain one for less than $400. It ranked with IBM's announcement of the System/360 a decade earlier as one of the most significant product announcements in computing. But what a difference a decade made: the Altair was a genuine personal computer.

H. Edward Roberts, the head of the company that designed the Altair, invented the personal computer in 1974. His invention had two parts. First was the Altair itself: a capable, inexpensive computer designed around the Intel 8080 microprocessor. The second, just as important, was a culture that made a place for a personal computer. Both parts were crucial, and although many others could claim to have done one or the other, only Roberts did both.

► ALTAIR

None of the previous kits had the impact of the Altair's announcement. Why? One reason is that it was designed and promoted as a capable minicomputer, as powerful as those offered by DEC or Data General. The magazine article, written by Ed Roberts and William Yates, makes this point over and over: '. . . a full-blown computer that can hold its own against sophisticated mini computers . . .'; '. . . not a "demonstrator" or a souped-up calculator . . .'; '. . . performance competes with current commercial minicomputers . . .' and so on.[34]

The physical appearance of the Altair computer suggests its minicomputer lineage. It looked like models of the Nova of PDP-8: a rectangular metal case, with a front panel of switches that allowed one to control the contents of internal registers, and small lights indicating the presence of a binary one or zero. (Figure 4.2). Inside the Altair's case, one found a machine built mainly of TTL integrated circuits (except for the microprocessor, which was a MOS device), packaged in Dual-in-Line Packages, soldered onto circuit boards. Signals and power traveled from one part of the machine to another on a common bus. The Altair used integrated circuits, not magnetic cores, for its primary memory. The *Popular Electronics* cover called the Altair the 'world's first minicomputer kit'; except for its use of a microprocessor, that accurately describes its physical construction and design.[35]

But the Altair was a lot cheaper. The magazine offered an Altair for under $400 as a kit. The magazine cover said that one could 'save over $1,000.' In fact, the cheapest PDP-8 cost several thousand dollars. Of course, a PDP-8 was a fully-assembled, operating computer that was considerably more

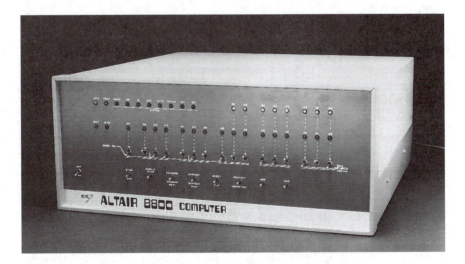

Figure 4.2   Altair 8800 (courtesy of Smithsonian Institution)

capable than the basic Altair. The low cost came mainly from its use of the Intel 8080 microprocessor. Intel was selling this chip for $360 each – almost what the whole Altair kit cost. MITS got them for only $75 by ordering them in quantity.[36] The 8080 had more instructions, was faster, and generally more capable than the 8008 that the Mark-8 and Scelbi-8 used. But it also permitted a simpler design since it required only six instead of 20 supporting chips to make a functional system. Two other key improvements over the 8008 were its ability to address up to 64K bytes of memory (vs. the 8008's 16K), and its use of main memory for the stack, which permitted essentially unlimited levels of subroutines instead of the 8008's seven levels.[37]

The 8080 processor was important to keeping the cost low, but it was only one factor in the Altair's success. Just as important was its bus architecture.[38] After building the prototype Altair, Roberts took some photographs of it and shipped it via Railway Express to the offices of *Popular Electronics* in New York. Railway Express, a vestige of an earlier American industrial revolution, was about to go bankrupt; they lost the package. The January issue described the prototype, with its light-coloured front panel and the words 'Altair 8800' on the upper left. The prototype's design used a set of four large circuit boards stacked on top of one another, with a wide ribbon cable carrying a 100-line bus from one board to another. By the time the first Altair kits were shipped to customers, Roberts had switched to a slightly larger, deep blue cabinet, and he had also made a crucial design change. Instead of using ribbon cable, he routed the bus along the cabinet's base, with a set of connectors that allowed one to plug in order cards as desired.[39]

The $400 kit came with only two cards to plug in to the bus: those two, plus a circuit board to control the front panel (and the power supply) made up the whole computer. The inside looked quite bare, in fact. But by laboriously soldering a set of wires to an expansion chassis, one could have a full set of slots onto which one could plug in a lot of cards. MITS was already designing cards for more memory, input/output, and other functions.

Following the tradition established by Digital Equipment Corporation and Data General, Roberts made the specifications of the bus public knowledge. That allowed others besides MITS to design and market cards for the Altair. That decision was as important to the Altair's success as its choice of an 8080 processor. The decision also led to the company's demise a few years later, since it allowed other companies to market at first compatible cards and later compatible computers to the Altair. But by then the floodgates had opened up. The open, bus architecture meant that if MITS was unable to deliver on its promise of making the Altair a serious machine (though it tried), other companies would step in. Although swamped with orders far beyond its expectations, MITS continued developing additional plug-in cards and peripheral equipment for its computers. So while it was true that for $400 one got very little, there was an ability to get the rest. Marketing the computer as a bare-bones kit offered a way for thousands of people to bootstrap their way into the computer age, at a pace that they, not a computer company, could control.

Assembling the Altair was much more difficult than assembling other electronics kits, such as those for home audio equipment sold by the Heath Company or Dynaco. Most customers looked to one another for support in finding the inevitable wiring errors and poorly soldered connections that they would make. The audience of electronic hobbyists, for whom the magazine article was aimed, compared the Altair not to the simpler Heathkits, but to building computer equipment from scratch. That alternative was almost impossible: not only was it hard to design such equipment, it was also hard to obtain the necessary chips. These devices were indeed cheap, but only if one purchased them in large quantities. Most semiconductor firms had no distribution channels set up for single unit sales, and if they did, they charged a high premium for the transaction. Seen in this light, customers felt, rightly, that they were getting an incredible bargain.

The limited capabilities of the basic $400 kit, plus the fact that because of the shipping loss no actual machine existed at the time the *Popular Electronics* article appeared, has led to the notion that it was a sham, a 'humbug,' not a serious product at all.[40] Although hobbyists first seized on the product, the Altair was designed and marketed as a serious computer to do the same kinds of things that a minicomputer could do. And nearly every person who bought one, after sending in their money and waiting for delivery, recognized that. The Altair was real.

MITS and the editors of *Popular Electronics* had figured out a way to bring the dramatic advances in integrated circuit production to individuals. The first customers were hobbyists, and the first thing they did with these machines, once they got them running, was play games.[41] This application was for the Altair as firing tables were to the ENIAC: important but hardly significant in relation to the phenomenon that it triggered. Among the first customers for the Altair were people with electronics skills who wanted to use it for 'serious' work. The *Popular Electronics* article proposed a list of 23 such applications, none of them games.[42] Because it took several years before MITS and others could supply peripheral equipment, memory, and software to make these practical, serious applications were rare at first. That, combined with the primitive capabilities of other machines like the Mark-8, led to the misleading assumption that the Altair was not a serious computer. But it was. Many of the proposed applications hinted at in the 1975 article were eventually implemented, and years later Altairs were found in a host of embedded systems just like their minicomputer cousins.[43]

The next three years, from January 1975 through the end of 1977, saw a burst of energy and creativity in computing that has almost no equal in its history. The Altair opened up the floodgates. Its shortcomings were clear to everyone. In its basic configuration one could do little more than program it to blink a pattern of lights on the front panel, and even that only by tediously flicking the toggle switches for each program step, then depositing that into a memory location, repeating for the next step, and so on – hopefully the power did not go off while this was going on – until one had the whole program (which had to be less than 256 bytes long!) stored in memory. Bruised fingers from the small toggle switches were the least of the frustrations this method of programming produced. But the bus architecture meant that other companies could design boards to remedy each of these shortcomings, or even design a copy of the Altair itself (as IMSAI did in 1976, whose machine was a close copy but with a more rugged power supply).

Although focused on their computer, the people at MITS and their hangers-on created a lot more. Selling a computer for less than $400 meant that the extensive support and infrastructure that mini and mainframe companies supplied had to come from elsewhere. For personal computer owners, it came from users' groups (following tradition set by the HP calculators), informal newsletters, commercial magazines, local clubs, conventions – even retail stores. All of these sprang up along with the Altair; many of them lived long after the last Altair computer itself was sold.

Other companies, beginning with Processor Technology, soon began offering plug-in boards that gave the machine more memory (MITS also offered boards, but they did not work well). Another board provided a way of connecting the machine to a Teletype. That allowed one's fingers to heal, but Teletypes were not easy to come by – an individual not affiliated with a corporation or university could usually only buy one second hand, and even then they were expensive. Before long, hobbyists-led small companies began offering ways of hooking up a television set and a keyboard. The board that connected to the Teletype sent data 'serially' – one bit at a time; another board was designed that sent out data in parallel, for connection to a line printer that minicomputers used; although like the Teletype these were also expensive and hard to come by.

The basic Altair lost its data when the power was shut off, but before long MITS designed an interface that put out data as audio tones, so that one could store programs on cheap audio cassettes. A group of MITS people, hobbyists, and others met in Kansas City in late 1975 and established a 'Kansas City standard' for the audio tones stored on cassettes, so that programs could be exchanged from one computer to another. But it was never universally adopted, and it did not work that well anyway. Some companies brought out inexpensive paper tape readers that did not require the purchase of Teletype. Others developed a tape cartridge like the old 8-track audio systems, which looped a piece of tape around and around. Cassette storage was slow and cumbersome – one usually had to record several copies of a program, and make several tries before successfully loading it into the computer. MITS and other companies soon began looking toward the 'floppy' disk – a medium invented at IBM for its system/370 mainframes. That would eventually prevail, but not until first of all the mainframe-style 8" drives gave way to less-expensive 5¼" drives specifically aimed at small systems, and more important, until control software and hardware were developed for disk storage.

▶ **SOFTWARE**

Equally important as the physical hardware of the Altair was the way that personal computer software emerged to serve it. Recall that BASIC was not invented for small computers. The Dartmouth version was compiled, not interpreted: that is, it did not execute a student's program until the entire program was typed in. But versions of BASIC quickly found their way onto PDP-11s, HP-2000s, and other small systems. These were 'interpreted': each line was executed, if possible, as it was typed in. More important was that these versions required very little memory compared to any other comparable languages. These modifications to BASIC disturbed Kemeney and Kurtz at Dartmouth: they abhorred changes others had made to the language in trying to shoehorn it onto systems less capable than their GE mainframe. [44] And even in its mainframe version, BASIC had severe limitations on the numbers and types of variables it allowed, for example. In the universities BASIC was disparaged as a toy language that fostered poor programming habits.

In the mid 1970s the standard programming language for serious applications was Fortran. Fortran was an old language and had many flaws that successive versions had only partially corrected. Still it was the language people used for serious programming. If, in 1974, one asked for a modern, concise, and well-designed language to replace Fortran, the answer would have been APL, an interactive language invented at IBM by Kenneth Iverson in the early 1960s. Or PL/1: IBM had thrown its resources toward this language, which it hoped would replace both Fortran and COBOL on its System/360 line. As mentioned below, Gary Kildall chose a subset of PL/1 to write a development kit for Intel's development kits. And a team within IBM designed a personal computer in 1973, the 'SCAMP,' that used APL (however the commercial version, the 5100, was too expensive to be a success). [45] But BASIC was easy to learn by novices, and it did have a track record of running on computers with very limited memory.

Roberts stated that he considered Fortran and APL, but he decided the Altair was to have BASIC.[46]

William Gates III was born in 1955, the year that work on the Fortran compiler began. He was a student at Harvard when the famous cover of *Popular Electronics* appeared describing the Altair. According to one biographer, his friend Paul Allen saw the magazine and showed it to Gates, and the two immediately decided that they would write a BASIC compiler for the machine.[47] Whether it was Gates's or Roberts's decision to go with BASIC for the Altair, BASIC it was.

In a newsletter sent out to Altair customers, Gates and Allen stated that a version of BASIC that required only 4K bytes of memory would be available beginning in June 1975. More powerful versions requiring more memory would be available soon after. The cost for those who also purchased Altair memory boards, was $60 for 4K BASIC, $75 for 8K, and $150 for 'extended' BASIC (requiring disk or other mass storage). If you wanted the language to run on another 8080-based system (they mentioned the Intel development kit), the cost was $500.[48]

Gates and Allen, with the help of Monte Davidoff, had not only written a version of BASIC that fitted into very little memory; they had written a BASIC with a lot of features and impressive performance. The language was true to its Dartmouth roots in that it was easy to learn and did not feel intimidating to the novice. It broke with those roots by providing avenues, via a 'USR' command, to subroutines written in machine language. These kept the language compact and gave their BASIC the performance of a compiled language like Fortran; while it remained an interactive, conversational language that made it easy to use. A programmer could even put bytes into or pull data out of specific memory locations, through the 'PEEK' and 'POKE' commands – something that would have caused havoc on the time-shared Dartmouth system. These extensions to BASIC were a key to Gates's and Allen's success; they reflected the influence of Digital Equipment Corporation modifications to BASIC for the PDP-11.

The developers of Microsoft BASIC were not formally trained in computer science or mathematics as were Kemeney and Kurtz, but they had the right sort of training to pull this off. Bill Gates's private school in Seattle was one of the first to have a General Electric time-sharing system available for its pupils, in 1968, when he was in the eighth grade. Not long after he got computer time on a much better system: a PDP-10 then being set up as a time-sharing utility by the Computer Center Corporation. A few years later he worked with a system of PDP-10s and PDP-11s that were being used to control hydroelectric power for the Bonneville Power Administration.[49]

Gates and Allen decided to develop a version of BASIC that incorporated the features that Digital Equipment Corporation had made available for their PDP-11.[50] By this time Gates was at Harvard. He did not have access to an 8080-based system, but he could use a PDP-10 at Harvard's computing center (named after Howard Aiken). With the help of fellow Harvard student Monte Davidoff, he developed a BASIC for the 8080 using the PDP-10. In early 1975 Paul Allen flew to Albuquerque and demonstrated it to Roberts and Yates. It worked. Soon after, MITS advertised its availability for the Altair.

Other hobbyists were also writing BASIC interpreters for the Altair and for the other small computers now beginning to flood the market. But none were as good as Gates's and Allen's, and it was not long before word of that got around. Roberts and his company had seemingly made one brilliant decision after another: the 8080 processor, the bus architecture, and now the right piece of software. But by mid-1975 Gates and Allen were not seeing it that way. Although Gates was still an undergraduate at Harvard, he and Allen cast their lot with MITS. Gates insists that he never became a MITS employee (although Allen was until 1976), and under the name 'Micro Soft', later 'Micro-Soft,' he and Allen retained the rights to their BASIC.[51] In a now-legendary 'Open Letter to Hobbyists,' widely distributed in early 1976, Gates complained about people making illicit copies of his BASIC, usually in the form of a roll of paper tape. (Figure 4.3). In it he claimed 'the value of the computer time we have used [to develop the language] exceeds $40,000.' But most users of it had not paid for it. If he and his programmers were not going to be paid, they would have little incentive to develop more software for personal computers. But the whole personal computing phenomenon was being put at risk: 'Nothing would please me more than to hire ten programmers and deluge the hobby market with good software.'[52]

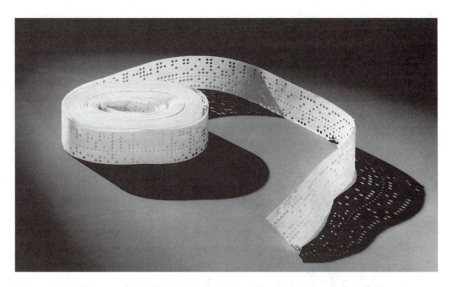

Figure 4.3   Microsoft BASIC on a Teletype tape (courtesy of Smithsonian Institution)

Some of the computer time he used to develop BASIC was on a PDP-10 at Harvard – a machine that students were not to use for commercial purposes although these distinctions were not as clear as they would be later. And the language itself was the invention of Kemeney and Kurtz of Dartmouth; the extended dialect that Microsoft based its version on was the result of many hours of programming by anonymous engineers from the Digital Equipment Corporation. Digital simply did not think of their software as something to sell; it was what you did to get people to buy your hardware.[53]

Bill Gates recognized that with the advent of cheap computers like the Altair, software could – and should – come to the fore as the principal driving agent in computing. And only by charging money for it – no matter that it had been free – could that happen. By 1978 his company, now called 'Microsoft,' had severed its relationship with MITS and was moving from Albuquerque to the Seattle suburb of Bellevue. (MITS itself had lost its identity, having been bought by Pertec in 1977.) Computers were indeed coming to 'the people,' as Stewart Brand predicted after his visits through the Bay area in 1972. But the driving force that brought them to the people was not the vision of a Utopia of shared and free information; it was the force of the free marketplace. Gates made good on his promise to 'hire ten programmers and deluge the hobby market with good software.'

▶ CP/M

System software for the Personal Computer came a little later, but it was just as crucial as BASIC to the phenomenon. Its creation was due to Gary Kildall.

It did not take long for people to realize that paper tape and audio cassette storage were inadequate. A floppy was different: although it was a rotating medium, one could treat it as if it had 'random' access: one did not have to run through the entire spool of tape to get at a specific piece of data. But to do that required tricky programming – which in the mainframe world had been known as a 'Disk Operating System,' or 'DOS.'[54]

Gary Kildall (1942–1994) had developed a programming system for Intel when the 8080 was announced. He worked on a mainframe, and the language PL/M was similar to IBM's PL/1. Kildall also wrote a small control program for its disk drive to make his job easier. 'It turned out that the operating system, which was called CP/M for Control Program for Micros, was useful, too, fortunately.[55] Kildall stated that PL/M was 'The Base for CP/M,' but the actual commands in CP/M are clearly derived from Digital Equipment's software.[56] Specifying the drive in use by a letter, the file names followed by a period and three-character extension, the 'DIR' (Directory) command, 'PIP,' and 'DDT' were among those carried over directly, and these are not a part of PL/1.[57] It was announced to the hobbyist community as 'similar to DECSYSTEM 10' in an article by Jim Warren in *Dr. Dobb's Journal of Computer Calisthenics and Orthodontia* [sic] in April 1976. Warren was clearly excited by CP/M, stating that it is 'well-designed, based on an easy-to-use operating system that has been around for a DECade. [sic]'[58] Suggested prices were well under $100, with a complete floppy system including drive and controller around $800 – not cheap but clearly a superior value compared to the alternatives of Teletype, cassette, or paper tape.

Gary Kildall and his wife, Dorothy McEwen, had slowly eased themselves into the business of making and selling commercial software. Selling software was something they did part time, while Kildall also drew a salary as an instructor at the Naval Postgraduate School in Monterey, California. But interest in CP/M picked up, and he found himself writing variations of it for a number of customers. The publicity in *Dr. Dobb's Journal* led to enough sales to convince him of the potential market for CP/M. In 1976 he quit his job and with Dorothy founded a company, Digital Research. The company enjoyed almost instant success in selling CP/M.[59]

The next year, 1977, he designed a version of CP/M that contained a significant departure from DEC's designs. IMSAI, the company that had built a successful 'clone' of the Altair, wanted a license to use CP/M for its products. Working with IMSAI employee Glen Ewing, Kildall rewrote CP/M so that only a small portion of it needed to be customized for the specifics of the IMSAI. The rest would be common code that would not have to be rewritten each time a new computer or disk drive came along. He called the specialized code the 'BIOS' – 'Basic Input/Output System.'[60]

This change in effect standardized the system software in the same way that the 100-pin Altair bus standardized hardware. IMSAI's computer system became a standard, with its rugged power supply, room for expansion with plenty of internal slots, external floppy drive, and CP/M. When the Altair standard evolved into a standard set by IBM after 1981, CP/M evolved into MS-DOS, sold by Microsoft. The arcane acronyms 'PIP' and

'DDT' disappeared, but in most other respects MS-DOS clearly showed its roots in the old DEC operating systems of the late 1960s and early 1970s.

► END OF THE PIONEERING PHASE, 1977

By 1977 the pieces were all in place for personal computing to come of age. The Altair's design shortcomings were corrected. Microsoft BASIC allowed programmers to write interesting, and for the first time, serious software for PCs. The ethic of charging money for this software gave an incentive to such programmers, although software piracy also became established. Many computers were being offered with BASIC supplied on a Read-Only-Memory, the manufacturer paying Microsoft a simple royalty fee. (When the start-up codes were also put into ROM, there was no longer a need for the front panel, with its array of lights and switches.) Five-and-a-quarter-inch floppy disk drives, controlled by CP/M, provided a way to develop and exchange software independent of the particular model one owned. Machines came with serial and parallel ports, and relatively-standard connections for printers, keyboard, and video monitors were becoming common. Finally, by 1977 there was a strong and healthy industry of publications, software companies, and support groups to bring the novice on board.

Three computers introduced that year completed the transition. The retail giant Radio Shack began offering its 'TRS-80' Model 1 in its stores, at prices starting at $400. It was a complete system, including a keyboard and monitor, with cassettes used for storage.[61] The Commodore PET, designed and sold by a company that had made calculators, also came complete with monitor, keyboard, and cassette player built in to a single box. It sold modestly in the U.S. and very well in Continental Europe where it became a standard for many years.

The third machine introduced in 1977 was the Apple II, created by two idealistic young men, Steve Jobs and Steve Wozniak, in a Silicon Valley 'garage.' The Apple II used a different microprocessor than the Altair, but in other ways it was the Altair's spiritual descendant. It came with a version of BASIC written by Microsoft. (A payment of $10,500 from Apple to Microsoft in August 1977, for part of the license fee, is said to have rescued Microsoft from insolvency at a critical moment of that company's history.)[62] It had a bus, which allowed Apple and other companies to expand the computer's capabilities and keep it viable into the 1980s. Among the cards in 1980 offered was a 'SoftCard,' from Microsoft, that allowed the Apple II to run CP/M. The Apple II was not quite a complete system in a box, but it went a long way toward the goal expressed in the hobbyist journals to have an 'appliance' computer: one that you could simply plug in and use with a minimum of set-up.

Who, then, really invented personal computing? The work of Jobs, Wozniak, and their fellow members of the Homebrew Computer Club came closest to Stewart Brand's prediction that computers would not only come 'to the people' but would also be embraced by those people as a friendly, non-threatening piece of technology (the Apple II) that would enrich their personal lives. Intel built the foundation for personal computing with its microprocessors. Digital Equipment Corporation provided the basic architecture, as well as the structure of the crucial disk operating system. Data General provided the model for the physical packaging. Hewlett-Packard opened up a consumer market for relatively expensive and complex digital electronic devices.

But Ed Roberts and his small group in Albuquerque deserve credit for engineering the convergence of technical and social forces. From the top came a mental notion of what personal computing ought to look like – a mental model based on mostly experimental and a few commercial time sharing systems running off expensive mainframes. Rising up from below was a profusion of powerful and inexpensive integrated circuits, which offered all the processing power of yesterday's mainframe. But the chip manufacturers did not really understand what to do with their products. Intel even built a computer using its chips – but marketed it not as a computer but as a development system for engineers to learn on.

The Altair, offered first as a $400 kit of parts and later enhanced with BASIC and the CP/M operating system, joined the two forces. It was unabashedly a personal product, sold to individuals for whatever use one wanted. It skillfully exploited the advantages, including low cost, of the microprocessor. It was not just a computer but a whole world that included users' clubs (one of which MITS founded itself), magazines, retail stores, trade fairs, and software libraries – all the other pieces of the mosaic that were needed to complete the picture. What started as a $400 kit of parts eventually drove the giants of the industry: IBM, DEC, Data General, UNIVAC, Burroughs, Wang, and Control Data – into years of red ink and shrinking revenues and even in some cases bankruptcy. The historian I. Bernard Cohen has pointed out that when one talks of a technological 'revolution' one often forgets that its political counterpart, especially the French Revolution, is measured as much by those who lose power and wealth as by those who gain it.[63] Given what has happened to these firms, perhaps the word 'revolutionary,' when applied to the invention of the personal computer is appropriate after all.

▶ **NOTES**

1 Stewart Brand, 'Spacewar: Fanatic Life and Symbolic Death Among the Computer Bums,' *Rolling Stone*, (Dec. 7, 1972), pp. 50–58.
2 Engelbart, in Adele Goldberg ed., *History of Personal Workstations*, (Reading, MA: Addison Wesley, 1988), p. 187.
3 Wiebe E. Bijker, Thomas P. Hughes, and Trevor Pinch, eds., *The Social Construction of Technological Systems*, (Cambridge: MIT Press, 1987).
4 See, for example, Paul Freiberger, *Fire in the Valley: the Making of the Personal Computer*, (Berkeley: Osborne/McGraw-Hill, 1984); but also Stan Augarten, *Bit By*

Bit: an Illustrated History of Computers. (New York: Ticknor and Fields, 1994), Chapter 9. There have also been numerous accounts of the history of specific machines and the companies or people who produced them; e.g. Apple, the IBM PC, Atari. These will be cited as those machines are discussed later in the text.

5 For example, Ernest Braun, and Stuart Macdonald, *Revolution in Miniature: the History and Impact of Semiconductor Electronics Re-explored in an Updated and Revised Second Edition* (Second Edition, Cambridge: Cambridge University Press, 1982); as well as the more popular Dirk Hanson, *The New Alchemists: Silicon Valley and the Microelectronics Revolution* (Boston: Little, Brown, 1982). As with the previous literature, there are also more detailed studies of the invention of specific circuits; these will be addressed later in the text.

6 The sudden emergence of the Internet phenomenon since 1992 has obviously made personal computing 'unstable,' but most personal computers in the mid-1990s continue to use a microprocessor that is a direct descendant of the Intel 8080 used in the Altair. The standard personal computer operating systems, as well as the physical packaging of computers, also trace their lineage back to that era.

7 C. Gordon Bell; J. Craig Mudge, and John McNamara, *Computer Engineering: a DEC View of Hardware Systems Design*, (Bedford, MA, Digital Press, 1978), Chapter 21.

8 See, e.g. advertisements in *Datamation* from that period. A Digital spokesperson called the PDP-6 'the first of what might be called a "personal" mainframe.' Jamie Pearson, *Digital at Work*, (Bedford, MA: Digital Press, 1992), pp. 54–55.

9 Gordon Moore, 'Microprocessors and Integrated Electronics Technology,' *Proceedings of the IEEE, 64* (June, 1976), pp. 837–841. One other exception to this rule has been in computer games, which I do not discuss in this paper.

10 Joseph Weizenbaum, *Computer Power and Human Reason: From Judgment to Calculation*, (San Francisco: W. H Freeman, 1976), Chapter 4.

11 '65-Notes' (Newsletter of the HP-65 Users' Club), vol. 2 #1 (January 1975), p. 7. HP-65 customers were overwhelmingly male; the newsletter made a special note of the first female member to join the Users' Club, a year after its founding.

12 Weizenbaum, p. 116.

13 Ted Nelson, *Computer Lib*, (South Bend, Indiana, Ted Nelson, 1974).

14 The 'von Neumann' argument came from the fact that most calculators, unlike general purpose computers, stored their programs in a memory deliberately kept separate from data. This was done to make the machine easier to use by non-specialists, but a common memory was felt to be a central 'defining' feature of a true computer.

15 'The Programmable Pocket Calculator Owner: Who Does He Think He Is?' 'HP-65 Notes,' Vol. 3, Number 6 (1976), p. 2.

16 '65 Notes,' vol. 2 #1 (1975), pp. 4–7.

17 Gordon E. Moore, 'Progress in Digital Integrated Electronics,' *Proc. Int. Electron Devices Meeting*, (Dec. 1975), pp. 11–13. Robert Noyce stated that Moore first noticed this trend in 1964: Noyce, 'Microelectronics,' *Sci. Am.* (Sept. 1977), pp. 63–69. Moore predicted that the rate would flatten out to a doubling every two years by 1980. That has led to some confusion in the popular press over what exactly is meant by 'Moore's Law.'

18 Clifford Barney, 'He Started MOS From Scratch,' *Electronics Week*, October 8, 1984, p. 64.

19 Carver Mead called the invention of the microprocessor a 'no-brainer,' in San Jose *Mercury News*, Dec. 2, 1990, p. 27; see also C. Gordon Bell, interview with the author, June 1992.

20 Thomas E. Kurtz, 'BASIC Session,' in Richard L. Wexelblatt, ed., *History of Programming Languages*, (New York: Academic Press, 1981), pp. 515–550.

21 John Kemeney, *Man and the Computer* (New York: Scribner's, 1972), p. vii.

22 Kemeney, op. cit., pp. 33, 35. Dartmouth at that time had no female students.

23 William Aspray and Bernard Williams, 'Arming American Scientists: NSF and the Provision of Scientific Computing,' *Annals of the History of Computing*, 16/4 (1994), pp. 60–74.

24 Frederick Brooks, Jr., *The Mythical Man-Month: Essays on Software Engineering*, (Reading, Mass.: Addison-Wesley, 1975).

25 Emerson Pugh, *Building IBM: Shaping an Industry and Its Technology* (Cambridge: MIT Press, 1995), pp. 295–6.

26 C. Gordon Bell, interview with the author.

27 Person, *Digital at Work*, pp. 64–65, 86; also C. Gordon Bell, interview with the author, June 1992. The problem with IBM's operating system development for the System/360 has been well-documented in Fred Brooks (1975).

28 Susan Douglas, 'Oppositional Uses of Technology and Corporate Competition: the Case of Radio Broadcasting,' in William Aspray, ed., *Technological Competitiveness*, (New York: IEEE, 1993), pp. 208–219.

29 The construction of the World Trade Center obliterated Radio Row, but by then integrated electronics was well underway. A single microprocessor and its associated support chips might contain more circuits than the entire contents of every stall on Radio Row.

30 *QST*, March 1974, p. 154.

31 Stan Veit, *Stan Veit's History of the Personal Computer*, (Asheville, NC: World-Comm, 1993), p. 11; also Thomas Haddock, *A Collector's Guide to Personal Computers*, (Florence, Alabama: Thomas Haddock, 1993), p. 20.

32 'Build the Mark-8, Your personal Minicomputer.' *Radio Electronics*, July 1974, cover, pp. 29–33.

33 NMAH Collections; also Steve Ditlea, ed., *Digital Deli*, (New York: Workman, 1984), p. 37.

34 H. Edward Roberts and William Yates, 'Exclusive! Altair 8800: the Most Powerful Minicomputer Project Ever Presented – can be Built for Under $400,' *Popular Electronics*, January 1975, cover, pp. 33–38.

35 Journalists soon began calling these machines 'microcomputers': an accurate but also ambiguous term, as it could imply two slightly different things: A microcomputer used a microprocessor, and minicomputers did not. That was true at the time, but eventually nearly every class of computer would use microprocessors. The other was that a microcomputer was smaller and/or cheaper than a minicomputer. The Altair was much cheaper but not that much smaller.

36 Intel Corporation, 'A Revolution in Progress,' (Intel, 1984), p. 14; also Stan Veit, *Stan Veit's History of the Personal Computer*, (Asheville, NC: WorldComm, 1993), p. 43; Veit stated that Roberts obtained chips from Intel that had cosmetic flaws, but Roberts and Intel both stated flatly that the 8080 chips used in the Altair were not defective in any way; see MITS, 'Computer Notes,' vol. 1 #3 (August 1975), p. 2 (National Museum of American History, Mims-Altair file).

37 Ibid., p. 14; also Barden, pp. 101–103.

38 Spelled 'buss' in the *Popular Electronics* article.

39 Veit (op. cit.) argues that it is to Railway Express's ineptitude that we owe the momentous decision to have a bus; others have stated that the decision came from Roberts's finding a supply of 100-slot connectors at an especially good price.

40 e.g. in Steven Manes and Paul Andrews, *Gates*, p. 64.

41 Jim Warren, Personal Computing: An Overview for Computer Professionals, *NCC Proceedings*, 46 (1977), pp. 493–498.

42 These included 'Multichannel data acquisition system,' 'Machine controller,' 'Automatic controller for heat, air conditioning, dehumidifying,' as well as 'Brain for a robot,' among others.

43 As an example, the author observed several Altair-based systems running experiments for the Agricultural Engineering Department at Clemson University, in the early 1980s. Over the years nearly all the circuit boards as well as the power supply of the Altairs had been replaced, but the systems remained housed in their original 'Altair' boxes.

44 e.g., the practice of allowing one to make the word 'LET' optional, and having more than one BASIC statement on a line, among others.

45 Jon Eklund, 'Personal Computers,' in Anthony Ralston and Edwin Reilley, eds., *Encyclopedia of Computer Science*, 3rd ed., (New York: van Nostrand Reinhold, 1993) pp. 460–463.

46 Forrest Mims, III, 'The Tenth Anniversary of the Altair 8800,' *Computers and Electronics*, January 1985, p. 62. Roberts's account has been disputed by others and remains controversial.

47 Stephen Manes and Paul Andrews, *Gates: How Microsoft's Mogul Reinvented an Industry, and Made Himself the Richest Man in America*, (New York: Doubleday, 1993), p. 63.

48 MITS Corporation, 'Computer Notes,' vol. 1 #2 (July, 1975), pp. 6–7.

49 Stephen Manes and Paul Andrews, *Gates: How Microsoft's Mogul Reinvented an Industry and Made Himself the Richest Man in America*, (New York: Doubleday 1993), Chapter 2, 3.

50 Ibid., pp. 71–72.

51 This, too, is a matter of great dispute – Roberts insists that MITS had the rights to BASIC. In a letter to 'Computer Notes' in April 1976, Gates stated that 'I am not a MITS employee,' but that was written after his rift with Roberts had grown deep. See also Stephen Manes and Paul Andrews, *Gates*.

52 MITS Corporation, 'Computer Notes,' Feb. 1976, p. 3. The open letter was distributed to many hobbyist publications and was widely read.

53 C. Gordon Bell, interview with the author, June 1992, Los Gatos, CA.

54 This term had been used, for example, with the IBM System/360 beginning in the late 1960s; see Emerson Pugh *IBM's 360 and Early 370 Systems* (Cambridge: MIT Press, 1991), Chapter 6.

55 Gary Kildall, 'Microcomputer Software Design – a Checkpoint,' *National Computer Conference* 44 (1975), pp. 99–106; also Kildall, quoted in Susan Lammers, (ed.) *Programmers at Work*, (Redmond, WA: Microsoft Press, 1989), p. 61.

56 Gary Kildall, 'CP/M: A Family of 8- and 16-Bit Operating Systems,' *Byte*, June 1981, pp. 216–229.

57 The above argument is based on PDP-10 and CP/M manuals in the author's possession, as well as conversations with Kip Crosby, to whom I am grateful for posting this question over an Internet discussion forum. One possible link between CP/M and IBM mainframe software is its unusual syntax: in CP/M a command is followed by the destination, then the source (the opposite of MS-DOS and UNIX). This is the way that IBM System/360 Assembler language works.

58 Jim C. Warren, 'First World on a Floppy-disc Operating System,' *Dr. Dobb's Journal . . .*, April 1976, p. 5.

59 Robert Slater, *Portraits in Silicon*, (Cambridge, MA: MIT Press, 1982), Chapter 23.

60 Ibid.; also Stan Veit (1993), p. 64; and Digital Research, 'An Introduction to CP/M Features and Facilities,' (1976), manual in the author's possession.
61 Kidwell (1994), p. 97; Viet (1993), p. 159–162.
62 Manes & Andrews, p. 111.
63 I. Bernard Cohen, *Revolution in Science*, (Cambridge: Harvard 1986).

5 ▶ Constructing a bridge

Eda Kranakis

The most immediate considerations for Finley [James Finley, born in Maryland in 1762, inventor of the modern suspension bridge with a flat roadway] were economic. Finley's effort to work out a suspension bridge design represented an entrepreneurial undertaking within the context of the patent system. In the United States, a patent gave an inventor a 17-year monopoly on the right to exploit an innovation. Finley's aim was to patent a complete design system and then earn profits from it by selling the right to build bridges from his plan to anyone who asked, for the price of a dollar per linear foot of span. This idea was not unique to Finley, and the patent system rapidly became the primary forum of competition among American bridge designers after its establishment in 1790. . . .*

This particular institutional-economic context, involving competition through the patent system, had several implications for Finley's design. First, to earn much profit, Finley's design would have to be sold, or reproduced, as widely as possible, and this made it essential for the design to be economical relative to its competitors, standardized, and universally applicable. It also had to be technically viable: if bridges built according to his plan consistently fell down, it would not be long before no new buyers could be found. Since Finley's concept of a level-roadway suspension bridge was so radically new, and since iron had only recently begun to be used for building bridges, Finley could not rely on experience or on craft know-how. If the design could be justified scientifically, however, this would provide a basis for its technical viability.

A more general factor that Finley had to consider was the undeveloped state of America's economy and material infrastructure. Expansion into

* For reasons of space, notes other than the sources of direct quotations are omitted from this extract.

new frontier areas created a need for more and better roads, sturdier bridges, and a better-developed canal system. At the same time, most communities, particularly frontier communities, had very limited economic resources with which to undertake such projects.

Fayette County [in Western Pennsylvania, where Finley was a farmer and county judge] was on a major route of travel to the west, and a disproportionate number of heavily laden vehicles consequently passed through the region, which heightened the need for good roads and bridges. Yet the population density was low, and Finley was conscious of the region's limited financial resources. . . .

In view of Finley's immediate knowledge about county and state finances, it is not hard to imagine that he would have felt it important to design bridges that would be economical to build and maintain. Of course, not all road and bridge projects depended on public funding. After 1790 more and more of these projects were sponsored by groups of businessmen and citizens for profit. But here again there was little willingness to spend more than was absolutely necessary. As I have already noted, a further reason for pursuing an economical design was that it promised to be an important advantage in the competition with other inventors who were also patenting and attempting to sell their bridge designs.

Just as important as the quest for economy was the need for technical simplicity. Finley realized that, to achieve widespread application, his design should not require the availability of builders with advanced knowledge of mathematics or with specialized technical qualifications (beyond general carpentry or blacksmithing skills). Unlike France, the United States at the time had very few formally trained engineers. It has been estimated that there were no more than 30 civil engineers in the entire country in 1815. Before the founding of West Point (in 1802) there was no institution in America that provided comprehensive training for civil engineers, and even West Point did not offer a solid course in the subject until 1816. What few engineers there were generally became involved with large-scale public works (such as canals) rather than with comparatively small enterprises like bridge building.

Under the circumstances (and particularly in sparsely settled rural and frontier areas, with their limited resources), bridge construction remained largely in the hands of local craftsmen. Although some craftsmen who specialized in bridge building were extraordinarily creative and capable, the average carpenter or blacksmith could not be expected to undertake very large or complex projects. For example, it could not be assumed that a local craftsman would have anything more than a rudimentary knowledge of arithmetic. As a consequence, for a design to be widely applicable it had to be relatively straightforward and uncomplicated to build. Finley took particular care to create a standard plan that could be adapted to a particular site with no need for any but the most trivial arithmetic or geometric calculations. His plan was supposed to 'enable any person to make a rough estimate for any particular case.'[1]

The net result of these conditions, and of Finley's response to them, was a design that can best be characterized as a standardized do-it-yourself plan. It was a distinctly utilitarian design, with no attempt to achieve aesthetic elegance or monumentality. In many respects it was an archetypical frontier technology. Yet it embodied significant experimental research, and some of the characteristics of the design plan that made it easy to follow were discovered by Finley through this research.

► FINLEY'S RESEARCH PROGRAM AND ITS ROOTS IN COMMON SENSE PHILOSOPHY

The aim of Finley's research was to learn more about the laws governing the behavior of suspension bridges. His methodology – which was to discover these laws through systematic experimentation – drew on the Scottish Enlightenment ideas that were being widely discussed at that time through-out America. Particularly influential were the Common Sense philosophers Frances Hutcheson, Thomas Reid, and Dugald Stewart, who created an intellectual synthesis that emphasized the use of observation, experiment, and inductive reasoning to expand man's understanding of himself and of the universe. They believed that applying inductive scientific methods would make it possible to uncover fundamental laws in any field, from agriculture to moral philosophy.

Finley adopted the empirico-inductive methodology advocated by the Common Sense philosophers and their American followers. In particular, he carried out inductive experiments with cables and pulleys in order to discover laws for proportioning suspension bridges so that they would withstand the forces acting on them. Most important, the main cables – which were, in Finley's words, 'the whole skeleton' of the bridge and its 'whole strength'[2] had to be able to resist tremendous tension.

An essential conceptual basis for Finley's experiments was the recognition that the tension in a cable supported at both ends (as in a suspension bridge) is not necessarily equal to the distributed load it supports (i.e., the weight of the roadway and the traffic upon it). The experiments he carried out helped him to determine quantitative relationships between these two parameters – load and tension – for a cable with a given sag/span ratio. (Figure 5.1 illustrates what is meant by sag and span.) The fundamental parameters of Finley's experiments were thus sag, span, tension, and load. The main procedure he followed was that of systematic parametric variation.

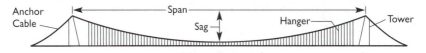

Figure 5.1 A schematic showing the principal elements of a suspension bridge

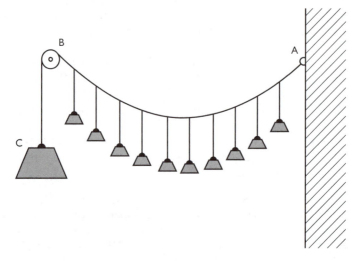

Figure 5.2 Finley's method of studying the behavior of cables bearing distributed loads, in particular the relationships among sag, span, load, and tension

One of the experimental setups Finley designed is illustrated schematically in Figure 5.2. One end of a cable was secured at A and the other end was extended over a pulley at B. Finley attached a weight to the free end of the cable, at C. He then loaded the portion of the cable between A and B uniformly along the horizontal (as ideally would be the case in a suspension bridge). With this setup, the span (i.e., the horizontal distance between A and B) remained constant. This left three variables: the tension in the cable, the load between A and B, and the sag of the cable. To vary the tension in the cable it was necessary to vary the weight at C, for with this experimental setup the weight at C determined the tension. (The cable's tension at C had to be equal to the weight at C, and because pulleys equalize tension the tension had to be the same on both sides of the pulley.) The sag of the cable changed when either the weight at C or the load between A and B was altered.

Finley systematically varied the weight at C and the load between A and B in order to determine the general relationships among the three variables. First, with the tension (i.e., the weight at C) held constant, Finley varied the load between A and B. He found that the sag of the cable increased as the load increased. Then, with the load between A and B held constant, Finley varied the weight at C. He found that the sag of the cable decreased as its tension increased. In other words, Finley found a direct relation between the cable's load and its sag, and an inverse relation between its tension and its sag.

Although these generalizations were significant, they were still qualitative and therefore not sufficient as guides in the design of suspension bridges. To gain quantitative information about the relations among sag, span, tension, and load, Finley devised a set of experiments that were ingenious and yet remarkably simple. He began once again with the experimental setup illustrated in Figure 5.2. First he attached a weight to the cable at point C (assume 10 pounds). He then distributed an equal amount uniformly along the horizontal between A and B (e.g., ten one-pound weights, evenly spaced). Finley knew that the tension in the cable between C and B was equal to the weight supported at C (that is, 10 pounds). And he knew that the tension in the cable was 10 pounds just on either side of the pulley, again because pulleys equalize tension. On the basis of symmetry, Finley concluded that the tension at A was equal to the tension at B and hence to the tension at C.

But if the tension in the cable was 10 pounds at A and at B, was it also 10 pounds everywhere in between? Since the cable formed a curve and therefore continually changed direction, might not the tension vary accordingly? Finley devised a clever experimental setup to investigate this question, illustrated schematically in Figure 5.3. A cable, loaded uniformly along the horizontal, passed over two pulleys at F and G so as to represent half of a suspension bridge's cable (the portion from the center to one end of the span). The ends of the cable were weighted at E and H. Finley discovered that, to maintain equilibrium, the load at E had to be greater than the load at H. This demonstrated that the tension of such a cable was least at the center and greatest at the supports. (The difference between the weights at E and H when equilibrium was established gave the variance in tension. Finley observed that the tension was 'about an eleventh less at the middle of the bridge than at the ends.'[3]) Within the context of the original problem, Finley now understood that the 10 pounds of tension at the supports A and B in Figure 5.2 represented the cable's maximum tension.

To reiterate: Finley had loaded a cable uniformly along the horizontal so that the total load was equal to the maximum tension in the cable. Under these circumstances he found the sag/span ratio to be approximately 1/6.5 or 1/7. Finley realized that this ratio provided an easy means of estimating cable tension: if the sag/span ratio of a suspension bridge were set by the designer at about 1/6.5 or 1/7, then the maximum overall tension that the cables would have to bear would simply be equal to the total load they supported (i.e., the weight of the bridge itself and the loads upon it, assumed to be uniformly distributed).

Finley went on to examine the conditions prevailing for sag/span ratios of 1/9, 1/14, and 1/30. He found that, for a given load (held constant), the tension in a cable increased as the cable's sag/span ratio decreased. For example, in a cable with a sag/span ratio of 1/9 the tension was 1.33 times the load. However, in a cable with a sag/span ratio of 1/14 the tension was 2.0 times the load.

Finley suggested that builders always adopt a sag/span ratio between 1/6 and 1/7, which would allow the amount of iron needed for the chains to be calculated directly from the load that the chains had to bear. In fact, adopting this ratio freed builders from having to distinguish between load and tension. Finley established a simple procedure for making the necessary calculations. First, the weight of the roadway was to be estimated. This required knowledge of the span and the width of the prospective bridge and

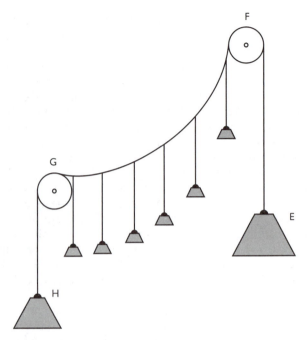

Figure 5.3 Finley's method for determining the variance in cable tension between middle of span and supports

data on the unit weight of the timber to be used. Finley provided an average value for the latter. Second, in order to reach an estimate for the total load the bridge cables had to support, the weight of the roadway was to be multiplied by 5 or 6. This factor was intended to allow for all the loads to which the bridge would be subjected (e.g., traffic and snow). It was also a safety factor – an allowance for weaknesses or flaws in the cables, or other imperfections. Finally, the total load calculated in the second step was to be used to calculate the size and the number of the chains needed for the bridge. This calculation required a value for the strength of iron; Finley used 60,000 pounds per square inch, which he got from a table in an article on the strength of materials written by John Robison for the third edition of the *Encyclopaedia Britannica* (1797). Finally, with a knowledge of the weight per linear foot of the chains (for which Finley again provided estimates), the builder could calculate the total weight of the iron needed to make the chains. . . .

▶ **NAVIER: AN ENGINEER-SCIENTIST**

The background and environment of Claude-Louis-Marie-Henri Navier [1785–1836] differed profoundly from Finley's. Whereas Finley lived in a rural community at the edge of civilization, Navier lived in a major European capital. Whereas Finley was an amateur inventor, Navier was a professionally trained engineer. And whereas Finley was an independent entrepreneur hoping to earn a profit, Navier was a salaried employee in a hierarchical bureaucracy: the Corps des Ponts et Chaussées, one of the national French engineering corps. In view of these important differences, it is not surprising that Navier's way of problematizing the suspension bridge – his response to the technology – was different from Finley's. The link between Navier's environment and the technology he created was indirect: his environment shaped his ideology, goals, and methods of practice, and these in turn influenced his manner of studying and designing suspension bridges. . . .

A variety of social and institutional mechanisms helped Navier gain access to, and encouraged him to become a part of, the French scientific community. His growing ties to this community can be seen in the way his research interests evolved, in the way he internalized important priorities of the community, and in the honors the community bestowed on him. These honors included, most notably, his nomination to the Société Philomathique in 1819 (a major stepping stone to the Académie des Sciences), his election to the mechanics section of the Académie des Sciences in 1824, and, in 1831, his appointment to what had been Cauchy's chair in analysis and mechanics at the Ecole Polytechnique. . . .

Navier's work on suspension bridges must be understood in the context of his dual ties to the engineering and mathematical communities. His concern with suspension bridge technology came about as a result of an official request by the Ponts et Chaussées administration that he visit Britain in order to study and report on the progress of suspension bridge construction there. Available evidence indicates that British interest in this technology was initiated by knowledge of Finley's work in the United States. Finley's contributions had become known in Britain by the autumn of 1811. The earliest British proposal for a suspension bridge for vehicle traffic, prepared by Thomas Telford in 1814, was for an iron-wire suspension bridge with a span of 1,000 feet, to be built at Runcorn, between Liverpool and London. This plan was never carried out, however. The first major suspension bridge to be completed in Britain was the Union Bridge, designed by Samuel Brown and constructed between August 1819 and July 1820.

Taking up the request of the Corps des Ponts et Chaussées, Navier became one of the principal agents for the diffusion of suspension bridge technology to France. He made two visits to Britain: one during the autumn of 1821 and one in the spring of 1823. While there, he read and collected available literature, talked with a number of engineers, and visited several suspension bridges completed or under construction. . . .

British engineers were strongly preoccupied with the problems of cable design and the strength of iron but were very little concerned with creating a comprehensive mathematical theory of suspension bridges. British engineers were not even consistent in the choice of the mathematical techniques they used to determine the tension in a loaded cable. Indeed, Telford and Brown did not use mathematical techniques at all; they used empirical, experimental techniques to determine cable tension. . . .

What did Navier do with all the information he gathered, and how did he subsequently direct his own research into this technology? In order to answer this question, and to understand how Navier's cultural environment shaped his response to the suspension bridge, it is important to recognize what he did *not* do with the information he acquired. Unlike James Finley and Samuel Brown, Navier never attempted to patent a specific design system. And unlike Finley, Brown, Telford, and others, Navier did not build and test models, nor did he do extensive experiments on the strength of iron or experiments on the relation between load and tension in a suspension bridge cable. Nor did Navier investigate new methods for designing cables – a matter that was of central importance to British engineers. Rather than pursue the best method of fabricating suspension bridge cables, he created mathematical descriptions of their curves and movements. . . .

▶ THEORIZING THE SUSPENSION BRIDGE

In physical terms, Navier's abstract and idealized model portrayed the suspension bridge as a perfectly flexible, inextensible, massless cable, supported at both ends and loaded uniformly along the horizontal so as to form a parabola. In mathematical terms, Navier's model amounted to a set of interrelated formulas for the cable's curve, tension, angle of inclination, sag, and length. The crucial parameters on which the equations depended were the cable's sag/span ratio and its load per unit of span. . . .

The most sophisticated part of Navier's theory from a mathematical standpoint was his treatment of cable vibrations due to impact loading. In a suspension bridge, vibrations from impact loading first affect the roadway and from there are transmitted through the hangers to the cables. Navier witnessed such vibrations when he examined Samuel Brown's Union Bridge:

> When vehicles cross [the bridge], different types of effects are produced: the deck sags; because the presence of the vehicle alters the distribution of the load supported by the chains, these chains must take a new shape . . . which conforms to the new state of equilibrium being established. The result of this situation is that the contour of the deck is continually modified throughout the duration of the crossing. . . .
>
> Apart from these changes of contour . . . which must be considered as a static effect resulting from the flexibility of the chains and the deck, dynamic effects are produced, some of which result equally from the flexibility of the construction, and others which are due to the elasticity of the materials. The little jolts [on the roadway] stemming from the movement of vehicles and horses . . . are transmitted by the hangers to the chains, which, during the passage of vehicles, always display a very

small horizontal swaying . . . such that these chains . . . are almost always in movement.[4]

Such vibrations were also described by Robert Stevenson in the article on suspension bridges cited by Navier. Referring to the Dryburgh Bridge, Stevenson called attention to 'the sudden impulses, or jerking motion of the load.'[5] British engineers did not attempt to analyze this phenomenon mathematically as did Navier, however.

Navier based his analysis on the assumption that a load, P, would fall directly onto a cable precisely at the middle of the span. In other words, roadway and hangers (and also towers and anchor cables) were abstracted out of the model. Navier justified this idealization on the grounds that his analysis would establish the theoretical upper limits for the effects of impact loading. A further advantage of this approach for Navier was that the problem became analogous to the vibrating string problem, a mathematical classic that had been investigated by a host of prestigious researchers, including Euler, d'Alembert, Lagrange, the Bernoullis, and others, up to and including Fourier.

Navier considered two types of vibrations: transverse oscillations (such as those produced when a violin string is plucked) and longitudinal vibrations (i.e., elastic expansions and contractions of the cables). To solve the problem of transverse oscillations, he worked out two partial differential equations that gave the cable's displacement as a function of time and position (along the x axis). One equation held for all points along the cable; the other held strictly for the midpoint, which carried the load P. It was one thing to derive these equations and quite another to solve them. A general methodology for solving them had eluded eighteenth-century mathematicians; it was worked out only in Navier's generation, largely through the efforts of Fourier, whose technique Navier explicitly followed. This technique enabled Navier to determine the desired solution by computing the constants of its trigonometric-series expansion. The resulting function occupied two full lines and comprised multiple sine and cosine expressions.

Owing to its mathematical complexity, the solution function Navier derived gave no immediate physical insight into the character of suspension bridge vibrations. Navier had to simplify and restructure the expression carefully in order to find out roughly how the frequency and the amplitude of the oscillations were related to bridge size. He discovered that both the frequency and the amplitude would decrease as the overall size of the bridge increased, with the sag/span ratio assumed to be constant. He found, in other words, that large bridges would be relatively less affected by vibrations than small ones. . . .

We can now draw some more detailed conclusions about the ways in which his research was socially shaped (beyond the mere fact that he chose to develop a mathematical theory at all). First, Navier's theory was socially shaped in the sense that, to create it, he employed a set of mathematical tools and techniques which were developed, taught, used, or promoted within his own institutional environment. These included basic mathematical tools, such as calculus, which in principle were known to every graduate of the Ecole Polytechnique (but which were not known to Finley, Telford, and Brown, for example). They also included more special-ized mathematical techniques and equations, such as the equation for belt friction and the general equation for the length of a curve and the equa-tion for the movement of a pendulum. Finally, they included advanced and specialized mathematical knowledge and techniques that were not widely known or used at that time outside of a small research community, of which Navier was a part. Prominent among the latter were the theory of elasticity and Fourier's technique for solving partial differential equations by means of trigonometric series expansions.

It might be argued that the mere use of a particular set of mathematical tools and techniques does not reflect any social shaping. But it does reflect such shaping in the sense that people most readily use the tools that are available, accepted, and relevant within their own environment. For ex-ample, in principle Finley could have read and used Lagrange's *Mécanique Analytique* in his research on suspension bridges, but it would have been unusual had he done so: that tool was not readily available to him, and it had no accepted place or particular relevance in early-nineteenth-century Fayette County. Of course, many creative individuals master tools and techniques that are foreign to their environments; yet it is extremely rare for an individual to reject all the tools provided within his or her own environment. Navier did adopt a new technique when he helped to introduce suspension bridge construction into the Corps des Ponts et Chaussées, yet he attempted to make this new and foreign technique more acceptable and relevant by evaluating it, by reinterpreting it, and by restruc-turing it on the basis of mathematical tools and techniques that were already accepted and respected within his community.

Navier's use of a particular set of mathematical tools reflects social shaping at another level as well. The Continental system of calculus and the style of Navier's mathematical analysis were very different from those used by British researchers. As Joan Richards has observed: 'The English clung to Newtonian fluxional notation, which was geometrical and also more constraining than the algebraical Leibnizian dy/dx symbols used on the Continent. Throughout the eighteenth century, English and Continen-tal calculus developed along these different lines and in virtual isolation from each other.'[6] Richards has argued convincingly that these differences, in the case of France and England, were linked to the very different insti-tutional structures in which mathematics was promoted and diffused in the two countries – structures that reflected very different philosophies con-cerning the nature of mathematics and its role in society.

Navier's theory was also socially shaped in the sense that it embodied many decisions about what to analyze and what to ignore that reflected goals and priorities stemming from or rooted in his environment. Why did Navier devote considerable attention to the problem of cable vibrations but completely ignore the question of the behavior of the deck, which was at least as important for builders? Why did he choose to analyze in detail the problem of cable slippage over the towers (taking into account friction, elasticity, and temperature variations) but, for the most part, to ignore the issue of cable design?

The evidence provided in the preceding sections has shown that Navier's decisions about what to analyze and in what depth were not simply dictated by 'the technology,' or 'the evidence,' or even just by the needs of builders. This is not to imply that these things were irrelevant. The way Navier mathematized his observations of vehicles crossing the Union Bridge and the way in which he classified different tower-cable systems show that his decisions *were* made partly on the basis of empirical evidence and technological precedent. But they were also made partly on the basis of who he wanted to impress, and what mathematical issues were currently of concern to the Parisian scientists with whom he was increasingly interacting.

Navier's decision to devote much attention to cable vibrations due to impact loading from traffic but only cursory attention to wind-induced oscillations provides an example. The difficulty of the wind problem cannot fully account for this decision, nor can it be accounted for by the empirical evidence available to Navier. Navier had clear evidence of the potential seriousness of wind-induced oscillations: direct evidence from the Dryburgh Bridge and indirect evidence from Brunel's bridges and from Brown's Trinity chain pier. (Brunel's two bridges and the Trinity chain pier had provisions to keep the roadways in place during windstorms.) Navier's empirical evidence concerning cable vibrations, in contrast, did not suggest such danger. Referring to the vibrations of the Union Bridge due to traffic, Navier commented that they 'in any event do not appear to jeopardize the solidity of the construction.'

Navier's decision to devote nearly thirty pages of difficult mathematics to the problem of cable vibrations but only four pages to the problem of wind becomes more understandable when we recall certain facts about his growing stature within the French scientific community. Navier, who had been elected to the Société Philomathique in 1819, was hoping in the early 1820s to achieve the next step: election to the Académie des Sciences. By 1820–21 Navier had begun to do important research on the mathematical theory of elasticity, which was one of the main interests of Parisian mathematicians. It should also be noted that Fourier became *secrétaire perpétuel* of the Academy of Sciences in 1822, which was the most powerful position in that organization. And Fourier's research on heat, published in *Mémoires de l'Académie des Sciences* in 1819 and 1820 and then as a book in 1822, introduced his new technique of using trigonometric series to solve partial differential equations. What better way to make an impact on Fourier and on mathematicians concerned with the theory of elasticity than to apply their theories and mathematical techniques to solve a new and difficult problem? In 1823 Navier even extracted a key portion of his cable-vibration analysis and published it separately in the *Bulletin de la Société Philomathique*, where it would be more visible to the mathematics community. . . .

► **THE PONT DES INVALIDES**

Although Navier's research on suspension bridges reflected his growing ties to the French mathematical community, he did not abandon his identity as a practicing engineer and builder. Not content simply to theorize, he wanted also to design and build suspension bridges. . . .

Navier's wish to build a monumental suspension bridge, and the ideas that guided his design, reveal goals and priorities very different from those that guided James Finley. Whereas Finley's plan was intended for mass application with little reliance on the skills of professional engineers, Navier's bridge was deliberately intended to serve as a monument to the glory of France and the Corps des Ponts et Chaussées. Whereas Finley patented his design system, the very concept of a patent was inimical to Navier's aims, since he wanted to build a unique design that would not be replicated. (Navier did not take out a patent in his entire career.)

Navier's attitude was shaped by the tradition of monumental architecture with which the Corps des Ponts et Chaussées was closely associated and which students encountered at both the Ecole Polytechnique and the Ecole des Ponts et Chaussées. A major public work, such as a bridge in the nation's capital, was expected to be a unique work of art that conveyed a sense of majesty and harmonized architecturally and decoratively with its surroundings. This attitude led many Corps engineers to manifest a certain disdain for what they regarded as the more petty objectives and the limited vision of entrepreneurs, who tended to seek out economical designs rather than majestic ones. As the director of the Corps put it in 1831: 'When it is a question of constructing a bridge in the heart of the capital and in the vicinity of other magnificent monuments, it is indispensable that the new construction be in harmony with its environment. There are special considerations of taste and of decoration which would make it difficult to leave the work to the entrepreneur, who, as we all know, always seeks that combination which costs him the least.'[7]

Navier was not only motivated by aesthetic considerations, however; he also intended the Pont des Invalides [his planned bridge over the river Seine in Paris] to be a vindication of his theoretical approach. In his report to Louis Becquey (Directeur Général of the Corps des Ponts et Chaussées) Navier had argued that mathematical analysis was essential in order to fully understand the practical potentials offered by suspension bridges: 'It was necessary . . . to make a thorough study of a type of construction which seemed to offer great advantages and about which almost nothing had yet been learned through time and experience; but this study would not have been possible without the progress made in mathematical analysis in recent times, and without the institutions by means of which those charged with the direction of public works are initiated into the most advanced ideas of mathematics.'[8] By means of the Pont des Invalides, Navier sought to carry this argument one step further and to give a material proof of the applicability of his theory to practice and, by extension, of the relevance of mathematical theory to engineering practice generally. Navier opposed empirical approaches to structural design because in his estimation they led to overbuilding. He believed that only on the basis of theory could a precise correlation between the design and the function of structural elements be achieved.

Accordingly, in Navier's plan for the Pont des Invalides, every possible element – cables, hangers, columns, floor beams, etc. – was designed and proportioned on the basis of theoretical analysis. For example, in his theory Navier had developed equations to determine the forces acting on the towers of a suspension bridge, tending either to crush or overturn them. He had shown that the overturning forces depended on the amount of friction between cables and towers, and he had applied the belt-friction equation to this problem. This equation assumed the top of a tower to be an arc of a circle, and consequently Navier actually designed the tower tops of the Pont des Invalides that way, so that the equation would remain valid in practice. Navier applied the equation to determine the upper limit of the overturning force that could develop in the Pont des Invalides, and through this calculation he found the minimum dimensions that would ensure stability of the columns. He also determined the vertical crushing forces that would act on the columns, and to ensure that these could be withstood he designed the columns with an internal network of iron reinforcing bars.

Navier also calculated the dimensions of the iron chains theoretically, using 14 kilograms per square millimeter as a stress limit for the cables so as to ensure that they would not be stressed beyond the yield point of wrought iron. James Finley had figured the maximum tension in the cables by multiplying the projected deck weight by 5 (assuming a 1/7 sag/span ratio). Navier's method was more intricate. First, he calculated the combined weight of the hangers and the deck, including the side rails. To this he added a value for the maximum live load the bridge would be required to bear, based on the assumption of the deck's being entirely covered with people standing shoulder to shoulder like 'troops marshalled for battle.' (He had calculated that this weight would be greater than if the bridge were entirely covered with cattle, cavalry, or the most heavily loaded wagons.) Then, using his formula for cable tension, he calculated the maximum tension that the total load would produce in the cables, and from this he determined the total cross-sectional area of iron needed to resist this tension.

▶ **FINLEY'S AND NAVIER'S BRIDGES COMPARED**

The technical features of Finley's and Navier's designs – their choices of hardware, and the technical means of interconnecting the various elements of the structures – simultaneously embodied their knowledge, their goals, and their ideas about how the structures would be built. The technical details of the designs embodied tradeoffs among these elements, and specific choices often undercut other goals or influenced other elements of the design. Finley's decision to adopt a sag/span ratio of 1/7 simplified the calculations that builders needed to carry out, and ensured that no iron would be 'wasted' to accommodate additional stresses due to greater tautness; yet that ratio sacrificed stability. Finley's simple tower design helped to make the bridge easy to build; but it effectively limited the length of the spans that could be built, and it led to greater reliance on multiple spans. Reliance on multiple spans, in turn, sacrificed stability. Finley's cable design, with only one link between hangers, saved iron and made the structure cheaper, but it necessitated the design of two types of 'keys' to connect the hangers to the cables.

Navier's decision to adopt a sag/span ratio of 1/10 for the Pont des Invalides helped to ensure greater stability. His compound cable enhanced the monumentality of the design but sacrificed economy and ease of construction. His method for determining hanger lengths enhanced re-liance on mathematical theory, but it too sacrificed economy and ease of construction (in the sense of requiring more highly skilled labor in the design-to-construction process in order to carry out the necessary math-ematical calculations). Navier's deck design attempted to achieve stability primarily through weight rather than stiffness. (The dead weight per linear foot of his bridge was greater than that of any previous suspension bridge.) Using weight to achieve stability tied the Invalides design – at least in Navier's view – more closely to the precepts of his theory. Yet it sacrificed economy, because the added weight increased the cost.

A fuller understanding of Finley's and Navier's different responses to the suspension bridge has to go beyond the mere existence of a time gap; it must take into account the very different environments in which the two individuals worked. Environments shape strategies. Individuals take stock of their environments in order to decide which strategies are most likely to bring success. This does not mean that environments are one-dimensional or all-knowable. An individual responds according to his or her perceptions of what are the most significant elements and characteristics of the environ-ment, and these perceptions may in turn be shaped by a host of factors (including personal psychology). But the environment places limits on the range of strategies that can be made to succeed.

How far would Navier have gotten if he had tried to do in Fayette County what he sought to do in Paris? How would he have mobilized the equivalent of a million francs to have a monumental suspension bridge built there? How would he have convinced the farmers who served as county commis-sioners to let him build such a bridge, or even to contribute to the cost of it? Where would he have found an engineer assistant who could compute hanger and link lengths to four decimal places, following a complex 14-page mathematical procedure? How many in Fayette County would have agreed that such a bridge should be built partly as a proof of the applic-ability of mathematical theory to constructive practice?

Finley and Navier attempted to structure their research and design efforts in relation to their environments. For the purposes of comparison, it is helpful to think of these environments in terms of three dimensions: an intellectual dimension, an economic dimension, and an organizational dimension.

With regard to the intellectual dimension, Finley adopted an empirico-inductive research methodology advocated by the Common Sense philosophy that was popular within his intellectual community. By following this methodology, Finley helped to ensure that his research findings would be understood and accepted by his peers. He would have had much less of a chance to convince a rural blacksmith or carpenter or county commissioner to build one of his bridges if he had given a proof of its technical adequacy based on the mathematics of Euler, Bernoulli, and Lagrange, which the majority of them had not read and did not care about.

In contrast, Navier followed a theoretico-deductive methodology that was respected within his intellectual community. By so doing, Navier helped to ensure that his research findings would be considered valuable and insightful by certain groups of his peers (and his superiors) in the scientific community and in the Corps des Ponts et Chaussées. If Navier had gone the route of Finley, Brown, and Telford, and had studied the equilibrium of loaded cables empirically, without reference to the mathematical contributions of his colleagues and predecessors, his work would probably have been scorned by the likes of Becquey, Prony, and Fourier.

Both Finley and Navier sought a deeper understanding of the character and behavior of suspension bridges, but their distinct research methodologies and intellectual perspectives led them to employ different tools to achieve this understanding. Finley used models, and experimental apparatus comprising cables, pulleys, and weights. He also sought understanding through encyclopedia articles on bridges and on the strength of iron, and through the construction (in 1801) of a 70-foot trial span over a stream close to Uniontown. Navier's tools were mainly mathematical: calculus, Fourier series, elasticity theory, mathematical statics, and a variety of specific equations such as the equation for belt friction. Navier also sought understanding through inspection of existing British bridges, through discussions with the builders of some of those bridges, and through published literature, not only on suspension bridges, but also on mathematics, theoretical mechanics, and elasticity theory.

Finley's research was closely tied to his design goals: he wanted knowledge that would enable him to come up with a workable yet simple and generic design. He therefore paid careful attention to such matters as the procedures smiths should follow to fabricate the links of the chain cables. Navier's research was not focused as much on design and hardware; it was wider ranging, it was often more abstract, and it was explicitly linked to research agendas that were of scientific rather than technological interest. Navier's analysis was also more comprehensive than Finley's. Finley could say nothing specific or quantitative about the effects of friction, cable elasticity, or temperature variations on the equilibrium of suspension bridges; Navier analyzed these issues in detail. Yet Finley had no special reason to seek out such knowledge. It made sense for him to seek out design solutions to minimize effects of vibration, expansion, and contraction, rather than to study the extent of the manifestations of the phenomena in extreme cases.

Navier's intellectual environment included an acceptance of monumentality as a worthwhile goal in structural engineering. This was not the only goal that was considered valid by his peers, but it was an important one: within the Corps des Ponts et Chaussées, monumental projects were the ones that received greatest respect and acclaim. It was quite reasonable for Navier to attempt to integrate the suspension bridge into this tradition. Finley, on the other hand, deliberately avoided monumentality, because it conflicted with the strict religious outlook that was prevalent within his environment, which equated monumentality with wastefulness, 'vanity,' 'contamination' of the mind, and 'idle elegance and show.' (These were expressions used by Finley to describe the nature and consequences of monumentality and to justify the non-monumental character of his design.) Finley also perceived that monumentality would not be a practical selling point in the many communities where bridges were needed but where money was scarce. In Navier's environment of the Corps des Ponts et Chaussées, however, monumentality was a good selling point.

Concerning the economic dimension of the environment, Finley adapted the suspension bridge to succeed in a competitive market against other bridge designs – particularly wooden arches and trusses. Since price was an important criterion, and since Finley had to 'sell' many bridges to earn a reasonable profit through royalties, he needed to do everything possible to minimize the construction price of his bridges. Navier's environment did not oblige him to compete in the marketplace at all. He was a salaried employee, and in principle it made no difference to his income whether he built one bridge or a hundred. Navier probably had more to gain by building one expensive, monumental bridge than a host of cheap ones. To ensure that he could build the monumental one, however, Navier had to convince his superiors that such a bridge was a worthwhile project. By presenting his project as a sophisticated application of theory that would establish a new technology in France on a monumental scale, Navier was striking chords that he knew his superiors – especially Becquey and Prony – would respond to.

Finley and Navier adapted the technical elements of their bridge designs to these different economic contexts. Whereas Finley's design choices show that he tried always to minimize costs, Navier's design choices reveal that keeping costs low was not an overwhelming priority for him. In his treatise, Navier had pointed out that the cost of a bridge was roughly proportional to the weight of the materials in it. Yet he did not attempt to minimize the weight of materials in the Pont des Invalides, but rather to maximize it (relatively speaking), so as to improve stability, despite the added cost. Finley sought to minimize weight in order to save money. Navier proposed to use heavy oak and iron for his bridge deck; Finley advised builders to use light pine. Finley sought stability through stiffness, Navier through weight. The latter was the more expensive option, at least in these instances.

Finley's use of short, multiple spans was also a money-saving choice. He saw the possibility of long-span suspension bridges, but he chose not to take up that possibility in his design. Shorter spans meant shorter, cheaper towers that could be made of wood. Shorter spans also made it possible to rely on the skills of local craftsmen, because shorter spans were easier to construct (unless the intermediate piers created special difficulties). Navier chose the long-span option to accord with his aim of monumentality, yet this choice resulted in a comparatively more expensive structure. Navier's use of compound cables to achieve a more monumental look also increased costs. The compound cables were more expensive because they were more difficult to construct. Navier had problems trying to adjust the cables in the sun – the job required that special canopies be built to shade the cables. Finally, Navier's design had technically unnecessary decorative features that were undoubtedly expensive, such as the reclining lions at the anchorages.

Finally, Finley and Navier adapted their designs to be built within different organizational structures and networks. Finley made use of the patent system, aiming to collect royalties. Navier worked within the administrative structures of the Corps des Ponts et Chaussées. These organizational structures involved different tradeoffs and compromises with regard to technical control over the building process, financial control (whether and how to finance the building process), and control over the supply of labor.

Because Finley used the patent system, permitting anyone to implement his design as long as they paid royalties, he often had no immediate involvement in the building of a bridge based on his design. This was an advantage in terms of his own career. Being a judge and a farmer, with local commitments in Uniontown, he probably did not wish to travel around the country building bridges. Yet the consequent lack of technical control became a significant problem for Finley, to the point of jeopardizing the reputation of his design system.

Using the patent system also meant that Finley generally could not exert direct control over financing. This organizational context gave him no authority to decide if a bridge should be financed by a local government, or through the creation of a corporation, or by some other means, or not at all. All Finley could do was advertise his bridge design, and otherwise encourage prospective builders, and then demand a royalty payment if a bridge were built. Finley also could not exert direct control over the supply of labor for the building process, and therefore he could not ensure that those doing the building would have specialized technical qualifications (apart from e.g. a general knowledge of carpentry).

Finley adapted his bridge design to this organizational context in several ways. First, he created a generic design, to which he added a step-by-step plan to guide builders through the process of adapting the general design to a specific location. Second, he attempted simultaneously to maximize the ease of construction and to minimize the need for technical expertise (both of which reduced costs), by creating a design that could be built by local craftsmen without any need for mathematics beyond arithmetic and without any need to differentiate between load and tension in a cable. The key here was the 1/7 sag/span ratio. Builders could likewise determine hanger and link lengths using a 'board fence' and a simple empirical technique.

Navier's tradeoffs were different. Working within the organizational context of the Corps des Ponts et Chaussées, Navier first had to accept design changes ordered by his superiors. Yet this context guaranteed that, once a design was agreed upon by the Corps, there would be strict technical control over the building process. This control was achieved by means of a legally binding set of technical and administrative specifications (the *cahier des charges*), by daily inspection of all work and materials by an assistant engineer, and by the agreement that made Navier the technical director of the project. Navier's context also guaranteed that he would have access to specified materials (e.g., stone of a certain type) and to a labor supply with the technical expertise he wanted. He could count on having an assistant with enough mathematical and technical knowledge to understand and apply his mathematical analysis, and to oversee materials, labor, and construction methods. Like Finley, however, Navier could not exert significant financial control over the project. Here he had to submit to the authority of the Corps' executive council, despite the fact that their decision to have the Pont des Invalides financed by private capital directly opposed his own wishes. The only influence Navier managed to exert on this score was to carry out a study to demonstrate that the bridge would bring an adequate return to investors. (Certainly this was an important way to attract capital, however.)

Navier adapted his design to the organizational context. His design presupposed the availability of highly skilled labor and the ability to control the technical details of the construction process. It is evident that Finley, in his organizational context, could never have hoped to succeed with the kind of anchorage used by Navier. Navier's anchorage required that the underground buttresses be placed where theory said they should be – the design presupposed that someone conversant with the theory would monitor the construction process to ensure that the buttresses were correctly built and positioned. Navier's design also required careful and systematic testing of links, and mathematical calculation of hanger and link lengths. The latter presupposed the availability of highly skilled labor and expert monitoring of the construction process.

Both Finley and Navier innovated, but their innovative work was not a transcendent quest for novelty; it was fundamentally structured by their respective environments. Finley created a new structural form – the suspension bridge with a level roadway – with the idea that it would provide a cheap means of spanning the many streams and rivers of a largely undeveloped country. He innovated to achieve a design that could be technically and economically successful in regions with little money, comparatively few inhabitants, and laborers having only general technical skills. Navier created a new mathematical theory that built on the mathematical contributions of his colleagues and his predecessors. He also innovated in adapting the suspension bridge to the classical French tradition of monumental architecture, and in working out a design that followed guidelines and precepts deduced from his theory.

In the long term, some of Navier's and Finley's innovations had considerable impact beyond their initial environments. Finley's idea of a level-roadway suspension bridge opened a realm of possibilities which builders have explored ever since. Navier's theoretical work – manifested not only in his theory of the suspension bridge but also in his teaching – brought more sophistication to structural theory and helped to make statics, mechanics, and the theory of elasticity more powerful tools in the hands of engineers.

In the short run, however, the innovative work of Finley and Navier did not prove entirely successful. Both designers found their efforts thwarted by forces beyond their control. The initial success of Finley's system – because it occurred within a competitive market environment – stimulated further competition. Other designers began to beat Finley at his own game with new wooden arch and truss designs that cut into the market for chain bridges. At the same time, his lack of control over the building process led to accidents which may have damaged public confidence in his design system. Navier was thwarted by forces both within and beyond the Corps des Ponts et Chaussées. Constructing the Pont des Invalides within that organizational context protected his project from outside technical and economic competition. Yet it also meant that the project was dependent on the Corps' continued protection. And when the failure of the anchorage led to financial disagreements, to public opposition, to embarrassment to the Corps, to a hostile resolution from the Paris municipal council, and to the threat of a lawsuit, the Corps withdrew its support and the Pont des Invalides was removed.

 NOTES

1 James Finley, 'A description of the patent chain bridge,' *Port Folio*, n.s. 3 (June 1810), p. 443.
2 Ibid., p. 452.
3 Ibid., p. 447.
4 C-L-M-H Navier, *Rapport à Monsieur Becquey et mémoire sur les pont suspendus* (Paris: Imprimerie Royale, 1823). A second edition of this work (Paris: Carilian-Goeury, 1830), published after the removal of Navier's Invalides Bridge, has different pagination and includes a 50-page appendix concerning the *Pont des Invalides* but otherwise corresponds to the first edition. The quotations in the text are on pp. 145–54 of the second edition.
5 Robert Stevenson, 'Description of bridges of suspension,' *Edinburgh Philosophical Journal* 5 (1821), April–October, p. 255.
6 Joan L. Richards, 'Rigor and clarity: Foundations of mathematics in France and England, 1800–1840,' *Science in Context* 4 (1991), no. 2, p. 298. Peter Barlow was an exception to this pattern; his 1817 *Essay on the Strength and Stress of Timber* (London: J. Taylor) does make use of Continental notation.
7 Quoted from James M. Oliver, *The Corps des Ponts et Chaussées, 1830–1848* (University of Missouri, 1967), pp. 175–76.
8 Navier, *Ponts suspendus* (second edition), p. 15.

6 ▶ Competing technologies and economic prediction

W. Brian Arthur

'Every steam carriage which passes along the street justifies
the confidence placed in it; and unless the objectionable
features of the petrol carriage can be removed, it is bound to
be driven from the road, to give place to its less objectionable
rival, the steam-driven vehicle of the day.'
(William Fletcher, *Steam Carriages and Traction Engines*,
1904, page ix.)

In 1890 there were three ways to power automobiles – steam, gasoline, and
electricity – and of these one was patently *inferior* to the other two: gasoline.
Yet today the entire automotive technology is based upon gasoline. It is
possible, of course, that gasoline possessed hidden engineering advantages
that were only slowly uncovered. But another, quite different explanation
can be put forward.

Very often, technologies show increasing returns to adoption – the more
they are adopted the more they are improved, and the more attractive they
become. Aircraft designs, for example, improve greatly in structural sound-
ness, maintenance costs, and payload capacity as they accumulate experience through actual airline operation. When two or more increasing-
returns technologies compete for adopters, insignificant 'chance' events
may give one of the technologies an initial adoption advantage. Then
more experience is gained with this technology and so it improves; it is
then further adopted, and in turn it further improves. Thus, the technology
that by 'chance' gets off to a good start may eventually 'corner the market'
of potential adopters, with the other technologies gradually being shut out.

Whether the automotive industry is locked-in to a gasoline technology by
historical small events magnified by increasing returns, or by the innate
superiority of gasoline engines, is a matter that would require careful
historical weighing of evidence together with detailed engineering analysis.
If we take the increasing-returns explanation as valid, however, we can see
in this example four key features of the dynamics of markets where increasing returns are present.

First, the technology that 'wins' a market does not necessarily have to be the 'best' or most efficient. In the case of the automobile, the steam (Rankine) cycle is thermodynamically more efficient than the gasoline (Otto) cycle. Given as much development as the gasoline engine has undergone over the last ninety years, it is quite possible that a steam engine could have been more economical. (There are several recent steam prototypes that achieve better fuel mileage and have lower exhaust emissions than current gasoline power sources.) In the dynamics of choice under increasing returns, even when individual choices are perfectly rational, there is a potential economic *inefficiency* of outcome.

Second, an industry (or economy) can get 'locked-in' to a technological path that is difficult to get away from. As more and more people choose one technology from a group of competing technologies, that technology becomes more attractive. The other technologies become 'frozen out' of the market and often disappear. To re-establish them, a widening change-over gap would then have to be closed. In cases with increasing returns, there is a potential *inflexibility* where ultimate 'market shares' cannot always be easily altered as a matter of policy.

Third, even with hindsight, the reasons why a particular technology came to be adopted are difficult to pinpoint. Exact causality is hard to ascribe. Where increasing returns are present, it is often a mistake to explain adoption by the 'superiority' of the technology, as is traditional. There is a *non-ergodicity**: historical 'small events' are not averaged out and 'forgotten' but may well decide the path of adoption shares.

Fourth, even if we know all the preferences and possibilities of those choosing, the outcome – the share of the market taken by each technology – is often impossible to predict in advance. If small events can decide the outcome, and if these are in some sense 'too small' for the economist's notice, then with increasing returns there is a *non-predictability*: knowledge of supply and demand usually does not suffice to predict theoretically the share of the market that each technology will take. Of course, with increasing returns we may be able to predict that one technology will come to dominate, we may be able to give odds on each, but we cannot with accuracy say *which* technology will dominate.

► DYNAMICS OF CHOICE UNDER INCREASING RETURNS

As one possible, simple model of an adoption process with increasing returns, imagine two technologies, A and B, competing with each other to fulfill a particular economic purpose. They compete in the sense that

---

* Editors' note: An ergodic process is one in which the initial state is, in the long run, irrelevant. 'Ergodicity' is a term from the mathematical theory of probability, in which a process involving probabilistic transitions between a set of states is described as 'ergodic' if the probabilities of the states tend, in the long run, to values that are independent of the state from which the process begins.

Figure 6.1 Stanley Steamer (reproduced from N. Taylor, *The Stanley Steamer and Other Steam Cars.* © 1981, Bellerphon Books, 36 Anacapa Street, Santa Barbara, California 93101, U.S.A.)

adoption of one will displace or preclude the adoption of the other technology.

Imagine manufacturers – economic agents – having to choose between the two technologies. Once he has chosen a technology, each agent stays with it and his payoff is not affected by future changes. The agents fall into two groups or types, R and S, with equal numbers in each type, but differing in the use to which they put the technologies. Let us say R-agents, initially at least, prefer technology A, and S-agents prefer B.

Now assume that payoff or returns to adopting A or B increase linearly (at a given rate) with the numbers who have chosen A or B respectively. And assume each agent's moment of choice is subject to small, but unknown, events, so that, to us as observers, choice order looks like a binary sequence of R- and S-agent types, with the probability that an R or an S stands in the nth position in line equal to one-half.

This is a well-defined, neoclassical model of choice: two types of agents choose between A and B, each agent demands one unit inelastically and the supply-cost (or returns) are known. The only unknown is the order in which the agents choose; this is subject to 'small events' below the notice of our model. What happens to the market share of the two technologies?

Initially at least, if an R-agent arrives at the 'adoption window' to make his choice he will adopt A; if an S-agent arrives he will adopt B. Thus the difference-in-adoptions between A and B moves up or down by one unit depending on whether the next adopter is an R or an S, that is, it moves up or down with probability one-half. This process is a simple gambler's-coin-toss random walk. There is only one complication. If, by 'chance' a large number of R-types cumulates in the line of choosers, A will then be heavily adopted and hence improved in payoff. In fact, if A gains a sufficient lead over B in adoptions it will pay S-types to switch over. Then both R- and S-types will be adopting A, and only A, from then on. The adoption process is locked-in to technology A. Similarly, if a sufficient number of S-types by

Figure 6.2 Clock by Paolo Uccello, Florence, Italy (Casa Editrice Giusti di Becocci)

Table 6.1 Properties of the three regimes

	Necessarily efficient	Necessarily flexible	Predictable	Ergodic
Constant returns	Yes	Yes	Yes	Yes
Diminishing returns	Yes	Yes	Yes	Yes
Increasing returns	No	No	No	No

'chance' arrives to adopt B over A, B will improve sufficiently to cause R-types to switch over. The process will then lock-in to B. Our random walk is really a random walk with absorbing barriers on each side, the barriers corresponding to the lead in adoption it takes for each agent-type to switch its choice.

All this is fine. We can now use the well-worked-out theory of random walks to 'prove' the properties I pointed to earlier. The important fact about a random walk with absorbing barriers is that absorption occurs eventually with certainty. Thus in the model I have described, the economy *must* lock-in to one of the two technologies, A or B. But *which* technology is not predictable in advance. Also, the order of choice of agents is not 'averaged away'; on the contrary, it decides the eventual market outcome. Thus the process is non-ergodic. Nor is it flexible. Standard policy measures of favoring one technology over another by tax or subsidy merely shift the barriers. But if the process has become locked-in, the leading technology is constantly improving, so that after a certain time any given boost to the payoff of the excluded technology will not be sufficient. Further, it is easy to construct examples in which this 'greedy algorithm' of each agent taking the technology that pays off best at his time of choice may miss high rewards to the future adoption and development of the excluded technology. Economic efficiency is not guaranteed.

These results are drastically altered in the standard textbook diminishing-returns case. Here technologies, as they become adopted, exert pressure on scarce resources, so that their returns fall with adoption. Hydroelectric power, for instance, becomes more expensive with increased use as the more suitable dam sites are taken up. It is easy to show that the market shares of technologies in the diminishing-returns case are governed by a random walk with reflecting barriers. Here the market for the two technologies is usually shared: the outcome is predictable, as it is the same regardless of the 'small events' sequence; it can always be changed as a matter of policy; and it is always economically efficient.

Where technologies remain the same in payoff regardless of the numbers of adopters – the constant-returns case – the dynamics are governed by a random walk without barriers. Table 6.1 summarizes the properties of the three contrasting regimes.

▶ IMPLICATIONS

There are several implications of the increasing-returns mechanism I have sketched out here. If this type of mechanism is valid, we would expect past history to contain a 'fossil record' of technologies that could have been as good as, or, given equal development, might have been better than, the technologies which were eventually adopted. One example is the direction of motion of the hands on the Uccello clock in the Cathedral in Florence, Italy. They turn anti-clockwise. The Uccello clock was constructed in 1433: it wasn't until after 1550 or so that the clockwise movement became standard.

We would also expect to see technologies which are patently inefficient but which we are 'stuck with'. The U.S. color television system, the driving-on-the-left convention in Britain (bad for car exporters) and the extreme longevity of the 1950s' programming language FORTRAN are examples. The 'standard' keyboard on typewriters is a case in point. Before 1873, early typewriters came with a variety of keyboard arrangements. But in that year, Christopher Scholes, together with his brother-in-law, a schoolteacher, designed a keyboard to overcome mechanical problems with sticking key-bars. The first six letters on the upper row of Scholes' keyboard were QWERTY. The Remington Sewing Machine Company of New York started mass-producing typewriters on the Scholes model – with the QWERTY keyboard. An international meeting in 1904 was supposed to decide on one keyboard among the many alternatives to become the standard. No agreement was reached, primarily because of opposition to any change from typing teachers. QWERTY keyboards are now used in all but 3 of the 45 nations with Roman alphabets and superior competitors to the QWERTY system – the Dvorak system and the Maltron system – have had trouble in gaining a footing.

Policy measures are generally straightforward in the diminishing-returns and constant-returns cases. Here it is usually best to leave the adoption process alone and let the market find its way to an efficient mix of technologies. But where competing technologies show increasing returns to adpotion, the 'fittest' of the technologies may not survive. The government may then need to step in, to encourage and protect infant technologies that, if sufficiently adopted and developed, may pay off handsomely. But there are difficulties. Eventual returns to a technology (think of solar energy, for example) are hard to ascertain; so that while there are obvious dangers and costs of missing out on a potentially superior technology, there are equally obvious costs to exploring large numbers of unknown techno-logical paths.

The argument here implies that we should be careful in interpreting economic history. We usually look for reasons why a predominant tech-nology was superior, and how this 'innate' superiority eventually led to adoption. But this line of reasoning is valid only for cases of constant and diminishing returns. Where technologies exist potentially in ever more improved designs, superiority becomes a function of adoption or use. To return to our gasoline versus steam engine example, it is quite possible that gasoline was indeed innately superior. The matter has never been settled. But it is equally possible that a series of small events at the turn of the century gave gasoline a temporary lead that subsequently proved unassailable. In the North American case, we can, among other small events, single out an 1895 horseless carriage competition sponsored by the Chicago *Times-Herald*. This was won by a gasoline-powered Duryea – one of only two cars to finish out of six starters – and has been cited as the possible inspiration for R. E. Olds to patent in 1896 a gasoline power source, which he subsequently mass-produced in the 'Curved-Dash Olds'. Gasoline thus overcame its slow start. Steam continued viable as an automotive power source until in 1914 there was an outbreak of hoof-and-mouth disease in North America. This led to the withdrawal of horse troughs – which is where steam cars could fill with water. It took the Stanley brothers about three years to develop a condenser and boiler system that did not need to be filled every thirty or forty miles. But by then it was too late. The steam engine never recovered. Where increasing returns are present, it is often the missing 'horseshoe nail' that decides the technological path that is followed.

I have argued that, with increasing returns, the later development of an industry or economy may depend on 'small events' beyond the resolution of an economic observer or his model. Similar arguments have been applied in the last decade to the theoretical possibility of accurate meteorological forecasting. It has been proven that an observational net of weather ships would theoretically have to be finer than the radius of the smallest eddy for weather developments to be forecastable; otherwise these 'small events' become amplified by inherent positive feedbacks into large uncertainties. Given the inevitable presence in the economy of increasing returns to adoption or to allocation, we can speculate that an econometric model that predicts perfectly accurately is not just a practical, but also a *theoretical*, impossibility.

▶ NOTE

Further development of Professor Arthur's argument can be found in W. Brian Arthur (1994) *Increasing Returns and Path Dependence in the Economy*. Ann Arbor, MI: University of Michigan Press.

The social construction of
technology

Ronald Kline and **Trevor Pinch**

Although Douglas, Fischer, and Nye have used the rubric of 'the social
construction of technology,'[1] we maintain that a specific model known as
SCOT (Social Construction of Technology), developed by Trevor Pinch and
Wiebe Bijker in the 1980s, has several advantages in analyzing users as
agents of technological change.[2] In SCOT, 'relevant social groups' who
play a role in the development of a technological artifact are defined as
those groups who share a meaning of the artifact. This meaning can then be
used to explain particular developmental paths. Typical groups might include engineers, advertisers, consumers, and so on. Such groups are not
static; newly emergent groups can also be identified. Although relevant
social groups share a meaning of the artifact, they may of course share
other properties of family resemblance, which also give them their group
characteristic.[3] Thus, some women users of bicycles who shared the meaning of the high-wheeler as an 'unsafe machine' also shared the family
resemblance that they were women.[4] SCOT emphasizes the 'interpretative
flexibility' of an artifact. Different social groups associate different meanings with artifacts leading to interpretative flexibility appearing over the
artifact. The same artifact can mean different things to different social
groups of users. For young men riding the bicycle for sporting purposes the
high-wheeler meant the 'macho machine' as opposed to the meaning given
to it by women and elderly men who wanted to use the bike for transport.
For this latter group, as already mentioned, the high-wheeler was the
'unsafe machine' (because of its habit of throwing people over the handle
bars – known as 'doing a header'). Such meanings can get embedded in new
artifacts, and developmental paths can be traced which reinforce this
meaning (e.g., placing even larger wheels on bicycles to enable them to go
even faster). Interpretative flexibility, however, does not continue forever.
'Closure' and stabilization occur, such that some artifacts appear to have

fewer problems and become increasingly the dominant form of the technology. This, it should be noted, may not result in all rivals vanishing, and
often two very different technologies can exist side by side (for example, jet
planes and propeller planes). Also this process of closure and stabilization
need not be final. New problems can emerge and interpretative flexibility
may reappear.

'Interpretative flexibility' distinguishes SCOT from other social constructivist approaches in the history of technology. SCOT underscores artifacts
and, in particular, their working as subject to radically different interpretations that are coextensive with social groups. This goes beyond saying that
technology is merely embedded in human affairs. SCOT focuses attention
upon what counts as a viable working artifact, and what indeed counts as a
satisfactory test of that artifact. Various case studies have shown how social
groups have contested workability and test results.[5] Such studies point to
the dangers of the analyst assuming a taken-for-granted bedrock of a
technical realm that sets the meaning of an artifact for all spaces, times,
and communities.

Although SCOT has been refined and developed over the last decade,
important weaknesses have appeared.[6] First, SCOT as originally conceived
dealt mainly with the design stage of technologies.[7] The notion of closure
was a little too rigid. What was missing was a sense of how and in what
circumstances the 'black box' of technology could be reopened as it was
taken up by different social groups.[8] Second, SCOT, as many commentators
have remarked, said little about the social structure and power relationships
within which technological development takes place.[9] A related concern is
the neglect of the reciprocal relationship between artifacts and social
groups. We agree that it is important to show not only how social groups
shape technology, but also how the identities of social groups are reconstituted in the process.

► **NOTES**

1 Susan J. Douglas, *Inventing American Broadcasting, 1899–1922* (Baltimore, 1987);
  Claude S. Fischer, *America Calling: A Social History of the Telephone to 1940*
  (Berkeley, 1992); Michele Martin, *'Hello Central?' Gender, Technology and Culture in
  the Formation of Telephone Systems* (Montreal, 1991); and David E. Nye, *Electrifying
  America: Social Meanings of a New Technology, 1880–1940* (Cambridge, Mass., 1990).
2 Trevor Pinch and Wiebe Bijker, 'The Social Construction of Facts and Artifacts,'
  Social Studies of Science 14 (1984): 399–441.
3 Wiebe Bijker, *Of Bicycles, Bakelite, and Bulbs: Towards a Theory of Sociotechnical
  Change* (Cambridge, Mass., 1995), has introduced the notion of a 'technological
  frame' to understand how individuals may deviate from the shared group
  meaning. Often one individual can partake in a number of different technological
  frames and can be weakly included in some frames and strongly included in
  others.

4 Langdom Winner, 'Upon Opening the Black Box and Finding It Empty: Social Constructivism and the Philosophy of Technology,' *Science, Technology, and Human Values* 18 (1993): 362–78, has criticized SCOT for an overly restrictive definition of a social group and for ignoring 'irrelevant' social groups. It is the possibility that groups share more than one family resemblance, which enables historians using SCOT to focus upon excluded or marginalized groups. Thus, on a priori grounds one might expect certain groups to be marginalized, e.g., women, African Americans, etc. Using this family resemblance property historians can analyze these neglected groups within the SCOT framework.

5 See for instance, Donald MacKenzie, 'From Kwajalein to Armageddon? Testing and the Social Construction of Missile Accuracy,' in *The Uses of Experiment*, ed. David Gooding, Trevor Pinch, and Simon Schaffer (Cambridge, 1989), pp. 409–35; Pinch, ' "Testing, One, Two, Three . . . Testing": Towards a Sociology of Testing,' *Science Technology, and Human Values* 18 (1993): 25–41.

6 For example, Boelie Elzen, 'Two Ultracentrifuges: A Comparative Study of the Social Construction of Artifacts,' *Social Studies of Science* 16 (1986): 621–62; Thomas Misa, 'Controversy and Closure in Technological Change: Constructing "Steel," ' in *Shaping Technology/Building Society: Studies in Sociotechnical Change*, ed. Wiebe Bijker and John Law (Cambridge, Mass., 1992), pp. 109–39.

7 H. Mackay and Gareth Gillespie, 'Extending the Social Shaping of Technology Approach: Ideology and Appropriation,' *Social Studies of Science* 22 (1992): 685–716.

8 It is important to realize how the term 'black box' is being used here. A technology that is black-boxed is one where design has stabilized. This does not mean it has literally to be treated as a black box, meaning that the inner workings are opaque to the user (although this may happen for some technologies and some users). Thus the Model T was a stabilized black box which was designed in such a way that it could easily be repaired.

9 Stewart Russell, 'The Social Construction of Artifacts: A Response to Pinch and Bijker,' *Social Studies of Science* 16 (1986): 331–46; Trevor J. Pinch and Wiebe E. Bijker, 'Science Relativism and the New Sociology of Technology: Reply to Russell', *Social Studies of Science* 16 (1986): 347–60.

8 ▶ Redefining the social link: from baboons to humans

Shirley Strum and Bruno Latour

▶ REDEFINING THE NOTION OF SOCIAL

Sciences of society currently subscribe to a paradigm in which 'society', although difficult to probe and to encompass, is something that can be the object of an ostensive definition [that is, a definition in which society is something that is 'there', that can be pointed to, so to speak]. The actors of society, even if the degree of activity granted them varies from one school of sociology to the next, are *inside* this larger society. Thus, social scientists recognize a difference of scale: the micro-level (that of the actors, members, participants) and a macro-level (that of society as a whole) (Knorr and Cicourel 1981). In the last two decades this ostensive definition of society has been challenged by ethnomethodology (Garfinkel 1967) [ethnomethodology is a view of sociology in which society is achieved through interaction, rather than society being the given, existing, structure within which interaction takes place] and by the sociology of science (especially of the social sciences (Law 1986) and the sociology of technology (Latour 1986b). In light of these studies, the conventional distinctions between micro- and macro-levels become less clear-cut and it is more difficult to accept a traditional definition of society. Instead, society is more compellingly seen as continually constructed or 'performed' by active social beings who violate 'levels' in the process of their 'work'.

The two positions, the ostensive and the performative model, differ in principle and in practice, with crucial consequences for how the social link is characterized. These two views can be summarized as follows:

Ostensive definition of the social link:

1 It is, *in principle*, possible to discover the typical properties of what holds a society together, properties which could explain the social link and its evolution, although *in practice*, it may be difficult to detect them.
2 These properties or elements are social. If other properties are included then the explanation of society is economic, biological, psychological, etc.
3 Social actors (whatever their size – micro or macro) are *in* the society as defined in #1. To the extent that they are active, their activity is restricted because they are only part of a larger society.
4 Because actors are in the society, they can be useful informants for scientists interested in discovering the principles of society. But because they are only *part* of society, even if they are 'aware', they can never see or know the whole picture.
5 With the proper methodology, social scientists can discover the principles of what holds society together, distinguishing between actors' beliefs and behavior. The picture of society as a whole, thus devised, is unavailable to the individual social actors who are within it.

According to the traditional paradigm, society exists, actors enter it adhering to rules and a structure that are already determined. The overall nature of the society is unknown and unknowable to the actors. Only scientists, standing outside of society, have the capacity to understand it and see it in its entirety.

Performative definition of the social link:

1 It is impossible, *in principle*, to establish properties which would be peculiar to life in society, although, *in practice*, it is possible to do so.
2 A variety of elements or properties contribute to the social link as defined by social actors. These are not restricted to the purely social and can include economic, biological, psychological, etc.
3 *In practice*, actors (no matter what their size, macro or micro) define, for themselves and for others, what society is, both its whole and its parts.
4 Actors 'performing' society know what is necessary for their success. This may include a knowledge of the parts and of the whole and of the difference between beliefs and behavior.
5 Social scientists raise the same questions as any other social actor and are themselves 'performing' society, no more and no less than nonscientists. They may, however, have different practical ways of enforcing their definition of what society is.

According to the performative view, society is constructed through the many efforts to define it; it is something achieved in practice by all actors, including scientists who themselves strive to define what society is. To use Garfinkel's expression (1967), social actors are transformed, in this view, from 'cultural dopes' to active achievers of society. This shifts the emphasis from looking for the social link in the *relations between actors* to focusing on *how* actors achieve this link in their search for what society is.

Going from the traditional to the performative framework creates two sets of inverse relationships, one that reveals a strange symmetry among all actors and another that points out a new asymmetry. The first inverse relationship is the following: the more active the actors, the less they differ from one another. This shift in definition is tantamount to saying that actors are full fledged social scientists researching what the society is, what holds it together and how it can be altered. The second inverse relationship is this: the more actors are seen to be equal, *in principle*, the more the *practical* differences between them become apparent in the means available to them to achieve society. Let us now see how we can apply these principles in the case of baboon societies. . . .

The trend [in studies of baboons] has been in the direction of granting baboons more social skill and more social awareness (Griffin 1981, 1984) than the sociobiological 'smart biology' argument allowed. These skills involve negotiating, testing, assessing and manipulating (Strum 1975a,b, 1981, 1982, 1983a,b,c, in press; Western and Strum 1983). A male baboon, motivated by his genes to maximize his reproductive success, cannot simply rely on his size, strength or dominance rank to get him what he wants. Even if dominance was sufficient, we are still left with the question: how do baboons know who is dominant or not? Is dominance a fact or an artefact? If it is an artefact, whose artefact is it – is it the observer's, who is searching for a society into which he can put the baboons? (Even in the classic dominance study, the investigator had to intervene by pairing males in contests over food, in order to 'discover' the dominance hierarchy.) Or is it a universal problem, one that both observer and baboon have to solve?

If baboons are constantly testing, trying to see who is allied with whom, who is leading whom, which strategies can further their goals, as recent evidence suggests, then both baboons and scientists are asking the same questions. And to the extent that baboons are constantly negotiating, the social link is transformed into a process of acquiring knowledge about 'what the society is'. To put it in a slightly different way, if we grant that baboons are not *entering into a stable structure* but rather negotiating what that structure will be, and monitoring and testing and pushing all other such negotiations, the variety of baboon society and its ill fit to a simple structure can be seen to be a result of the 'performative' question. The evidence is more striking in reverse. If there was a structure to be entered, why all this behavior geared to testing, negotiating and monitoring . . .?

We can summarize the baboon data and argument as follows: first, the traditional, ostensive, definition of baboon society has been unable to accommodate the variety of data on baboon social life. As a result, some information has been treated as 'data' and other information as discrepancies to be ignored or explained away. Second, more recent studies demonstrate that baboons invest a great deal of time in negotiating, testing, monitoring and interfering with each other.

A performative definition of society allows us to integrate both sets of 'facts'. Under this definition, baboons would not be seen as being *in* a group. Instead they would be seen as striving to define the society and the groups in which they exist, the structure and the boundaries. They would not be seen as being *in* a hierarchy, rather they would be ordering their social world by their very activity. In such a view, shifting or stable hierarchies might develop not as one of the principles of an overarching society into which baboons must fit, but as the provisional outcome of their search for some basis of predictable interactions. Rather than entering an alliance system, baboons performing society would be testing the availability and solidity of alliances without knowing for certain, in advance, which relationships will hold and which will break. In short, performative baboons are social players actively negotiating and renegotiating what their society is and what it will be.

▶ SOCIAL COMPLEXITY AND SOCIAL COMPLICATION

When we transform baboons into active performers of their society does this put them on a par with humans? The performative paradigm suggests an important distinction. What differs is the *practical* means that actors have to enforce their version of society or to organize others on a larger scale, thereby putting into practice their own individual version of what society is.

If actors have only themselves, only their bodies as resources, the task of building stable societies will be difficult. This is probably the case with baboons. They try to decide who is a member of the group, what are the relevant units of the group that have to be considered, what is the nature of the interaction of these other units, etc., but they have no simple or simplifying means to decide these issues or to separate out one at a time to focus upon (see references above). Age, gender, and perhaps kinship can be taken as givens in most interactions. To the extent that dominance systems are linked to kinship, dominance rank may also be a given (Chapais and Schulman 1980; Hausfater *et al.* 1982). But even age, kinship and kinship-linked dominance may be the object of negotiation at critical points (Altmann in 1980; Cheney 1977; Chepko-Sade and Sade 1979; Popp and DeVore 1979; Trivers 1972; Walters 1981; Wasser 1982; Wasser and Barash 1981). A profusion of other variables impinge simultaneously. This is the definition of *complexity*, 'to simultaneously embrace a multitude of objects'. As far as baboons are concerned they assimilate a variety of factors all at once.

For the rest of our discussion we will consider that baboons live in *complex societies* and have complex sociality. When they construct and repair their social order, they do so only with limited resources, their bodies, their social skills and whatever social strategies they can construct. A baboon is, in our view, the ideal case of the *competent member* portrayed by ethnomethodologists, a social actor having difficulty negotiating one factor at a time, constantly subject to the interference of others with similar problems. These limited resources make possible only limited social stability.

Greater stability is acquired only with additional resources; something besides what is encoded in bodies and attainable through social skills is

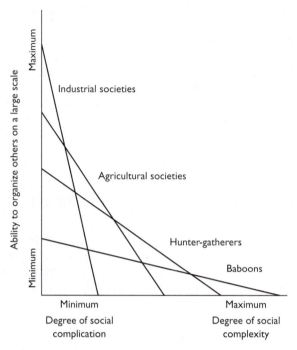

Figure 8.1 Complexity versus complication: the trade-off

needed. Material resources and symbols can be used to enforce or reinforce a particular view of 'what society is' and permit social life to shift away from complexity to what we will call *complication*. Something is 'complicated' when it is made of a succession of simple operations. Computers are the archetype of a complicated structure where tasks are achieved by the machine doing a series of simple steps. We suggest that the shift from complexity to complication is the crucial *practical* distinction between types of social life.

To understand this point better, we might look at what baboon-watchers do in order to understand baboon social life. First, individuals are identified and named, and the composition of the group is determined by age, sex and kinship, and perhaps also dominance rankings. Items of behavior are identified, defined and coded. Then attention is consciously focused on a subset of individuals, times, and activities, among the variety of interactions that occur simultaneously. Of course we could interpret this procedure as merely a rigorous way of getting at the social structure that exists and informs baboon societies. This interpretation of the scientific work fits nicely with the ostensive definition of society. In our view, however, the work that human observers do in order to understand baboon societies is the very same process that makes human societies different from baboon ones. Modern scientific observers replace a complexity of shifting, often fuzzy and continuous behaviors, relationships and meanings with a complicated array of simple, symbolic, clear-cut items. It is an enormous task of simplification.

How does the shift from social complexity to social complication happen? Figure 8.1 illustrates how we imagine this progression. The first line represents a baboon-like society in which socialness is complex, by our use of that term, and society is complex but not complicated because individuals are unable to organize others on a large scale. The intensity of their social negotiation reflects their relative powerlessness to enforce their version of society on others, or to make it stick as a stable, lasting version.

The second line positions hypothetical hunter-gatherers who are rich in material and symbolic means to use in constructing society compared to baboons, although impoverished by comparison with modern industrial societies. Here language, symbols, and material objects can be used to simplify the task of ascertaining and negotiating the nature of the social order. Bodies continue their social strategies in the performance of society, but on a larger, more durable, less complex scale. Material resources and the symbolic innovations related to language allow individuals to influence and have more power over others thereby determining the nature of the social order.

Line 3 represents agricultural societies where even more resources can be brought to bear in creating the social bond. In fact, the social bond can be maintained in the relative absence of the individuals. These societies are more complicated and more powerful than hunter-gatherer groups and the performance of society is possible on a larger scale because negotiations at each step are much less complex.

Modern industrial societies are depicted by the fourth line on the diagram. Here individuals are able to organize and 'mobilize' others on a grand scale. According to our scheme, the skills in an industrial society are those of simplification making social tasks *less complex* rather than making them more complex by comparison with other human and animal societies. By holding a variety of factors constant and sequentially negotiating one variable at a time, a stable *complicated* structure is created. Through extra-somatic resources employed in the process of social complication, units like multinational corporations, states and nations can be constituted (Latour 1987). The trend as we have sketched it, is from complex sociality, as found among baboons, to complicated sociality as found among humans. Starting with individuals who have little power to affect others, or enforce their version of society, or make a lasting social order, we encounter a situation where individuals employ more and more material and 'extra-social' means to simplify social negotiations. This gives them the ability to organize others on a large scale, even when those others are not physically present. By using additional new resources, social actors can make weak and renegotiable associations, like alliances between male baboons, into strong and unbreakable units (Callon and Latour 1981; Latour 1986a).

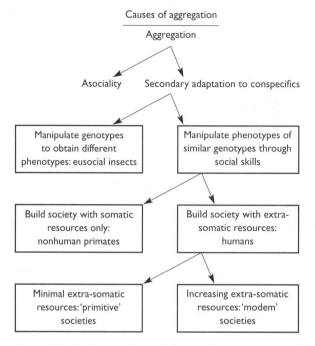

Figure 8.2 The evolution of the performative social bond

► **THE EVOLUTION OF THE PERFORMATIVE SOCIAL BOND**

Our use of a performative framework produces two important permuta-
tions. Firstly, it grants full activity to all social participants. Individually and
together they create society and, in theory, they are all equal. But, secondly,
new asymmetries are introduced when we consider what practical means
actors have to enforce their own definition of the social bond and to
organize others according to individual views of what society is.

The performative framework we are advocating, in effect, gives back to
the word 'social' its original meaning of association. Using this definition
we can compare the *practical* ways in which organisms achieve societies.
Figure 8.2 summarizes our views about the possible evolution of the
performative social bond. We focus on the types of resources that actors
have with which to create society and to associate, but we do not restrict the
idea of 'resources' in any sense.

Aggregations of conspecifics is the first meaning of social in various
accounts of the origin of society (see Latour and Strum 1986 and references
included there). However most accounts fail to distinguish between this
aggregation and the origin of social skills. Once aggregation occurs, whatever its cause (e.g. Alcock 1975; Hamilton 1971), two different strategies are
possible in our model. The first is for the actor to depart, fleeing others as

soon as possible. This option generates asocial animals who exist alone except for brief reproductive interludes and temporary associations.

The second option is of greater interest. If the aggregated individual is not going to flee, he or she must adapt to a new environment of conspecifics. This is the meaning of social most common in the animal behavior litera-ture: to modify one's behavior in order to live in close proximity to others of the same species. Acquiring the skill to create society and hold it together is then a SECONDARY adaptation to an environment made up, in large part, of conspecifics. In order not to be exploited by their new social environ-ment, individuals must become smarter at manipulating and maneuvering around each other.

Once the social option has been chosen, two other possibilities appear. In the first, it is the genotypes that are modified until they are socially distinct. Insect societies are an example where the actors' own bodies are irreversibly molded. In the second possibility we find a different meaning of social. In this case the genotypes produce similar phenotypes. These phenotypes are then manipulated by the ever-increasing social skills of individuals. This option also branches into two alternatives.

Baboons provide an example of the first. Social skills are necessary to enroll others in the actor's definition of what society is. But baboons have only 'soft tools' and can build only 'soft' societies. They have nothing more to convince and enlist others in their definition than their bodies, their intelligence and a history of interactions built up over time. This is a *complex* task and only socially 'smart' and skillful individuals may hope to be successful in baboon society.

The second possibility is to acquire additional means of defining and strengthening the social bond. Here we have the human case where the creation of society uses material resources and symbols to simplify the task. Social interactions become more *complicated* but not more complex. Much of the skill necessary to achieve society in the other, baboon-like, option now resides in the creation of symbolic and material bonds. The result is that actors rather than appearing to create society, now appear to be inserted into a material society that overpowers them (the traditional paradigm discussed earlier).

For human societies there is an additional branching: 'primitive' societies are created with a minimal amount of material resources; increasing such resources produces 'modern' societies. Thus technology becomes one way of solving the problem of building society on a larger scale. In this sense even modern technology is social. It represents a further resource in the mobilization of individuals in the performance of society.

To summarize our theoretical model, once individuals are aggregated and choose not to avoid each other, there must be a secondary adaptation to a new competitive environment of conspecifics. Two strategies are possible: manipulate the genotypes to obtain different phenotypes (eusocial insects) or manipulate the phenotypes of similar genotypes through increasing social skills. Similar bodies adapting to social life have, themselves, two possibilities: build the society using only social skills (nonhuman primates) or utilize additional material resources and symbols, as necessary, to define the social bond (human societies). In the human step different types of societies are created depending upon the extent of new resources that are used.

 POLITICS

What relevance does our exploration of the meanings of social have for politics? . . .

The thrust of our argument is to draw a closer parallel between what we call 'social' and what has been defined as political. These efforts do not erase the significant differences between ants, baboons and, for instance, the technocrats of the Pentagon. Rather they highlight the source of those differences in a new way: the resources used and the practical work required in mobilizing them. In our definition of resources, genes, power, language, capital, and technology, for instance, are all seen as strategic means of enhancing one's influence over others in increasingly more durable ways. Politics is not one realm of action separated from the others. Politics, in our view, is what allows many heterogeneous resources to be woven together into a social link that becomes increasingly harder and harder to break.

REFERENCES

Alcock, J. (1975) *Animal Behavior: an evolutionary approach*. Sunderland Mass.: Sinnuer.

Altmann, J. (1980) *Baboon Mothers and Infants*. Cambridge: Harvard University Press.

Callon, M. and B. Latour (1981) Unscrewing the big Leviathan: how actors macro-structure reality and how sociologists help them to do so. In K. Knorr-Cetina and A. V. Cicourel (eds) *Advances in Social Theory and Methodology: Toward an Integration of Micro- and Macro-Sociologies*. Boston: Routledge and Kegan Paul, pp. 277–303.

Chapais, B. and S. Schulman (1980) An evolutionary model of female dominance relations in primates. *J. Theoret. Biol.* 82: 47–89.

Cheney, D. (1977) The acquisition of rank and the development of reciprocal alliances among free-ranging immature baboons. *Behav. Ecol. Sociobiol.* 2: 303–18.

Chepko-Sade, B. and D. Sade (1979) Patterns of group splitting within matrilineal kinship groups. *Behav. Ecol. Sociobiol.* 5: 67–86.

Garfinkel, H. (1967) *Studies in Ethnomethodology*. New Jersey: Prentice-Hall.

Griffin, D. (1981) *The Question of Animal Awareness*, 2nd edition. New York: Rockefeller University Press.

Griffin, D. (1984) *Animal Thinking*. Cambridge: Harvard University Press.

Hamilton, W. D. (1971) Geometry for the selfish herd. *J. Theoret. Biol.* 31: 295–311.

Hausfater, G., J. Altmann, and S. Altmann (1982) Long-term consistency of dominance relations among female baboons. *Science* 217: 752–755.

Knorr, K. and A. Cicourel (1981) *Advances in Social Theory and Methodology: Towards an Integration of Micro and Macro Sociologies*. Boston, MA: Routledge and Kegan Paul.

Latour, B. (1986a) Visualization and cognition: thinking with eyes and hands. *Knowledge and Society Studies: Past and Present*, Vol. 6: 1–40.

Latour, B. (1986b) The powers of association. In John Law (ed.). *Power Action and belief: a New Sociology of Knowledge. Soc. Rev.* Monograph: 264–80.

Latour, B. (1987) *Science in Action*. Milton Keynes: Open University Press and Cambridge Mass., Harvard University Press.

Latour, B. and S. Strum (1986) Human social origins: please tell us another story. *J. Soc. Biol. Struct.* 9: 169–187.

Popp, J. and I. DeVore (1979) Aggressive competition and social dominance theory: synopsis. In D. Hamburg and E. McCown (eds) *The Great Apes*. Menlo Park: W. A. Benjamin.

Strum, S. (1975a) Life with the Pumphouse Gang. *Nat. Geo.* 147: 672–91.

Strum, S. (1975b) Primate predation: interim report on the development of a tradition in a troop of olive baboons. *Science* 187: 755–57.

Strum, S. (1981) Processes and products of change: baboon predatory behavior at Gilgil, Kenya. In G. Teleki and R. Harding (eds). *Omnivorous Primates*. New York: Columbia University Press.

Strum, S. (1982) Agonistic dominance in male baboons: an alternative view. *Int. J. Primatol.* 3: 175–202.

Strum, S. (1983a) Why males use infants. In D. Taub (ed.). *Primate Paternalism*. New York: Van Nostrand Reinhold.

Strum, S. (1983b) Use of females by male olive baboons. *Amer. J. Primatol.* 5: 93–109.

Strum, S. (1983c) Baboon cues for eating meat. *J. Human Evol.* 12: 327–36.

Strum, S. (in press) Are there alternatives to aggression in baboon society? In H. Steklis and R. Harding (eds). Sherwood Washburn Festschrift.

Trivers, R. (1972) Parent-offspring conflict. *Am. Zool.* 14: 249–64.

Walters, J. (1981) Inferring kinship from behaviour: maternity determinations in yellow baboons. *Anim. Behav.* 1981: 126–136.

Wasser, S. (1982) Reciprocity and the trade-off between associate quality and relatedness. *Amer. Natural.* 119: 720–31.

Wasser, S. and D. Barash (1981) The 'selfish' allowmother. *Ethol. Sociobiol.* 2: 91–3.

Western, J. D. and S. Strum (1983) Sex, kinship, and the evolution of social manipulation. *Ethol. Sociobiol.* 4: 19–28.

9 ▶ Caught in the wheels: the high cost of being a female cog in the male machinery of engineering

Cynthia Cockburn

Every self-respecting man knows that 'women are no good with machinery', they are 'hopeless at technical things'. After all, the facts speak for themselves. Women don't fiddle about inside TV sets, keep an oscilloscope in the garage, or aspire to fly a Tornado jet. Every self-respecting woman, however, feels there is something fishy about these facts. We know that women make good, competent and enthusiastic technicians and engineers – most of us have met one or two. Just as many women as men reach maturity with a bent for calculation, problem-solving, design and construction. Why, then, do most of us finish up using it to interpret knitting patterns and construct patchwork quilts? Have we chosen or were we pushed? And if we chose – what exactly was the choice we were offered?

I recently bought a book of encylopedic scope on the history of machinery and technical invention. Working through the index I have given up at P (Ptolemy IV) without finding a woman among the inventors. No doubt we women too have had our technologies, but they are not in our history books. Ours are not the technologies that soared into capitalist profitability in the eighteenth and nineteenth centuries. . . .

▶ 'EQUAL CHANCES'?

People concerned with equal opportunities study the statistics [of women in engineering] and seek to help women get their 'fair share'. With this in mind, the Equal Opportunities Commission, for instance, has funded several projects aiming to understand women's disadvantages in the technical field and to encourage girls into technology. It has sponsored several career-opportunities courses for young women. It has funded a project in which women are working with teachers and pupils in schools to bring girls into technical subjects. . . .

Most initiatives to date have been based on the philosophy that women must be 'given more confidence'. We are continuing, the argument goes, to *fail* to make our mark in occupations that are seen to be challenging and rewarding. But are women really such weeds? I would suggest that women know very well both where we are unwelcome and what we are rejecting. We are not failing, we are on strike.

▶ TECHNOLOGY: NEITHER NEUTER NOR NEUTRAL

The prevailing belief concerning technology is that it is neutral, mankind's heritage, equally available and relevant to us all. All that women have to do is to reach out and grasp it. The 'man' in mankind, however, is no slip of the tongue. Technology is far from neutral. This should not be a difficult concept to Marxists, who are, after all, used to understanding that our technology is capitalist technology and bears the marks and serves the purposes of the class that owns it. It needs only a little further broadening of the mind to understand that our industrial technology also has the imprint and the limitations that come of being both the social property and one of the formative processes of men. Industrial, commercial, military technologies are masculine in a very historical and material sense. They cannot readily be used in a feminine, nor even a sexless, mode. Women are not merely failing to enter technology. On the one hand we are being repelled, and on the other we are refusing.

Men's greater average physical stature and strength are often cited as a reason for men's preponderance in engineering occupations. Yet it is not self-evident that they should be all male. Many machines, from the lever to the mill, have been developed precisely to *substitute* for human physical strength. The masculinity of technology, men's proprietorial grasp of machinery, has to be seen as a product of social rather than biological history.

To this we have to relate the male appropriation of the whole concept of 'work'. The paradigm worker, model for both employer and trade unions, is, by men's design, a man. In this context, women's physiology is seen, not as a worker's norm, but as defective. Men scorn the fact, for instance, that some women need a work load that can be modified or varied to take account of their monthly cycle or of pregnancy. Women feel guilty if they display any 'weakness' in this respect. Yet it could have been held to be to working class advantage, as well as to the advantage of women, to make our collective demands to the employer on the basis of the weakest common factor. That this has never been the case can lead to only one conclusion:

work, whatever it may mean to the capitalist, is also an important sphere for men in which to establish and maintain power over women as a sex.

▶ ALL THE MANLY VIRTUES

If the norm for the industrial worker in general is male, that for the qualified engineer, the skilled technician, most certainly is. Engineering represents everything that is defined as manly – the propensity to control and manipulate nature; the celebration of muscle and machine in action upon raw materials; the tolerance of, even pleasure in, dirt, *viz*, grease, swarf and metal shavings. (It is worth remembering that this is not the only kind of dirt, however. Men are popularly seen as having a natural aversion to that other kind of dirt, human faeces, blood, vomit, with which women are supposed to feel more ability to cope.) Technical work involves the acceptance of physical risk – exposure to frequent accidents, cuts, contusions. It affords free movement round and about its object, in contrast to the physical confinement of much women's work. It implies control – designing solutions to physical problems, making energy work for you. The all-male workshop fosters and develops masculine patterns of relationships, it is the home of camaraderie based on the exchange of anecdote and slander concerning women.

Add to this that engineering as we know it is firmly embedded in the

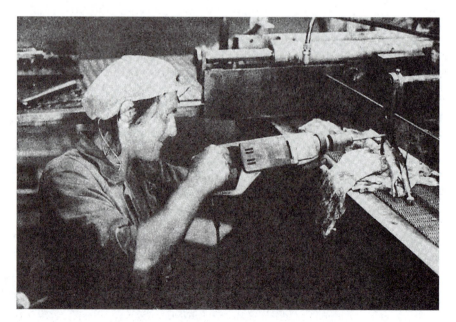

Figure 9.1 Translating energy into work . . .

capitalist business world and the state. The atmosphere is competitive, it is about performance: your firm's machine against the next, your country's weapons against those of its rival. Capitalist industry and contemporary technology both express and embody values that have on the one hand developed out of patriarchy, and on the other have developed to make patriarchy what it is in modern society. The relations surrounding technology continually renew and extend male hegemony over the rest of us. The growth of industrial technology has to be seen as part and parcel of the historical development of gender difference. It has been formative in the growth of class relations. But it has also been part of what has made males into 'men' and females into 'girls'.

WOMEN'S VALUES

If engineering occupations have developed as a heartland of male hegemony, it is hardly surprising that female incursions into this domain don't occur easily or painlessly. What women and women's work have come to mean, both to ourselves and to men, is something quite different. It is accepting rather than defying physical and social limitations. While 'men's work' means single-mindedly pitting everything you have, in the army, in the mines, on the high seas, 'women's work' means refusing to let go of your other self. Men's work is predicated upon someone looking after them. We look after ourselves. Women's work means staying at home with sick children if need be. It means carrying human preoccupations into the job: nursing, teaching, social work. It seems as though for women more than for men the social purpose of work is important.

To emphasise this *difference* is not to say that women and men are born this way, immutable. It is to recognise that over hundreds and possibly thousands of years society has constructed gender difference, gender complementarity, and continues to confirm and elaborate it every day, in work as well as outside work. And much in the feminine gender is good. Women (and indeed many men) value it above masculinity. We do not want to have to abandon a concern with feelings and people and purpose in order to take on technology. We don't want to exchange the society of women for the society of men, to become a kind of de-sexed satellite of a male world.

Of course, men treasure their unique possession of technology. Many skilled trade unions have deliberately kept women out of traditionally male jobs. Individual men resist and resent the intrusion of women into work that is comparable to their own. Managements too, unless it suits them to do otherwise, often recruit personnel into the existing sexually-segregated pattern. Many discourage women from applying for technical jobs and some blatantly discriminate against them.

However, several engineering firms in which I have done interviews claim to be short of skilled engineers and more than willing to employ women. There is no 'discrimination' here, they claim. And I am sure that, broadly

Figure 9.2 The masculinity of technology: an end result (photo Mike Abrahams, by permission of Network)

speaking, they are telling the truth. But women, they say, are just not coming forward. Now and then one is offered a job and turns it down. The fact is that, while a handful of women are determined enough to persevere, the great majority are turning aside from each gateway as it is reached. . . .

▶ **MOVING FORWARD**

Women are caught up in a contradiction over engineering, indeed over all technology. I have heard women arguing: should we keep our hands clean, keep well away from men's technology and run the risk of typecasting ourselves as whimsical earth mothers? Or do we need fire to fight fire, at the risk of burning our hands?

We cannot move forward into the male industrial field without great individual cost, and cost to the women's movement by a continual drain of strong and able women into mere competition with men and collaboration with capital. Yet we cannot leave things as they are. Why? Because, like it or not, we now live in a world in which power lies in the economic ownership of these technical forces of production (and of distribution, reproduction and war) and in the practical control of these things. We cannot continue to be the passive objects of some technologies (at the receiving end of medical and military technologies, for instance, that we should be questioning or resisting), and the manipulated and exploited operators of others (type-writers, washing machines). We have to learn technical skills. If we are to learn, we have to get in there. It cannot be done at a distance.

Besides, the cultural barrier that is erected between women and tech-nology is all too closely related to other physical taboos that confine and limit us. When you see a woman take a set of spanners and approach a car, you suddenly become aware of the manifold informal pressures against women in public places using their bodies in the way men do: getting dirty and sweaty, climbing up things, lying on the floor, spreading their legs, exerting muscular force. Learning to understand and use tools that work metal and wood, to translate geometry into motion and energy into work, these are things that we cannot do without if we are to stop being the world's victims. We have to get our bodies out of their cocoons, and this involves overcoming a certain physical cowardice and reticence that is the bad side of our gendered character.

▶ ON OUR TERMS

Is there a feminist and socialist strategy for getting us into the technical world without getting us hurt or doing harm to others? It seems to me we need a firm grasp, first and foremost, of a theory that recognises *systemic* male dominance and the part of technology within it. Only this can help us make informed choices. And when it comes to those choices, at a practical level?

The first thing, surely, is that we should not deny our social values. We should work our way into technology along paths that make sense for us. If, as seems likely, we can do nothing but hurt ourselves and others by breaking into aerospace and ballistics, well, let us leave those fads to the lads, and opt consciously instead for work in industrial sectors that seem to us to be more humanly useful and malleable: engineering for construction perhaps, or for the media, domestic equipment. Women seem to be taking this course naturally. Look for instance at the disproportionate number of women physicists who choose to work in medical physics in the health service. Media resources officers in education, who deal with video and printing, are often women.

A second route is to recognise the alien nature of most business enterprise and opt for self-organised collectives. There are already many small co-operatives, mixed or women only, in printing and building, for instance. They afford a way of separating out the already difficult relations of technology from those of capitalist employment.

▶ ## ORGANISING INDEPENDENCE

In my view, by far the most effective principle evolved to date is separate, woman-only organisation. It enables us to learn (teach each other) without being put down. Provide schoolgirls with separate facilities and the boys won't be able to grab the computer and bully the girls off the console. Provide young women with all-women courses so that they can gain the experience to make an informed choice about an engineering career. We need to demand a massive increase in resources from the state, from industry, from industrial training boards, for women-run, women-only initiatives. Everywhere we have tried it, from women's caucuses to Greenham Common [the women's peace camp at a cruise missile base], autonomy works wonders for our feelings and our strength. We need, before all else, a great expansion of the autonomous sphere in technology.

Nonetheless, we have already learned that while autonomy is necessary, it is also not the whole answer. It only works in some situations and therefore it only gets us so far. What occurs in engineering is very similar to what has long been noted by women trying to make an impact within male-dominated trade unions. It is what some Italian feminists have called the contradiction between 'mutilation and marginalisation'. It is perhaps our essential contradiction as feminists. We begin by standing on the edge of the action, we are marginal. Then we seek equality and make our choice between failure or being a pseudo-man: mutilation. Then we assert our difference and organise autonomous women's groups, committees, demonstrations. This feels great. But over time it is clear just how the male machinery, of trade unions, of left politics, of factory and workshop, the world in short, grinds on regardless. Still we are marginal.

There is a limit to what we can earn and what technologies we can handle in women's collectives. There are limits to the impact we can have and what we can learn. It is a choice more available to middle-class than to working-class women. So we also need devices to take us into the sphere of male and capitalist work. We can invent mechanisms that help us to avoid isolation. Progressive teachers and employers must be pressed to ensure that space is made not just for the token woman, but for groups of women at one time. For women supervisors and managers too. We can ourselves form women's groups alongside work or study to help each other talk through and deal with our minority situation.

WOMAN-LED, CAN IT WORK?

There is one further step. Just what do we mean by equal representation of men and women in an occupation? It will never be exactly 50% of each sex. The norm today is a preponderance of men, a handful of women. We should start to visualise women-led situations. How impossible it seems to imagine a technical training course, a workshop, an engineering plant where women are simply in a majority and in positions that can influence the relations of work and the mode of control. Where do we ever see men under women's tutelage, lads learning from women, men obliged to do things women's way? That is more revolutionary than autonomy itself.

If we made this one of our goals, there would be some interesting consequences. The sex/gender struggle in the workplace would change subtly, the balance tipped a little in favour of women. Men's reluctance to participate on such terms would be glaringly exposed. We would see them for what they are: the original separatists.

Of course there are many problems. Such a strategy is impossible until enough women exist, trained and experienced, ready and willing, to enter such situations. Besides, male dominance is not based merely on numbers. It is systemic, every one man is backed up by men's organisation, wealth and ideology in society as a whole. Nonetheless, playing the numbers game can be a start. Where positive discrimination in favour of women is both legal and attainable, we should remember perhaps that between the strategy of the sacrificial token woman and the strategy of women-only, there exists another option: putting men into a minority and releasing women's creativeness in a context within which men may learn and things may change.

 ## ACKNOWLEDGEMENT

Thanks for advice received from Liz Allen, Mary Clemmey, Heather Hunt, Anne Phillips and Elaine Sinclair.

10 ▶ Making 'white' people white

Richard Dyer

The photographic media and, *a fortiori*, movie lighting assume, privilege and construct whiteness. The apparatus was developed with white people in mind and habitual use and instruction continue in the same vein, so much so that photographing non-white people is typically construed as a problem.

All technologies work within material parameters that cannot be wished away. Human skin does have different colours which reflect light differently. Methods of calculating this differ, but the degree of difference registered is roughly the same: Millerson (1972: 31), discussing colour television, gives light skin 43 per cent light reflectance and dark skin 29 per cent; Malkiewicz (1986: 53) states that 'a Caucasian face has about 35 per cent reflectance but a black face reflects less than 16 per cent'. This creates problems if shooting very light and very dark people in the same frame. Writing in *Scientific American* in 1921, Frederick Mills, 'electrical illuminating engineer at the Lasky Studios', noted that

> when there are two persons in [a] scene, possibly a star and a leading player, if one has a dark make-up and the other a light, much care must be exercised in so regulating the light that it neither 'burns up' the light make-up nor is of insufficient strength to light up the dark make-up.
> (1921: 148)

The problem is memorably attested in a racial context in school photos where either the black pupils' faces look like blobs or the white pupils have theirs bleached out.

The technology at one's disposal also sets limits. The chemistry of different stocks registers shades and colours differently. Cameras offer varying degrees of flexibility with regard to exposure (affecting their ability to take a wide lightness/darkness range). Different kinds of lighting have different colours and degrees of warmth, with concomitant effects on different skins. However, what is at one's disposal is not all that could exist. Stocks, cameras and lighting were developed taking the white face as the touchstone. The resultant apparatus came to be seen as fixed and inevitable, existing independently of the fact that it was humanly constructed. It may be – certainly was – true that photo and film apparatuses have seemed to work better with light-skinned peoples, but that is because they were made that way, not because they could be no other way.

All this is complicated still further by the habitual practices and uses of the apparatus. Certain exposures and lighting set-ups, as well as make-ups and developing processes, have become established as normal. They are constituted as the way to use the medium. Anything else becomes a departure from the norm, or even a problem. In practice, such normality is white. . . .

Innovation in the photographic media has generally taken the human face as its touchstone, and the white face as the norm of that. The very early experimenters did not take the face as subject at all, but once they and their followers turned to portraits, and especially once photographic portraiture replaced painted portraits in popularity (from the 1840s on), the issue of the 'right' technology (apparatus, consumables, practice) focused on the face and, given the clientele, the white face. Experiment with, for instance, the chemistry of photographic stock, aperture size, length of development and artificial light all proceeded on the assumption that what had to be got right was the look of the white face. This is where the big money lay, in the everyday practices of professional portraiture and amateur snapshots. By the time of film (some sixty years after the first photographs), technologies and practices were already well established. Film borrowed these, gradually and selectively, carrying forward the assumptions that had gone into them. In turn, film history involves many refinements, variations and inno-vations, always keeping the white face central as a touchstone and occasion-ally revealing this quite explicitly, when it is not implicit within such terms as 'beauty', 'glamour' and 'truthfulness'. Let me provide some instances of this.

The interactions of film stock, lighting and make-up illustrate the as-sumption of the white face at various points in film history. Film stock repeatedly failed to get the whiteness of the white face. The earliest stock, orthochromatic, was insensitive to red and yellow, rendering both colours dark. Charles Handley, looking back in 1954, noted that with ortho-chromatic stock, 'even a reasonably light-red object would photograph black' (1967: 121). White skin is reasonably light-red. Fashion in make-up also had to be guarded against, as noted in one of the standard manuals of the era, Carl Louis Gregory's *Condensed Course in Motion Picture Photography* (1920):

Be very sparing in the use of lip rouge. Remember that red photographs black and that a heavy application of rouge shows an unnaturally black mouth on the screen.

(316)

Yellow also posed problems. One derived from theatrical practices of make-up, against which Gregory inveighs in a passage of remarkable racial resonance:

> Another myth that numerous actors entertain is the yellow grease-paint theory. Nobody can explain why a performer should make-up in chinese yellow. . . . The objections to yellow are that it is non-actinic and if the actor happens to step out of the rays of the arcs for a moment or if he is shaded from the distinct force of the light by another actor, his face photographs BLACK instantly.
>
> (ibid.: 317, emphasis in original)

The solution to these problems was a 'dreadful white make-up' (actress Geraldine Farrar, interviewed in Brownlow 1968: 418) worn under carbon arc lights so hot that they made the make-up run, involving endless retouching. . . .

Colour brought with it a new set of problems, explored in Brian Winston's article on the invention of 'colour film that more readily photographs Caucasians than other human types' (1985: 106). Winston argues that at each stage the search for a colour film stock (including the development process, crucial to the subtractive systems that have proved most workable) was guided by how it rendered white flesh tones. Not long after the introduction of colour in the mid-1930s, the cinematographer Joseph Valentine commented that 'perhaps the most important single factor in dramatic cinematography is the relation between the colour sensitivity of an emulsion and the reproduction of pleasing flesh tones' (1939: 54). Winston looks at one such example of the search for 'pleasing flesh tones' in researches undertaken by Kodak in the early 1950s. A series of prints of 'a young lady' were prepared and submitted to a panel, and a report observed:

> Optimum reproduction of skin colour is not 'exact' reproduction . . . 'exact reproduction' is rejected almost unanimously as 'beefy'. On the other hand, when the print of highest acceptance is masked and compared with the original subject, it seems quite pale.
>
> (David L. MacAdam 1951, quoted in Winston 1985: 120)

As noted above, white skin is taken as a norm but what that means in terms of colour is determined not by how it is but by how, as Winston puts it, it is 'preferred – a whiter shade of white' (ibid.: 121). Characteristically too, it is a woman's skin which provides the litmus test. . . .

A last example of the operation of the white face as a control on media technology comes from professional television production in the USA. In the late 1970s the WGBH Educational Foundation and the 3M Corporation developed a special television signal, to be recorded on videotape, for the purpose of evaluating tapes. This signal, known as 'skin', was of a pale orange colour and was intended to duplicate the appearance on a television set of white skin. The process of scanning was known as 'skinning'. Operatives would watch the blank pale orange screen produced by tapes prerecorded with the 'skin' signal, making notes whenever a visible defect appeared. The fewer defects, the greater the value of the tape (reckoned in several hundreds of dollars) and thus when and by whom it was used. The whole process centred on blank images representing nothing, and yet founded in the most explicit way on a particular human flesh colour.

► **REFERENCES**

Brownlow, Kevin (1968) *The Parade's Gone By*. London: Secker & Warburg.
Handley, C. W. (1967) 'History of Motion-Picture Studio Lighting' in Fielding 1967: 120–4. (First published in the *Journal of the Society of Motion Picture and Television Engineers*, October 1954.)
Malkiewicz, Kris (1986) *Film Lighting*. New York: Prentice-Hall.
Millerson, Gerald (1972) *The Technique of Lighting for Television and Motion Pictures*, London: Focal Press.
Mills, Frederick S. (1921) 'Film Lighting as a Fine Art: Explaining Why the Fireplace Glows and Why Films Stars Wear Halos', *Scientific American* 124: 148, 157–8.
Valentine, Joseph (1939) 'Make-up and Set Painting Aid New Film', *American Cinematographer* February: 54–6, 85.
Winston, Brian (1985) 'A Whole Technology of Dyeing: A Note on Ideology and the Apparatus of the Chromatic Moving Image', *Daedalus* 114(4): 105–23.

PART TWO

▶ The technology of production

▶ Introduction

▶ TECHNOLOGICAL DETERMINISM AND PRODUCTION

As we approach the twenty-first century, many social commentators are promoting the view that our societies have moved beyond the stage of being based upon industrial production and that we are living in a post-industrial age. Changes to the production process, which have seen diminishing numbers of people employed in manufacturing and manual work in many Western countries, have fostered a belief that production is no longer the organizing principle of contemporary society. The focus has shifted to information, consumption, culture and lifestyle.

This section reasserts the importance of production in the contemporary world. Production has not disappeared, but is being carried out in strikingly novel forms on an increasingly global basis. Much low skilled assembly-line work has moved off shore to the Third World and it is predominantly performed by women rather than men. The quintessential product and symbol of the new age, the computer, is often manufactured in precisely this fashion. In the West there has indeed been a major shift of employment from factory work to service industries and office work, but much of this white-collar work is an integral part of the production process and one in which technology plays a crucial role. Even those areas of white-collar work with little direct connection to production, such as universities, have become much more mechanized in recent years: in the West, every lecturer now has a computer on his or her desk, and soon every student will have to have one too (Beynon 1992).

The role of technology in the economy and society is indeed central to contemporary social theory, but theorists' understanding of that role is often simplistic. The three paradigmatic theories of the transformation that Western societies are undergoing – the theories of the information society, post-Fordism and postmodernity – contain, at their core, claims about technological change and its social impact (Kumar 1995). There are strong echoes of the earlier 'post-industrial society' thesis and its tendency to adopt a technologically determinist stance. Much emphasis is placed on major new clusters of scientific and technological innovations, particularly the widespread use of information technology, and the convergence of ways of life around the globe. The increased automation of production and the intensified use of the computer are said to be revolutionizing the character of work and the structure of the workforce. At the same time, leisure, education, family relationships and personal identities are seen as moulded by the pressures exerted and opportunities arising from the new technical forces.

The proponents of the *information society* would have us believe that information and communication technologies are relegating manufacturing technology to a subordinate role. Capital and labour are replaced as the key resources of society by information and knowledge. Exponents of *post-Fordism* also identify a new era in capitalist production. Flexible forms of work organization and customized production are replacing mass production. In this model, unskilled labour is rendered obsolete as a premium is now placed on the involvement of highly skilled workers. Theories of *postmodernity*, on the other hand, are hardly concerned with production at all. Here, the issue is not so much with transformation of production but with aesthetics, language, representations and culture. When postmodern theories talk about current economic and social arrangements, they tend to presuppose that technological change has led to the dominant form of work becoming information- or knowledge-based work.

These writings share a view of technological change as an external force which overpowers pre-existing forms of social differentiation. Technological determinism is also strongly entrenched in the sociology of work and organizations. We are still taught that industrial society developed inexorably from a spate of technical inventions during the period known as the industrial revolution. These techniques were, it is implied, discovered in a 'natural' process guided at most only by the desire for greater industrial efficiency. They were seen as having definite effects, good and bad. The good effect was the great increase of material wealth they make possible. The bad effects – or at least the effects about which there is greatest ambivalence – were and are the 'degradation' of work on the one hand, and the elimination of work altogether on the other.

The new technology *required*, it is suggested, the reduction of much industrial work to the performance of mindless, repetitive tasks:

> For many [workers] – though by no means for all – the introduction of machinery implied for the first time a complete separation from the means of production; the worker became a 'hand'. On almost all, however, the machine imposed a new discipline. No longer could the spinner turn her wheel and the weaver throw his shuttle at home, free

of supervision, both in their own good time. Now the work had to be done in a factory, at a pace set by tireless, inanimate equipment.

(Landes 1969: 43)

Sacrificing creative and interesting work was seen as the necessary price to be paid for the material wealth generated by modern industry. The drive to formulate the most efficient production process is still seen as one that inevitably subordinates workers' agency to the momentum of the machine.

Unfortunately, while the repetitive, machine-paced work apparently demanded by the assembly line was the subject of much critical commentary in the first three quarters of the twentieth century, the forms of white-collar service work that have replaced it seldom elicit attention. For example, one of the fastest-expanding types of work is the 'call centre', in which as many as 2000 workers – mostly women – are employed by banks, insurance companies, airlines and other 'postindustrial' companies to answer telephone inquiries. Computer systems monitor how long workers spend on each call, and supervisors can listen in on what is said. 'Walkaway' – time spent not answering calls – is logged and has to be explained (see, for example, Wainwright 1998). Workers doing this quick-fire, monotonous work do not even have the assembly-line worker's 'luxury' of disaffected daydreaming: they must speak what are often standard, formulaic phrases with an engaged, friendly tone of voice.

Technologically determinist understandings of production have important consequences for practical politics. Politicians of all persuasions insist that we must embrace technological change in the name of progress and economic advance. Most workers share the general belief that the forces of technology and the allegedly consequent organization of work are immutable. Those workers who resist their replacement by machines are dismissed as Luddites and blamed for the retention of obsolete machinery. So in the name of technological progress crucial and far-reaching decisions are made and presented as inevitable.

But what if this technological determinism is wrong? What if the 'effects' of technology on work are in fact (deliberately or even accidentally) built into the design of workplace technology, and the technology could actually be designed differently? This would change radically the questions we have to ask. For example, it is difficult to see unemployment as an unfortunate, if temporary, by-product of technological progress once we start to realize that much technology is expressly *designed* to eliminate human labour.

The contributions in this section provide an analysis of the development and introduction of particular technologies of production. They make the argument that our technology of production is in many ways the *result* of our social relations. Industrial innovation is a product of a historically specific activity carried out by social groups for particular purposes: 'behind the technology that affects social relations lie the very same social relations' (Noble 1979: 19). Here we attempt to go not only behind but inside the 'black box' of technology to see how technical choices are simultaneously social through and through.

Our first extract deals with feudal society. The extract is from Marc

Bloch's essay on the advent of the watermill and its eventual but partial triumph over the handmill in medieval Europe. One might presume that choosing the best method of milling grain is simply a matter of technical efficiency. Bloch however shows that social relations shaped technical choice in that peasants' and lords' different places in feudal social relations led them to prefer different ways of milling grain. Centralized watermilling enabled feudal overlords to exact dues on grain being milled, and its introduction met with vigorous opposition. Domestic handmills were impossible to monitor, and dues could thus be evaded. Despite considerable, and sometimes violent, suppression, the handmill survived in some peasant areas until the nineteenth century. The history of the watermill is thus a long history of struggle between different social groups, whose different place in the social order made them prefer different technologies.

▶ LABOUR PROCESS THEORY AND BEYOND

The next set of readings deal with capitalist production, where the writings of Karl Marx have been fundamental. Many industrial sociologists have been influenced by a reading of Marx which cast him as a technological determinist. For example, Robert Blauner (1964) argued that Marx attributed social change primarily to transformations in the forces of production, which he took to be synonymous with technology. In his classic reader on industrial sociology, *Industrial Man*, Tom Burns (1969: 35) entitled the section on Marx: 'Technology as the prime mover of industrialization and social change'. According to this interpretation, which was shared by many Marxists, the development of technology and productivity is the motor force of history; new machinery and equipment revolutionize production and transform society. In this technicist version of Marxism, of which Engels was an early exponent, the productive methods of capitalism form the basis for socialism, and technology itself is beyond class struggle (see Reinfelder's 1980 survey article on 'technicism' in twentieth-century Marxism).

Here we take a different view. There is much in Marx's writing to suggest that he regarded capitalist-worker relations as a major factor affecting the technology of production (see MacKenzie 1984). We have included the pages of *Capital* in which Marx draws on authors such as Ure to show how capitalists shape technology with class struggle in mind. The revival of interest in Marx's contribution to the study of technology, particularly in relation to work, was inspired by the publication of Harry Braverman's *Labor and Monopoly Capital* (1974). That work restored Marx's critique of technology and the division of labour to the centre of his analysis of the process of capitalist development. The reproduction and expansion of capital require the subordination of labour and, according to Braverman, this is achieved by the deskilling and homogenization of the working class.

Braverman's argument cannot be accepted as it stands. Few books have elicited a greater body of critical response – for a useful introduction to this literature, see the collection edited by Wood (1982). Braverman, it has been argued convincingly, both fails to grasp the full nature of Marx's analysis and suggests a simple tendency towards the deskilling of work that is belied by the complexity of reality. Nevertheless, Braverman's work has been crucial to modern debate about workplace technology, and accordingly we reprint some key pages from his analysis of machinery.

The justified criticism of Braverman's oversimplifications should not blind us to the real connections between technology and control over labour. Especially in the early phases of capitalist development, employers were painfully aware of the magnitude of the problems of work discipline they faced. The new industrial workforce had to be wrenched from its old working habits. The rhythms of agricultural production determined by the seasons and the hours of daylight were incompatible with the demands of regular factory hours and the tyranny of new time discipline (Pollard 1965; Thompson 1967; Marglin 1976). This traumatic shift was met with fierce resistance. Machinery was used by the owners and managers of capital as an important weapon in the battle for control over production. Even those who do not place themselves in any Marxist tradition have expressed considerable agreement with Marx's characterization of the introduction of machinery. Nathan Rosenberg (1976: 117) notes that: ' [t]he apparent recalcitrance of nineteenth-century English labor, especially skilled labor, in accepting the discipline and the terms of factory employment provided an inducement to technical change', and gives a list of particular innovations in which this process can be identified.

We move into the twentieth century with a study of machine tool automation, which exemplifies the strengths of the labour process approach. David Noble found that a major goal of automation was to secure managerial power, in this case by shifting control from the shop-floor to the centralized office. The most interesting aspect of Noble's work is, however, that he goes beyond this kind of observation to a detailed discussion of the technique of automatic control. He shows that automation did not have to proceed in the way it did. The form of automation was the result of deliberate selection.

Noble suggests that it was at least in part the social relations of production that tipped the balance in the choice of technology.[1] Machining was automated using the technique of numerical control. But there was also a prototype for a technique of automation called record playback. With record playback the control of the machine remains with the machinist, whereas numerical control appeared to offer a means of dispensing with these skilled workers. Managements, argues Noble, decided to take advantage of this possibility. So, he claims, the technical choices made can be understood only by paying attention to the conflictual relations of production within which machining takes place.

However, the intentions underlying technological design were not necessarily realized. The introduction of numerical control onto the shop floor did not simply shift control to management. It was met with fierce resistance from the workforce. At the same time, management found that it needed to retain skilled machinists to operate the new machines effectively. Consequently, management was never able to gain complete control over production. In reality, machines do not run themselves. This contradiction of capitalist production has not been eclipsed: 'It is still dependent upon the work force to turn a profit' (Noble 1979: 43).

But class conflict is not the whole story. A major flaw in the Marxist analysis of technology is its exclusive focus on capital-labour relations. An understanding of technical change as based on the social relations of production must include an account of divisions *within* the working class. A critical division, frequently omitted from the category of social relations of production, is that between men and women. In any discussion of technical innovation, the sex of the workforce and gender relations in the workplace are of fundamental importance. This is integral to Cynthia Cockburn's account of the history of typesetting technology in Britain. We discover how employers' desire to deskill the workforce underpinned the development of the new technology, and how its application took place in a context of intense struggles by print workers to retain their craft monopoly over the job. Rather than resisting mechanization, the male compositors (typesetters) fought instead to retain sole rights to the new equipment. Their success entailed the exclusion of unskilled women from the trade.

Gendered preferences for the design of technology are clearly exemplified in Cockburn's discussion of the nineteenth-century rival to the Linotype machine, the Hattersley typesetter. Compositors hated technical systems such as the Hattersley typesetter that separated the jobs of composing and distribution (putting the letters back in boxes). It had a separate mechanism for distributing type, designed for use by girls. The separation of the type-setting (seen as skilled) and distribution (seen as unskilled) was devised as a means of reducing overall labour costs. Compositors feared that employers would try to expand this use of cheaper, unskilled labour once it got a foothold into the composing room.

The Linotype machine on the other hand did not represent the destruction but merely the mechanization of the compositors' setting skills as a whole. The key aspect of this successful machine was that it eliminated distribution as a task – since letters were formed anew each time by the action of brass mould on molten metal. After the type was used, it was simply melted down ready to be reused. Although it automated their work, male compositors actually welcomed the Linotype machine because it did not depend for its success on the employment of child or female labour. On the contrary, by cutting out the task of distribution, it stopped any possible inroads that boys and women might make into the trade. Thus, perhaps in deference to the organizational strength of the union, the Linotype company and employers adopted a technology that was beneficial to the union men.

There is another way in which the technology of production reflects male power. It is often said that men are naturally stronger and therefore more suited to certain types of work. Against this, feminists argue convincingly that distinctions of skill between women's and men's work have as much to do with job control and wage levels as they have to do with technical competencies (see Wajcman 1991b). Understandably, feminists have downplayed the material realities of physical strength, because women's exclusion from occupations has so often been legitimated in biological terms. However Cockburn argues that this formulation underemphasizes the material realities of physical power and with them the tangible factors in skill. Men, having been reared to physical advantage, make full use of this in the selective design of tools and machinery to match the strengths they have cultivated (as Cockburn concedes, biology may give men a 'first step on the ladder', but any innate difference is magnified socially). Machinery is designed by men with men in mind. Industrial technology thus reflects male power as well as capitalist domination.

The next extract by Robert Thomas further broadens the notion of the social relations of production. He argues that divisions *within* management, and the pressures of professional self-interest, are also an important element in the process of technical innovation. In studying the development of the flexible machining system, Thomas found that choosing between technologies is not simply a matter of rational, economic or technical calculations. Rather, choices were profoundly influenced by the personal ambitions of divisional managers and engineers, and by their efforts to shape the organizational context within which the technology would be used. While middle management wanted to control shop-floor supervisors, the engineers were keen to enhance their careers by doing 'real engineering'. Top management were largely oblivious to these perceptions and interests.

In the second part of the extract, '*The limits of social choice*', Thomas takes issue with both strategic choice and labour process theories of technological change. The strategic choice perspective restricts our attention to decisions made by the higher echelons of management. As Thomas demonstrates, decisions made at lower levels of the organization have a considerable influence over the framing of alternatives open to top management. Similarly, some labour process theory views technical choice as little more than an extension of managerial will. It assumes that owners and managers share the common goal of control over labour. But, as we have seen, for managers the object of control is as likely to be other managers as it is workers. Thomas also claims that the role of engineers has been overlooked by labour process theorists. This claim is not accurate: Noble, for example, devotes considerable attention to the activity of engineers (Noble 1977, 1984). Rather, the point is that, instead of seeing engineers as subordinate to capital, as many labour process theorists have done, Thomas attributes a more autonomous role to them. Engineers have their own designs on technology, which are as likely to result in upgrading workers as in their deskilling. Thus while power relations in organizations shape the design and deployment of new technology, the multiplicity of competing interests called into play are not reducible to class conflict.

DECONSTRUCTING THE DESIGNER/USER DIVIDE

Our next reading extends the discussion of industrial technology to non-manual work. The introduction of information technologies into offices has been the focus of much feminist research, mainly because secretaries almost everywhere are women. Word processors were intended to raise typing productivity and reduce the number of office jobs by automating and rationalizing the work of female typists (Barker and Downing 1980; Webster 1990). They were therefore designed to resemble typewriters so as to make them more familiar to their principal users. The development of word-processing software reflected designers' assumptions about the gendered characteristics of the prospective users, their level of intelligence and their ability to learn by doing.

There is a tendency in some feminist literature to assume that gender is a fixed and unitary phenomenon embedded in information technology. Against this, Jeanette Hofmann argues that gendered conceptions of users are fluid and subject to a variety of interpretations. Comparing three kinds of word processing system, she suggests that the design of specific software was informed by utterly contrary ideas of women. Whereas dedicated word processing systems treated female operators as eternal beginners, programs like WordStar and WordPerfect assumed that typists would gain technical expertise in handling them. A comparison between the user interfaces of these writing programs illustrates that the designers of technical artifacts can attribute different characteristics to the same target group, that is, female secretaries.

Subsequently, however, these constructions of the sex and skill of the user were redrawn. A third type of approach targeted at professionals and managers (implicitly taken predominantly to be men) was developed in the 1970s at the Xerox's Palo Alto Research Center (Xerox PARC). This was the 'Star' computer, which enjoys the dubious claim to fame of having laid the commercially unsuccessful cornerstone for what went on to become a model of commercial and conceptual success: the famous Apple-Windows graphical user interface. The story of how Steven Jobs of Apple Corporation derived the idea for the Apple computer after seeing the Alto, a predecessor of the Star, at Xerox PARC is dramatically portrayed in Smith and Alexander (1988).

The primary objective of the Alto and Star projects, as they bore upon wordprocessing, was not to enhance typing productivity but to establish a fundamentally new conception of computer-assisted writing. The aim was to create a user interface for the user who could afford to have no interest in computers, the dilettante knowledge worker. The resulting personal computer system involved the first word processing program for non-expert users, with text production, along with other computing tasks, organized by means of pictorial symbols, or 'icons', and an inputting device simple enough for a child – the hand-held 'mouse'. The basics of the Star system could be learned in less than an hour. The scientists at Xerox PARC designed, built and used a complete system of hardware and software that fundamentally altered the nature of the human-computer interface itself. As that interface became pervasive, via Apple and then Microsoft's Windows, typing as a distinct occupation began to lose its function. Gendered approaches to interface design have been replaced by the image of a universal user. Although sexual divisions within the office shaped the early evolution of word processing technology, the technology has in turn undermined some of these long-established gendered arrangements.

Such an analysis is not without its problems. In assuming that the image of the user is crucial to the design of the word processing software, Hofmann may not give sufficient weight to the constraints imposed by the limited material capacity of the early dedicated word processing machines. An alternative scenario, that the designers had developed particular kinds of machine and then conceived of users, might also account for the contradictory images of secretaries. Indeed, the significance of the gendered user in the development of the technology is called into question by Hofmann's own argument about the universal character of the Apple-Windows graphical interface. We are drawn to the conclusion that the technology and the potential users were mutually shaped.

In our discussion so far, innovation has been represented as an activity restricted to engineers and computer scientists in research and development. This linear model of innovation treats technologies as coherent, finished products that are supplied in unproblematic forms to meet user needs. It is increasingly recognized, however, that implementation, the process of getting technologies to work as commercially successful operating systems, is a key phase in the overall development of technology. Here, James Fleck uses a study of the introduction of computer-aided production management to demonstrate that innovation and diffusion are not separate processes. His concept of 'innofusion' captures the idea that significant new developments are forged *during* implementation within user organizations.

Fleck's analysis highlights the importance of the local expertise and experience of users. Familiarity with a firm, its markets, its production and administrative practices is often embodied in tacit skills and procedures built up over years. In order to create functioning technical systems, generic technology knowledge must be combined with this local, practical knowledge. It follows from this that links between supplier and user firms are crucial. Implementation of new systems is often a joint process – involving supplier and user in collaborative development and application of technologies to specific requirements. Understanding how people work within the context of their organization is then a prerequisite for technical success. It is a process of mutual adaptation between organization and technology.

In this section we have been exploring how social relations are embodied in the actual design of artifacts and in the technologies of production. But how might technology be redesigned with alternative priorities in mind? The 1970s witnessed a number of political projects to develop technologies which would provide both enriching work and socially useful products. Ideas about ecologically appropriate and human-centred technologies encapsulated in books such as Schumacher's *Small is Beautiful* (1973) and Illich's *Tools for Conviviality* (1975) gained popular currency. At the same time, experiments in industrial democracy prompted a profound questioning of the nature of technology itself. They recognized that trade union intervention into the design of both technologies and working practices could influence the quality of jobs and services, and build more progressive values into the resulting artifacts. The Lucas Aerospace Alternative Plan, for example, was an attempt by a workforce in Britain to shift production from military to 'socially useful' products, with particular application in Third World countries. The UTOPIA project in Sweden, along with others in Scandinavia, emphasized participatory design methods, doing systems design *with* rather than *for* users.[2]

Inspiring in their own way, projects such as the Lucas Alternative Plan and UTOPIA were focused mainly on skilled, and thus male, manufacturing workers. The limitations of this approach have become painfully apparent against the backdrop of a 'post-industrial' landscape. The new forms of 'flexible' production have prompted volumes of writing from the academic left, but a politics of intervention in these production processes has been notably absent. Feminists too have been more active in relation to the environment, the military and medical technologies than in the politics of production.

A notable exception is Lucy Suchman who practises her feminist politics 'on the inside' of a multinational corporation (at the Xerox Palo Alto Research Center, where the Star computer discussed by Hofmann was developed). Suchman's key concern is to develop new methods for designing computer systems which both recognize and integrate the hidden knowledge and expertise of women at work.[3] Like Fleck, Suchman stresses the innovatory character of the implementation process which depends for its success on the local expertise of users. Conventional methods presume that the abstract knowledge of computer scientists practised at a distance from the sites of technologies-in-use creates effective technologies. As a result, Suchman argues, women's participation as inventors is obscured and denied. A recognition that the design process is complete only when the computer system is implemented and used is much more likely to include women within its framework. In real working environments, users are constantly required to interpret technologies with reference to their local circumstances. By ensuring that the skills of women workers are highly valued, feminist methods of systems development give women a central role in technical production and design.

Of course, we must be realistic about the constraints upon a feminist politics of production technology, desirable though that is. The operations of the market and of the processes of lock-in discussed in Part One of this book have powerful effects. Standard hardware (computers built around the pervasive Intel microprocessors), standard, easily-tailored software (spreadsheets, databases, etc.) and the now standard graphical user interface (derived from the Xerox Star, as noted above) offer so many advantages that local considerations are increasingly unlikely to override them. Nevertheless, standard components, their shape determined in the remote technological metropolis, can still be configured in flexible, locally appropriate ways.[4]

Marxism, as we have emphasized, had many blind spots, and – rooted as it was in the experience of masculinized industrial production – it is in many ways conceptually ill-equipped to tackle the problems and the opportunities of the new world of globalized, information-saturated, partially feminized production. Yet, at its best, Marxism offered both a diagnosis of what needed changing and of the real-world constraints upon the possibility of such changes. Neither modern feminism, nor contemporary social theory (dominated as it is by postmodernism), have an account of the technology of production that does an equivalent job. We live in new times, but still await a new social theory of production, and a new politics of technology, to match those new times.

► **NOTES**

1 The primary issue to be raised with respect to Noble's account is whether he deals adequately with the possibility that numerical control was favoured for cost and profit reasons, rather than for control over labour as such. See his discussion in note 10 on pp. 174–5. Noble has published a book-length version of his work on machine tools, and interested readers should consult this as well as the extract that forms Chapter 14: see Noble (1984).
2 For a review of human-centred projects, including Lucas and UTOPIA, see Hales (1993) and Pain *et al.* (1993).
3 For an overview of feminist approaches to systems design, and a discussion of their limitations, see Webster (1996, Ch. 6).
4 We are grateful to Robin Williams for helpful discussion of this point. See Williams (1997).

11 ▶ The watermill and feudal authority

Marc Bloch

It is in England especially that the war of wind and water against human muscle is seen in its clearest light.

Manorial rights were not an institution native to England. The Norman conquerors had imported them from the continent as one of the principal elements in the 'manorial' system which after the almost total dispossession of Saxon aristocracy they methodically established, by superimposing it on what remained of a much looser form of dependence. It is true that in England the system of seignorial monopolies always remained less complete than on the other shore of the Channel. But manorial rights over the mill were generally introduced, though not without resistance. Opposition was all the more passionate in this country because by reason of its remoteness from Mediterranean influences, and the strong impress of German and Scandinavian civilisation, the watermill, though familiar from the end of the 11th century on the great estates, only won its way very slowly among the middle classes. It is characteristic that among the privileges granted to English burgesses there should frequently figure this clause, totally unknown in French and German urban charters, allowing the use of the handmill, or more rarely the horse-driven mill. This was the state of affairs at Newcastle, Cardiff and Tewkesbury during the 12th century, and in London even in the middle of the 14th century.[1] But the pressure of rich and powerful communities was needed to overcome this tolerance. 'The men shall not be allowed to possess any handmills' – such was the clause inserted by the canons of Embsay in Yorkshire between 1120 and 1151, in a charter in which a noble lady made over to them a certain watermill. It expressed their own attitude and that of all lords of rivers and manors.[2] There were occasions when milling stones were seized by the lord's officials in the very houses of the owners and broken in pieces; there were insurrections on the part of housewives; there were law-suits which grimly pursued their endless and fruitless course, leaving the tenants always the losers. The chronicles and monastic cartularies of the 13th and 14th centuries are full of the noise of these quarrels.[3] At St Albans they assumed the scale of a veritable milling epic.

In this small Hertfordshire town, to which the monks who were lords of the place obstinately refused to give any privileges, the example of neighbouring citizens stirred up – in the words of the monastic chronicler – a particularly 'indomitable tenantry'. They were a collection of artisans rather than peasants, and it was not only the dues for milling grain or malt and the miller's exactions that they sought to avoid by milling at home. The drapers among them also claimed, in defiance of the lord's fulling-mill the right to set up their own fulling stocks for the pressing of material, at least for the coarser kinds of cloth, for it was generally considered at this period that fine materials must be fulled under foot. The first quarrel broke out in 1274, accompanied by the usual incidents. Millstones and lengths of cloth were confiscated; there was mutual violence perpetrated by the lord's officials and by tenants; a league was formed among the inhabitants, who clubbed together to establish a common purse to maintain their cause at law, whilst monks, barefoot before the High Altar, were chanting penitential psalms; attempts were made by the women to win the queen over to their side, but the abbot had taken the precaution of having her smuggled into the monastery by a secret entrance. Finally there were lengthy proceedings before the royal court, ending inevitably in the defeat of the recalcitrant party, who sought to appease their offended lord by the gift of five fine barrels of wine. Another incident took place in 1314. Then in 1326 the citizens demanded a charter which should contain among other clauses the right to domestic milling. This led to an open insurrection, in which the monastery was twice besieged. Final agreement was only reached under pressure from the king, but it left the problem of the lord's monopolies unresolved. Taking advantage of this uncertainty, the inhabitants soon had anything up to eighty handmills working in their homes. But in 1331 a new abbot – Richard II, the terrible leprous abbot – entered the lists. He won the day by going to law. From all over the town the millstones were brought in to the monastery, and the monks paved their parlours with them, like so many trophies. But when in 1381 the great insurrection of the common people broke out in England and Wat Tyler and John Ball emerged as leaders, the people of St Albans were infected by the same fever and attacked the abbey. They destroyed the notorious paved floor, the monument to their former humiliation, and as the stones were doubtless no longer any use for grinding, they broke them up and each took fragments of them as a sign of victory and solidarity, 'as the faithful do on Sundays with the holy bread'. The deed of liberation which they extorted from the monks recognised their freedom to maintain 'handmills' in every home. The insurrection however proved to be like a blaze of straw that soon burns itself out. When it had collapsed all over England the Charter of St Albans and all the other extorted privileges were annulled by royal statute. But was this the end of a struggle that had lasted more than a century? Far from it. The chronicler, as he draws to the close of his story, has to admit that for malting at any rate the detestable handmills have come into action again and have been again forbidden.[4]

They were destined to give humble service throughout the length and breadth of England for a long time to come. It is true that the narratives we possess hardly make any further mention of them. But here and there a manorial 'record' allows us to lift the veil for a moment. The Tudors had long ago succeeded the Plantagenets when in 1547 the people of the royal manor of Kingsthorpe obtained recognition for their right to grind at least a certain quantity of their grain at home. At Bury in Lancashire, it was not until the restoration of the Stuarts that the lord of the manor succeeded in suppressing competition with his own mill; and even so the obstinacy of those representing the parish made the suit drag on till 1713. This was only seventy-three years before the first large-scale flour-mill was opened in London.[5]

In short, when the iron and coal age opened, the ancient prehistoric tools had nowhere yielded altogether to the 'engines' which for so many centuries had also relied upon the inanimate forces of wind and water. There is therefore nothing surprising about the survival in working order of handmills in Ireland, Scotland, the Shetlands, Norway, East Prussia, and nearly everywhere in Slav territory, right up to the end of the 19th century; and even perhaps in our own day they have not altogether ceased to function. For these regions, situated as they are on the fringe of the West, had in all respects long been faithfully wedded to a fairly rudimentary degree of mechanisation. Prussian villagers were still grinding grain in 1896 according to the elementary methods of their ancestors, and felt obliged, like them, to hide from strangers as they did so – as though the lord's monopolies were still in existence. In these actions we can recognise, not only the dim potency of tradition, but also the fact that in the North winter frosts were not very favourable to the use of running water-power; moreover, in the Shetlands, Norway, Scotland, and even Ireland, there was no seignorial authority comparable to that prevailing in France. But even in the heart of our own civilisation, a more searching enquiry would no doubt reveal more than one scattered example of a similar survival. The Breton handmills, whose history might well tempt a scholar more familiar with the province than I am, were surely given something of a new lease of life by the suppression of the seignorial regime. In the second half of the 19th century Lamprecht observed near the tributaries of the Moselle, whose waters had turned some of the earliest millwheels in Gaul, material traces of the relatively recent practice by which, with the aid of human power, corn was crushed between two revolving stones.[6]

▶ NOTES

1 Ballard, *British Borough Charters*, 1916. To the list of towns quoted above should be added Bristol, although the text of its charter is less precise. For London cf. *Liber*

Albus, edit. Riley, p. 74. In Wales, about the 12th century, corn was still commonly hand-ground by women of servile status (*Ancient Laws of Wales*, Vol. II, p. 7, c. 17).

2 Bennett and Elton, *History of Corn Milling*, London, 1898–1904, Vol. I, p. 211, and a fascimile as frontispiece to the volume (for the date, cf. *Victoria County Histories*, Yorkshire, Vol. III, p. 195). See also, in Edward III's reign, the regulation of the Chester mills (Bennett, Vol. I, p. 212).

3 *Chronicum Petroburgense*, edit. Stapleton (Camden Society, 1849), p. 67 (1284); E. A. Fuller, *Cirencester: the manor and the town* in Trans. Arch. Soc. *Bristol & Glos*, Vol. IX, 1884–85, pp. 311ff (1306–1307); Glos, *Registra quorumdam abbatum monasterii S. Albani*, edit. Riley (*Rolls Series*), Vol. I, pp. 109ff. (1499: a rising by the women provoked by the prohibition of horse-mills); in 1381, the people of Watford, like those of St. Albans, had extorted from the abbey a charter recognising their right to use handmills; see Th. Walsingham, *Gesta S. Albani*, Vol. III, p. 325.

4 Thomas Walsingham, *Gesta S. Albani*, edit. Riley (*Rolls Series*), Vol. I, pp. 410ff; Vol. II, p. 149, 158, 287ff.; Vol. III, pp. 286ff.; 360–361, 367–371. Cf. the *Historia Anglicana* by the same author, edit. Riley (*Rolls Series*), p. 475. For the events of 1381, see C. Oman, *The Great Revolt of 1381*, pp. 91ff.

5 Bennett and Elton, Vol. I, p. 224; P. Mantoux, *La Révolution industrielle*, 1905, p. 341.

6 Bennett and Elton, Vol. I, p. 167; *Revue archéologique*, 1900, I, p. 35; *Zeitschrift für Ethnologie*, Vol. XXII, 1890, p. 607, and Vol. XXVIII, 1896, p. 372; Lamprecht, *Deutsche Wirtschaftsgeschichte*, Vol. I, p. 585. Research in provincial areas would undoubtedly shed much new light on the life-span of hand-milling. It is in this way that A. Demont (*Le blé dans les traditions artésiennes*, in *Revue de folklore français*, 1935, p. 49) discovered that a handmill was found in 1856 'right in the middle of the belfry of the church at Hermaville. The millstones showed signs of long use. The oak framework was a timber construction in 18th-century style.' Cf. also *Bull. Soc. Archeol. Limousin*, Vol. CXXLV, p. 47 and LXXV, p. 58.

12 ▶ The machine versus the worker

Karl Marx

The instrument of labour strikes down the labourer. This direct antagonism between the two comes out most strongly, whenever newly introduced machinery competes with handicrafts or manufactures, handed down from former times. But even in Modern Industry the continual improvement of machinery, and the development of the automatic system, has an analogous effect. 'The object of improved machinery is to diminish manual labour, to provide for the performance of a process or the completion of a link in a manufacture by the aid of an iron instead of the human apparatus.'[1] 'The adaptation of power to machinery heretofore moved by hand, is almost of daily occurrence . . . the minor improvements in machinery having for their object economy of power, the production of better work, the turning off more work in the same time, or in supplying the place of a child, a female, or a man, are constant, and although sometimes apparently of no great moment, have somewhat important results.'[2] 'Whenever a process requires peculiar dexterity and steadiness of hand, it is withdrawn, as soon as possible, from the cunning workman, who is prone to irregularities of many kinds, and it is placed in charge of a peculiar mechanism, so self-regulating that a child can superintend it.'[3] . . .

But machinery not only acts as a competitor who gets the better of the workman, and is constantly on the point of making him superfluous. It is also a power inimical to him, and as such capital proclaims it from the roof tops and as such makes use of it. It is the most powerful weapon for repressing strikes, those periodical revolts of the working-class against the autocracy of capital.[4] According to Gaskell, the steam-engine was from the very first an antagonist of human power, an antagonist that enabled the capitalist to tread under foot the growing claims of the workmen, who threatened the newly born factory system with a crisis.[5] It would be possible to write quite a history of the inventions, made since 1830, for the sole purpose of supplying capital with weapons against the revolts of the working-class. At the head of these in importance stands the self-acting mule, because it opened up a new epoch in the automatic system.[6]

Nasmyth, the inventor of the steam-hammer, gives the following evidence before the Trades' Union Commission, with regard to the improvements made by him in machinery and introduced in consequence of the widespread and long strikes of the engineers in 1851. 'The characteristic feature of our modern mechanical improvements, is the introduction of self-acting tool machinery. What every mechanical workman has now to do, and what every boy can do, is not to work himself but to superintend the beautiful labour of the machine. The whole class of workmen that depend exclusively on their skill, is now done away with. Formerly, I employed four boys to every mechanic. Thanks to these new mechanical combinations, I have reduced the number of grown-up men from 1,500 to 750. The result was a considerable increase in my profits.'

Ure says of a machine used in calico printing: 'At length capitalists sought deliverance from this intolerable bondage' (namely the, in their eyes, burdensome terms of their contracts with the workmen) 'in the resources of science, and were speedily re-instated in their legitimate rule, that of the head over the inferior members.' Speaking of an invention for dressing warps: 'Then the combined malcontents, who fancied themselves impregnably intrenched behind the old lines of division of labour, found their flanks turned and their defences rendered useless by the new mechanical tactics, and were obliged to surrender at discretion.' With regard to the invention of the self-acting mule, he says: 'A creation destined to restore order among the industrious classes. . . . This invention confirms the great doctrine already propounded, that when capital enlists science into her service, the refractory hand of labour will always be taught docility.'[7]

▶ NOTES

1 'Rep. Insp. Fact. for 31st October, 1858,' p. 43.
2 'Rep. Insp. Fact. for 31st October, 1856,' p. 15.
3 Andrew Ure, *The Philosophy of Manufactures*, London, 1835, p. 19. 'The great advantage of the machinery employed in brick-making consists in this, that the employer is made entirely independent of skilled labourers.' ('Ch. Empl. Comm. V. Report,' Lond., 1866, p. 130, n. 46.)
4 'The relation of master and man in the blown-flint bottle trades amounts to a chronic strike.' Hence the impetus given to the manufacture of pressed glass, in which the chief operations are done by machinery. One firm in Newcastle, who formerly produced 350,000 lbs. of blown-flint glass, now produces in its place 3,000,500 lbs. of pressed glass. ('Ch. Empl. Comm., Fourth Rep.' 1865, pp. 262–263).
5 Gaskell, 'The Manufacturing Population of England,' London, 1833, pp. 3, 4.
6 Fairbairn discovered several very important applications of machinery to the construction of machines, in consequence of strikes in his own workshops.
7 Ure, 1. c., pp. 368–370.

13 ▶ Technology and capitalist control

Harry Braverman

The evolution of machinery from its primitive forms, in which simple rigid frames replace the hand as guides for the motion of the tool, to those modern complexes in which the *entire process* is guided from start to finish by not only mechanical but also electrical, chemical, and other physical forces – this evolution may thus be described as an increase in human control over the action of tools. These tools are controlled, in their activities, as extensions of the human organs of work, including the sensory organs, and this feat is accomplished by an increasing human understanding of the properties of matter – in other words, by the growth of the scientific command of physical principles. The study and understanding of nature has, at its primary manifestation in human civilization, the increasing control by humans over labor processes by means of machines and machine systems.

But the control of humans over the labor process, thus far understood, is nothing more than an abstraction. This abstraction must acquire concrete form in the social setting in which machinery is being developed. And this social setting is, and has been from the beginnings of the development of machinery in its modern forms, one in which humanity is sharply divided, and nowhere more sharply divided than in the labor process itself. The mass of humanity is subjected to the labor process for the purposes of those who control it rather than for any general purposes of 'humanity' as such. In thus acquiring concrete form, the control of humans over the labor process turns into its opposite and becomes the control of the labor process over the mass of humans. Machinery comes into the world not as the servant of 'humanity,' but as the instrument of those to whom the accumulation of capital gives the *ownership* of the machines. The capacity of humans to control the labor process through machinery is seized upon by management from the beginning of capitalism as the *prime means whereby production may be controlled not by the direct producer but by the owners and representatives of capital.* Thus, in addition to its technical function of increasing the productivity of labor – which would be a mark of machinery under any social system – machinery also has in the capitalist system the function of divesting the mass of workers of their control over their own labor. It is ironic that this feat is accomplished by taking advantage of that great human advance represented by the technical and scientific developments that increase human control over the labor process. It is even more ironic that this appears perfectly 'natural' to the minds of those who, subjected to two centuries of this fetishism of capital, actually see the machine as an alien force which subjugates humanity!

The evolution of machinery represents an expansion of human capacities, an increase of human control over environment through the ability to elicit from instruments of production an increasing range and exactitude of response. But it is in the nature of machinery, and a corollary of technical development, that the control over the machine need no longer be vested in its immediate operator. This possibility is seized upon by the capitalist mode of production and utilized to the fullest extent. What was mere *technical possibility* has become, since the Industrial Revolution, an *inevitability* that devastates with the force of a natural calamity, although there is nothing more 'natural' about it than any other form of the organization of labor. Before the human capacity to control machinery can be transformed into its opposite, a series of special conditions must be met which have nothing to do with the physical character of the machine. The machine must be the property not of the producer, nor of the associated producers, but of an alien power. The interests of the two must be antagonistic. The manner in which labor is deployed around the machinery – from the labor required to design, build, repair, and control it to the labor required to feed and operate it – must be dictated not by the human needs of the producers but by the special needs of those who own both the machine and the labor power, and whose interest it is to bring these two together in a special way. Along with these conditions, a social evolution must take place which parallels the physical evolution of machinery: a step-by-step creation of a 'labor force' in place of self-directed human labor; that is to say, a working population conforming to the needs of this social organization of labor, in which knowledge of the machine becomes a specialized and segregated trait, while among the mass of the working population there grows only ignorance, incapacity, and thus a fitness for machine servitude. In this way the remarkable development of machinery becomes, for most of the working population, the source not of freedom but of enslavement, not of mastery but of helplessness, and not of the broadening of the horizon of labor but of the confinement of the worker within a blind round of servile duties in which the machine appears as the embodiment of science and the worker as little or nothing. But this is no more a technical necessity of machinery than appetite is, in the ironic words of Ambrose Bierce, 'an instinct thoughtfully implanted by Providence as a solution to the labor question.'

Machinery offers to management the opportunity to do by wholly mechanical means that which it had previously attempted to do by organizational and disciplinary means. The fact that many machines may be paced and controlled according to centralized decisions, and that these controls may thus be in the hands of management, removed from the site of production to the office – these technical possibilities are of just as great interest to management as the fact that the machine multiplies the productivity of labor.[1] It is not always necessary, for this purpose, that the machine be a well-developed or sophisticated example of its kind. The moving conveyor, when used for an assembly line, though it is an exceedingly primitive piece of machinery, answers perfectly to the needs of capital in the organization of work which may not be otherwise mechanized. Its pace is in the hands of management, and is determined by a mechanical device the construction of which could hardly be simpler but one which enables management to seize upon the single essential control element of the process.

▶ **NOTES**

1 'One great advantage which we may derive from machinery,' wrote Babbage, 'is from the check which it affords against the inattention, the idleness, or the dishonesty of human agents.' Charles Babbage, *On the Economy of Machinery and Manufactures* (1832; rept. ed., New York, 1963), p. 54.

14 ▶ Social choice in machine design: the case of automatically controlled machine tools

David F. Noble

 ## THE TECHNOLOGY: AUTOMATICALLY CONTROLLED MACHINE TOOLS

The focus here is numerically controlled machine tools, a particular production technology of relatively recent vintage. According to many observers, the advent of this new technology has produced something of a revolution in manufacturing, a revolution which, among other things, is leading to increased concentration in the metalworking industry and to a reorganization of the production process in the direction of greater managerial control. These changes in the horizontal and vertical relations of production are seen to follow logically and inevitably from the introduction of the new technology. 'We will see some companies die, but I think we will see other companies grow very rapidly,' a sanguine president of Data Systems Corporation opined (Stephanz 1971). Less sanguine are the owners of the vast majority of the smaller metalworking firms which, in 1971, constituted 83 percent of the industry; they have been less able to adopt the new technology because of the very high initial expense of the hardware, and the overheads and difficulties associated with the software (ibid.). In addition, within the larger, better endowed shops, where the technology has been introduced, another change in social relations has been taking place. Earl Lundgren, a sociologist who surveyed these shops in the late 1960s, observed a dramatic transfer of planning and control from the shop floor to the office (1969).

For the technological determinist, the story is pretty much told: numerical control leads to industrial concentration and greater managerial control over the production process. The social analyst, having identified the cause, has only to describe the inevitable effects. For the critical observer, however, the problem has merely been defined. This new technology was developed under the auspices of management within the large metalworking firms. Is it just a coincidence that the technology tends to strengthen the market position of these firms and enhance managerial authority in the shop? Why did this new technology take the form that it did, a form which seems to have rendered it accessible only to some firms, and why only this technology? Is there any other way to automate machine tools, a technology, for example, which would lend itself less to managerial control? To answer these questions, let us take a closer look at the technology.

A machine tool (for instance, a lathe or milling machine) is a machine used to cut away surplus material from a piece of metal in order to produce a part with the desired shape, size, and finish. Machine tools are really the guts of machine-based industry because they are the means whereby all machinery, including the machine tools themselves, are made. The machine tool has traditionally been operated by a machinist who transmits his skill and purpose to the machine by means of cranks, levers, and handles. Feedback is achieved through hands, ears, and eyes. Throughout the nineteenth century, technical advances in machining developed by innovative machinists built some intelligence into the machine tools themselves – automatic feeds, stops, throw-out dogs, mechanical cams – making them partially 'self-acting.' These mechanical devices relieved the machinist of certain manual tasks, but he retained control over the operation of the machine. Together with elaborate tooling – fixtures for holding the workpiece in the proper cutting position and jigs for guiding the path of the cutting tool – these design innovations made it possible for less skilled operators to use the machines to cut parts after they had been properly 'set up' by more skilled men; but the source of the intelligence was still the skilled machinist on the floor.

The 1930s and 1940s saw the development of tracer technology. Here patterns, or templates, were traced by a hydraulic or electronic sensing device which then conveyed the information to a cutting tool which reproduced the pattern in the workpiece. Tracer technology made possible elaborate contour cutting, but it was only a partial form of automation: for instance, different templates were needed for different surfaces on the same workpiece. With the war-spurred development of a whole host of new sensing and measuring devices, as well as precision servomotors which made possible the accurate control of mechanical motion, people began to think about the possibility of completely automating contour machining.

Automating a machine tool is different from automating, say, automotive manufacturing equipment, which is single-purpose, fixed automation, and cost-effective only if high demand makes possible a high product volume. Machine tools are general purpose, versatile machines, used primarily for small batch, low volume production of parts. The challenge of automating machine tools, then, was to render them self-acting while retaining their versatility. The solution was to develop a mechanism that translated electrical signals into machine motion and a medium (film, lines on paper, magnetic or punched paper tape, punched cards) on which the information could be stored and from which the signals could be reproduced.

The automating of machine tools, then, involves two separate processes. You need tape-reading and machine controls, a means of transmitting information from the medium to the machine to make the tables and cutting tool move as desired, and you need a means of getting the information on the medium, the tape, in the first place. The real challenge was the latter. Machine controls were just another step in a known direction, an extension of gunfire control technology developed during the war. The tape preparation was something new. The first viable solution was 'record playback,' a system developed in 1946–1947 by General Electric, Gisholt, and a few smaller firms.[1] It involved having a machinist make a part while the motions of the machine under his command were recorded on magnetic tape. After the first piece was made, identical parts could be made automatically by playing back the tape and reproducing the machine motions. John Diebold, a management consultant and one of the first people to write about 'flexible automation,' heralded record-playback as 'no small achievement . . . it means that automatic operation of machine tools is possible for the job shop – normally the last place in which anyone would expect even partial automation' (1952: 88). But record-playback enjoyed only a brief existence, for reasons we shall explore. (It was nevertheless immortalized as the inspiration for Kurt Vonnegut's *Player Piano*. Vonnegut was a publicist at GE at the time and saw the record-playback lathe which he describes in the novel.)

The second solution to the medium-preparation problem was 'numerical control' (N/C), a name coined by MIT engineers William Pease and James McDonough. Although some trace its history back to the Jacquard loom of 1804, N/C was in fact of more recent vintage; the brainchild of John Parsons, an air force subcontractor in Michigan who manufactured rotor blades for Sikorsky and Bell helicopters. In 1949 Parsons successfully sold the air force on his ideas, and then contracted out most of the research work to the Servomechanisms Laboratory at MIT; three years later the first numerically controlled machine tool, a vertical milling machine, was demonstrated and widely publicized.

Record-playback was, in reality, a multiplier of skill, simply a means of obtaining repeatability. The intelligence of production still came from the machinist who made the tape by producing the first part. Numerical control, however, was based upon an entirely different philosophy of manufacturing. The specifications for a part – the information contained in an engineering blueprint – are first broken down into a mathematical representation of the part, then into a mathematical description of the desired path of the cutting tool along up to five axes, and finally into hundreds or thousands of discrete instructions, translated for economy into a numerical code, which is read and translated into electrical signals for the machine controls. The N/C tape, in short, is a means of formally circumventing the role of the machinist as the source of the intelligence of production. This new approach to machining was heralded by the National Commission on Technology, Automation, and Economic Progress as 'probably the most significant development in manufacturing since the introduction of the moving assembly line' (Lynn *et al.* 1966: 89).

 CHOICE IN DESIGN: HORIZONTAL RELATIONS OF PRODUCTION

This short history of the automation of machine tools describes the evolution of new technology as if it were simply a technical, and thus logical, development. Hence it tells us very little about why the technology took the form that it did, why N/C was developed while record-playback was not, or why N/C as it was designed proved difficult for the metalworking industry as a whole to absorb. Answers to questions such as these require a closer look at the social context in which the N/C technology was developed. In this section we will look at the ways in which the design of the N/C technology reflected the horizontal relations of production, those between firms. In the following section, we will explore why N/C was chosen over record-playback by looking at the vertical relations of production, those between labor and management.

To begin with, we must examine the nature of the machine-tool industry itself. This tiny industry which produces capital goods for the nation's manufacturers is a boom or bust industry that is very sensitive to fluctuations in the business cycle, experiencing an exaggerated impact of good times – when everybody buys new equipment – and bad times – when nobody buys. Moreover, there is an emphasis on the production of 'special' machines, essentially custom-made for users. These two factors explain much of the cost of machine tools: manufacturers devote their attention to the requirements of the larger users so that they can cash in on the demand for high-performance specialized machinery, which is very expensive due to high labor costs and the relatively inefficient low-volume production methods (see Rosenberg 1963; Wagoner 1968; Brown and Rosenberg 1961; Melman 1959). The development of N/C exaggerated these tendencies. John Parsons conceived of the new technology while trying to figure out a way of cutting the difficult contours of helicopter rotor blade templates to close tolerances; since he was using a computer to calculate the points for drilling holes (which were then filed together to make the contour) he began to think of having the computer control the actual positioning of the drill itself. He extended this idea to three-axis milling when he examined the specification for a wing panel for a new combat fighter. The new high performance, high-speed aircraft demanded a great deal of difficult and expensive machining to produce airfoils (wing surfaces, jet engine blades), integrally stiffened wing sections for greater tensile strength and less weight, and variable thickness skins. Parsons took his idea, christened 'Cardomatic' after the IBM cards he used, to Wright Patterson Air Force Base and convinced people at the Air Material Command that the air force should underwrite the development of this potent new technology. When Parsons got the contract, he subcontracted with MIT's Servomechanism Laboratory, which had experience in gunfire control systems.[2] Between the signing of the initial contract in 1949 and 1959, when the air force ceased its formal support for the development of software, the military spent at least $62 million on the research, development, and transfer of N/C. Up until 1953, the air force and MIT mounted a large campaign to interest machine-tool builders and the aircraft industry in the new technology, but only one company, Giddings and Lewis, was sufficiently interested to put their own money into it. Then, in 1955, N/C promoters succeeded in having the specifications in the Air Material Command budget allocation for the stockpiling of machine tools changed from tracer-controlled machines to N/C machines. At that time, the only fully N/C machine in existence was in the Servomechanism Lab. The air force undertook to pay for the purchase, installation, and maintenance of over 100 N/C machines in factories of prime subcontractors; the contractors, aircraft manufacturers, and their suppliers would also be paid to learn to use the new technology. In short, the air force created a market for N/C. Not surprisingly, machine-tool builders got into action, and research and development expenditure in the industry multiplied eight-fold between 1951 and 1957.

The point is that what made N/C possible – massive air force support – also helped determine the shape the technology would take. While criteria for the design of machinery normally includes cost to the user, here this was not a major consideration; machine-tool builders were simply competing to meet performance and 'competence' specifications for government-funded users in the aircraft industry. They had little concern with cost effectiveness and absolutely no incentive to produce less expensive machinery for the commercial market.

But the development of the machinery itself is only part of the story; there was also the separate evolution of the software. Here, too, air force requirements dictated the shape of the technology. At the outset, no one fully appreciated the difficulty of getting the intelligence of production on tape, least of all the MIT engineers on the N/C project, few of whom had had any machining experience before becoming involved in the project. Although they were primarily control engineers and mathematicians, they had sufficient hubris to believe that they could readily synthesize the skill of a machinist. It did not take them long to discover their error. Once it was clear that tape preparation was the stumbling block to N/C's economic viability, programming became the major focus of the project. The first programs were prepared manually, a tedious, time-consuming operation performed by graduate students, but thereafter efforts were made to enlist the aid of Whirlwind, MIT's first digital computer. The earliest programs were essentially subroutines for particular geometric surfaces which were compiled by an executive program. In 1956, after MIT had received another air force contract for software development, a young engineer and mathematician named Douglas Ross came up with a new approach to programming. Rather than treating each separate problem with a separate subroutine, the new system, called APT (Automatically Programmed Tools), was essentially a skeleton program – a 'systematized solution,' as it was called – for moving a cutting tool through space; this skeleton was to be 'fleshed out' for every particular application. The APT system was flexible and fundamental; equally important, it met air force specifications that the language must have a capacity for up to five-axis control. The air force loved APT because of its flexibility; it seemed to allow for rapid mobilization, for rapid design change, and for interchangeability between machines within a plant, between users and vendors, and between contractors and subcontractors throughout the country (presumably of 'strategic importance' in case of enemy attack). With these ends in mind, the air force pushed for standardization of the APT system and the Air Material Command cooperated with the Aircraft Industries Association Committee on Numerical Control to make APT the industry standard, the machine tool and control manufacturers followed suit, developing 'postprocessors' to adapt each particular system for use with APT.

Before long the APT computer language had become the industry standard, despite initial resistance within aircraft company plants. Many of these companies had developed their own languages to program their N/C equipment, and these in-house languages, while less flexible than APT, were nevertheless proven, relatively simple to use, and suited to the needs of the company. APT was something else entirely. For all its advantages – indeed, because of them – the APT system had decided disadvantages. The more fundamental a system is, the more cumbersome it is, and the more complex it is, the more skilled a programmer must be, and the bigger a computer must be to handle the larger amount of information. In addition, the greater the amount of information, the greater the chance for error. But initial resistance was overcome by higher level management, who had come to believe it necessary to learn how to use the new system 'for business reasons' (cost-plus contracts with the air force). The exclusive use of APT was enforced. Thus began what Douglas Ross himself has described as 'the tremendous turmoil of practicalities of the APT system development'; the system remained 'erratic and unreliable,' and a major headache for the aircraft industry for a long time.

The standardization of APT, at the behest of the air force, had two other interrelated consequences. First, it inhibited for a decade the development of alternative, simpler languages, such as the strictly numerical language NUFORM (created by A. S. Thomas, Inc.), which might have rendered contour programming more accessible to smaller shops. Second, it forced those who ventured into N/C into a dependence on those who controlled the development of APT,[3] on large computers and mathematically sophisticated programmers. The aircraft companies, for all their headaches, could afford to grapple with APT because of the air force subsidy, but commercial users were not so lucky. Companies that wanted military contracts were compelled to adopt the APT system, and those who could not afford the system, with its training requirements, its computer demands, and its headaches, were thus deprived of government jobs. The point here is that the software system which became the de facto standard in industry had been designed with a user, the air force, in mind. As Ross explained, 'the universal factor throughout the design process is the economics involved. The advantage to be derived from a given aspect of the language must be balanced against the difficulties in incorporating that aspect into a complete and working system' (Ross 1978: 13). APT served the air force and the aircraft industry well, but at the expense of less endowed competitors.

▶ CHOICE IN DESIGN: VERTICAL RELATIONS OF PRODUCTION

Thus far we have talked only about the form of N/C, its hardware and software, and how these reflected the horizontal relations of production. But what about the precursor to N/C, record-playback? Here was a technology that was apparently perfectly suited to the small shop: tapes could be prepared by recording the motions of a machine tool, guided by a machinist or a tracer template, without programmers, mathematics, languages, or computers.[4] Yet this technology was abandoned in favor of N/C by the aircraft industry and by the control manufacturers. Small firms never saw it. The Gisholt system, designed by Hans Trechsel to be fully accessible to machinists on the floor, was shelved once that company was bought by Giddings and Lewis, one of the major N/C manufacturers. The GE record-playback system was never really marketed since demonstrations of the system for potential customers in the machine-tool and aircraft companies elicited little enthusiasm. Giddings and Lewis did in fact purchase a record-playback control for a large profile 'skin mill' at Lockheed but switched over to a modified N/C System before regular production got underway. GE's magnetic tape control system, the most popular system in the 1950s and 1960s, was initially described in sales literature as having a 'record-playback option,' but mention of this feature soon disappeared from the manuals, even though the system retained the record-playback capacity.[5]

Why was there so little interest in this technology? The answer to this question is complicated. First, air force performance specifications for four- and five-axis machining of complex parts, often out of difficult materials, were simply beyond the capacity of either record-playback or manual methods. In terms of expected cost reductions, moreover, neither of these methods appeared to make possible as much of a reduction in the manufacturing and storage costs of jigs, fixtures, and templates as did N/C. Along the same lines, N/C also promised to reduce more significantly the labor costs for toolmakers, machinists, and patternmakers. And, of course, the very large air force subsidization of N/C technology lured most manufacturers and users to where the action was. Yet there were still other, less practical, reasons for the adoption of N/C and the abandonment of record-playback, reasons that have more to do with the ideology of engineering than with economic calculations. However useful as a production technology, record-playback was considered quaint from the start, especially with the advent of N/C. N/C was always more than a technology for cutting metals, especially in the eyes of its MIT designers, who knew little about metalcutting: it was a symbol of the computer age, of mathematical elegance, of power, order, and predictability, of continuous flow, of remote control, of the automatic factory. Record-playback, on the other hand, however much it represented a significant advance on manual methods, retained a vestige of traditional human skills; as such, in the eyes of the future (and engineers always confuse the present and the future) it was obsolete.

The drive for total automation which N/C represented, like the drive to substitute capital for labor, is not always altogether rational. This is not to say that the profit motive is insignificant – hardly. But economic explanations are not the whole story, especially in cases where ample government financing renders cost-minimization less of an imperative. Here the ideology of control emerges most clearly as a motivating force, an ideology in which the distrust of the human agency is paramount, in which human judgment is construed as 'human error.' But this ideology is itself a reflection of something else: the reality of the capitalist mode of production. The distrust of human beings by engineers is a manifestation of capital's distrust of labor. The elimination of human error and uncertainty is the engineering expression of capital's attempt to minimize its dependence upon labor by increasing its control over production. The ideology of engineering, in short, mirrors the antagonistic social relations of capitalist production. Insofar as the design of machinery, like machine tools, is informed by this ideology, it reflects the social relations of production.[6] Here we will emphasize this aspect of the explanation – why N/C was developed and record-playback was not – primarily because it is the aspect most often left out of such stories.

Ever since the nineteenth century, labor-intensive machine shops have been a bastion of skilled labor and the locus of considerable shop-floor struggle. Frederick Taylor introduced his system of scientific management in part to try to put a stop to what he called 'systematic soldiering' (now called 'pacing'). Workers practiced pacing for many reasons: to keep some time for themselves, to exercise authority over their own work, to avoid killing 'gravy' piece-rate jobs by overproducing and risking a rate cut, to stretch out available work for fear of layoffs, to exercise their creativity and ingenuity in order to 'make out' on 'stinkers' (poorly rated jobs), and, of course, to express hostility to management (see articles by Roy; Mathewson 1969). Aside from collective cooperation and labor-prescribed norms of behavior, the chief vehicle available to machinists for achieving shop-floor control over production was their control over the machines. Machining is not a handicraft skill but a machine-based skill; the possession of this skill, together with control over the speeds, feeds, and motions of the machines, enables machinists alone to produce finished parts to tolerance (Montgomery 1976b). But the very same skills and shop-floor control that made production possible also make pacing possible. Taylor therefore tried to eliminate soldiering by changing the process of production itself, transferring skills from the hands of machinists to the handbooks of management; this, he thought, would enable management, not labor, to prescribe the details of production tasks. He was not altogether successful. For one thing, there is still no absolute science of metalcutting and methods engineers, time-study people, and Method Time Measurement (MTM) specialists – however much they may have changed the formal processes of machine-shop practice – have not succeeded in putting a stop to shop-floor control over production.[7]

Thus, when sociologist Donald Roy went to work in a machine shop in the 1940s, he found pacing alive and well. He recounts an incident that demonstrates how traditional patterns of authority rather than scientific management still reigned supreme:

> 'I want 25 or 30 of those by 11 o'clock,' Steve the superintendent said sharply, a couple of minutes after the 7:15 whistle blew. I [Roy] smiled at him agreeably. 'I mean it,' said Steve, half smiling himself, as McCann and Smith, who were standing near us, laughed aloud. Steve had to grin in spite of himself and walked away. 'What he wants and what he is going to get are two different things,' said McCann. (1953: 513)

Thirty years later, sociologist Michael Burawoy returned to the same shop and concluded, in his own study of shop-floor relations, that 'in a machine shop, the nature of the relationship of workers to their machines rules out coercion as a means of extracting surplus' (1976).

This was the larger context in which the automation of machine tools took place; it should be seen, therefore, as a further managerial attempt to wrest control over production from the shop-floor work force. As Peter Drucker once observed, 'What is today called automation is conceptually a logical extension of Taylor's scientific management' (1967: 26). Thus it is not surprising that when Parsons began to develop his N/C 'Cardomatic' system, he took care not to tell the union (the UAW) in his shop in Traverse City about his exciting new venture. At GE (Schenectady), a decade of work-stoppages over layoffs, rate cuts, speed-ups, and the replacement of machin-ists with less skilled apprentices and women during the war, culminated in 1946 in the biggest strike in the company's history, led by machinists in the United Electrical Workers (UE) and bitterly opposed by the GE Engineers' Association. GE's machine-tool automation project, launched by these engineers soon afterward, was secret, and although the project had strong management support, publicist Vonnegut recalled, with characteristic understatement, that 'they wanted no publicity this time.'[8]

During the first decade of machine-tool automation development, the aircraft industry – the major user of automatic machine tools – also ex-perienced serious labor trouble as the machinists and auto workers com-peted to organize the plants. The postwar depression had created discontent among workers faced with layoffs, company claims of inability to pay, and massive downward reclassifications (Allen and Schneider 1956). Major strikes took place at Boeing, Bell Aircraft (Parsons' prime contractor), Mc-Donnell Douglas, Wright Aeronautical, GE (Evandale) (jet engines), North American Aviation, and Republic Aircraft. It is not difficult, then, to explain the popularity among management and technical men of a November 1946

Figures 14.1, 14.2 From skilled craftworker to buttonpusher? (photo by permission of the Australian Metal Workers Union)

Fortune article entitled 'Machines Without Men. 'Surveying the technological fruits of the war (sensing and measuring devices, servomechanisms, computers, etc.), two Canadian physicists promised that 'these devices are not subject to any human limitations. They do not mind working around the clock. They never feel hunger or fatigue. They are always satisfied with working conditions, and never demand higher wages based on the company's ability to pay.' In short, 'they cause much less trouble than humans doing comparable work' (Leaver and Brown 1946: 203).

One of the people who was inspired by this article was Lowell Holmes, the young electrical engineer who directed the GE automation project. However, in record-playback, he developed a system for replacing machinists that ultimately retained machinist and shop-floor control over production because of the method of tape preparation.[9] This 'defect' was recognized immediately by those who attended the demonstration of the system; they showed little interest in the technology. 'Give us something that will do what we say, not what we do,' one of them said. The defects of record-playback were conceptual, not technical; the system simply did not meet the needs of the larger firms for managerial control over production. N/C did. 'Managers like N/C because it means they can sit in their offices, write down what they want, and give it to someone and say, "do it,"' the chief GE consulting engineer on both the record-playback and N/C projects explained. 'With N/C there is no need to get your hands dirty, or argue' (personal interview). Another consulting engineer, head of the Industrial Applications Group which served as intermediary between the research department and sales department at GE (Schenectady) and a key figure in the development of both technologies, explained the shift from record-playback to N/C: 'Look, with record-playback the control of the machine remains with the machinist – control of feeds, speeds, number of cuts, output; with N/C there is a shift of control to management. Management is no longer dependent upon the operator and can thus optimize the use of their machines. With N/C, control over the process is placed firmly in the hands of management – and why shouldn't we have it?' (personal interview). It is no wonder that at GE, N/C was often referred to as a management system, not as a technology of cutting metals.

Numerical control dovetailed nicely with larger efforts to computerize company operations, which also entailed concentrating the intelligence of manufacturing in a centralized office. In the intensely anti-Communist 1950s, moreover, as one former machine-tool design engineer has suggested, N/C looked like a solution to security problems, enabling management to remove blueprints from the floor so that subversives and spies couldn't get their hands on them. N/C also appeared to minimize the need for costly tooling and it made possible the cutting of complex shapes that defied manual and tracer methods, and reduced actual chip-cutting time. Equally important, however, N/C replaced problematic time-study methods with 'tape time' – using the time it takes to run a cycle as the base for calculating rates – replaced troublesome skilled machinists with more tractable 'button-pushers,' and eliminated once and for all the problem of pacing. If, with hindsight, N/C seems to have led to organizational changes in the factory, changes which enhanced managerial control over production, it is because the technology was chosen, in part, for just that purpose.[10]

► REALITY ON THE SHOP FLOOR

Although the evolution of a technology follows from the social choices that inform it, choices which mirror the social relations of production, it would be an error to assume that in having exposed the choices, we can simply deduce the rest of reality from them. Reality cannot be extrapolated from the intentions that underlie the technology any more than from the technology itself.[11] Desire is not identical to satisfaction. . . .

The introduction of N/C was not uneventful, especially in plants where the machinists' unions had a long history. Work stoppages and strikes over rates for the new machines were common in the 1960s, as they still are today. There are also less overt indications that management dreams of automatic machinery and a docile, disciplined work force but they have tended to remain just that. Here we will examine briefly three of management's expectations: the use of 'tape time' to set rates; the deskilling of machine operators; and the elimination of pacing. . . .

[In the extract chosen, only the second of these points is discussed.]

In reality, N/C machines do not run by themselves – as the United Electrical Workers argued in its 1960 *Guide to Automation*, the new equipment, like the old, requires a spectrum of manual intervention and careful attention to detail, depending upon the machine, the product, and so on.

The deskilling of machine operators has, on the whole, not taken place as expected, for two reasons. First, as mentioned earlier, the assigning of labor grades and thus rates to the new machinery was, and is, a hotly contested and unresolved issue in union shops. Second, in union and nonunion shops alike, the determination of skill requirements for N/C must take into account the actual degree of automation and reliability of the machinery. Management has thus had to have people on the machines who know what they are doing simply because the machines (and programming) are not totally reliable; they do not run by themselves and produce good finished parts. Also, the machinery is still very expensive (even without microprocessors) and thus so is a machine smash-up. Hence, while it is true that many manufacturers initially tried to put unskilled people on the new equipment, they rather quickly saw their error and upgraded the classification. (In some places the most skilled people were put on the N/C machines and given a premium but the lower formal classifications were retained, presumably in the hope that someday the skill requirements would actually drop to match the classification – and the union would be decertified.) The point is that the intelligence of production has neither been built entirely into the machinery nor been taken off the shop floor. It remains in the possession of the work force.

NOTES

1 The discussion of the record-playback technology is based upon extensive inter-views and correspondence with the engineers who participated in the projects at General Electric (Schenectady) and Gisholt (Madison, Wisconsin), and the trade journal and technical literature.

2 This brief history of the origins of N/C is based upon interviews with Parsons and MIT personnel, as well as the use of Parsons' personal files and the project records of the Servomechanism Laboratory.

3 The air-force funded development of APT was centered initially at MIT. In 1961 the effort was shifted to the Illinois Institute of Technology Research Institute (IITRI) where it has been carried on under the direction of a consortium composed of the air force, the Aircraft Industries Association (AIA), and major manufacturers of machine tools and electronic controls. Membership in the consortium has always been expensive, beyond the financial means of the vast majority of firms in the metalworking industry. APT system use, therefore, has tended to be restricted to those who enjoyed privileged access to information about the system's development. Moreover, the APT system has been treated as proprietary information within user plants; programmers have had to sign out for manuals and have been forbidden from taking them home or talking about their contents with people outside the company.

4 Technically, record-playback was as reliable as N/C, if not more so – since all the programming was done at the machine, errors could be eliminated during the programming process, before production began. Moreover, it could be used to reproduce parts to within a tolerance of a thousandth of an inch, just like N/C. (It is a common mistake to assume that if an N/C control system generates discrete pulses corresponding to increments of half a thousandth, the machine can produce parts to within the same tolerances. In reality, the limits of accuracy are set by the machine itself – not to mention the weather – rather than by the electrical signals.)

5 This history is based upon interviews with Hans Trechsel, designer of Gisholt's 'Factrol' system, and interviews and correspondence with participating engineer-ing and sales personnel at GE (Schenectady), as well as articles in various en-gineering and trade journals.

6 It could be argued that control in the capitalist mode of production is not an independent factor (a manifestation of class conflict), but merely a means to an economic end (the accumulation of capital). Technology introduced to increase managerial control over the work force and eliminate pacing is, in this view, introduced simply to increase profits. Such reductionism, which collapses control and class questions into economistic ones, renders impossible any expla-nation of technological development in terms of social relations or any careful distinction between productive technology which directly increases output per person-hour and technology which does so only indirectly by reducing worker resistance or restriction of output. Finally, it makes it hard to distinguish a technology that reduces pacing from a gun in the service of union-busting company agents; both investments ultimately have the same effect and the economic results look the same on the balance sheet. As Jeremy Brecher reminds us, 'The critical historian must go behind the economic category of cost-minimization to discover the social relations that it embodies (and conceals)' (1978).

7 The setting of rates on jobs in machine shops is still more of a guess than a

scientific determination. This fact is not lost on machinists, as their typical descriptions of the methods-men suggests: 'They ask their wives, they don't know; they ask their children, they don't know; so they ask their friends.' Of course, this apparent and acknowledged lack of scientific certainty comes into play during bargaining sessions over rates, when 'fairness' and power, not science, determine the outcome.

8 Kurt Vonnegut, letter to author, February 1977.

9 The fact that record-playback lends itself to shop-floor control of production more readily than N/C is borne out by a study of N/C in the United Kingdom done by Erik Christiansen in 1968. Only in those cases where record-playback or plugboard controls were in use (he found six British-made record-playback jig borers) did the machinist keep the same pay scale as with conventional equipment and retain control over the entire machining process. In Christiansen's words, record-playback (and plugboard programming) 'mean that the shop floor retains control of the work cycle through the skill of the man who first programmed the machine' (1968: 27, 31).

10 The cost effectiveness of N/C depends upon many factors, including training costs, programming costs, computer costs, and the like, beyond mere time saved in actual chip-cutting or reduction in direct labor costs. The MIT staff who conducted the early studies on the economics of N/C focused on the savings in cutting time and waxed eloquent about the new revolution. At the same time, however, they warned that the key to the economic viability of N/C was a reduction in programming (software) costs. Machine-tool company salesmen were not disposed to emphasize these potential drawbacks, though, and numerous users went bankrupt because they believed what they were told. In the early days, however, most users were buffered against such tragedy by state subsidy. Today, potential users are somewhat more cautious, and machine-tool builders are more restrained in their advertising, tempering their promise of economic success with qualifiers about proper use, the right lot and batch size, sufficient training, etc.

For the independent investigator, it is extremely difficult to assess the economic viability of such a technology. There are many reasons for this. First, the data is rarely available or accessible. Whatever the motivation – technical fascination, keeping up with competitors, etc. – the purchase of new capital equipment must be justified in economic terms. But justifications are not too difficult to come by if the item is desired enough by the right people. They are self-interested anticipations and thus usually optimistic ones. More important, firms rarely conduct postaudits on their purchases, to see if their justifications were warranted. Nobody wants to document his errors and if the machinery is fixed in its foundation, that is where it will stay, whatever a postaudit reveals; you learn to live with it. The point here is that the economics of capital equipment is not nearly so tidy as economists would sometimes have us believe. The invisible hand has to do quite a bit of sweeping up after the fact.

If the data does exist, it is very difficult to get a hold of. Companies have a proprietary interest in the information and are wary about disclosing it for fear of revealing (and thus jeopardizing) their position vis-à-vis labor unions (wages), competitors (prices), and government (regulations and taxes). Moreover, the data, if it were accessible, is not all tabulated and in a drawer somewhere. It is distributed among departments, with separate budgets, and the costs to one are the hidden costs to the others. Also, there is every reason to believe that the data that does exist is self-serving information provided by each operating unit to enhance its position in the firm. And, finally, there is the tricky question of how 'viability' is defined in the first place. Sometimes, machines make money for a company whether they were used productively or not.

The purpose of this aside is to emphasize the fact that 'bottom-line' explanations for complex historical developments, like the introduction of new capital equipment, are never in themselves sufficient, nor necessarily to be trusted. If a company wants to introduce something new, it must justify it in terms of making a profit. This is not to say, however, that profit making was its real (or, if so, its only) motive or that a profit was ever made. In the case of automation, steps are taken less out of careful calculation than on the faith that it is always good to replace labor with capital, a faith kindled deep in the soul of manufacturing engineers and managers (as economist Michael Piore, among others, has shown. See, for example, Piore 1968). Thus, automation is driven forward, not simply by the profit motive, but by the ideology of automation itself, which reflects the social relations of production.

11 This is an error that Braverman tended to make in discussing N/C.

REFERENCES

Allen, Arthur, and Schneider, Betty (1956) *Industrial Relations in the California Aircraft Industry*. Berkeley: Institute of Industrial Relations, University of California.

Braverman, Harry (1974) *Labor and Monopoly Capital: The Degradation of Work in the Twentieth Century*. New York: Monthly Review Press.

Brecher, Jeremy (April 1978) 'Beyond Technological Determinism: Some Comments.' Talk presented at the Organization of American Historians Convention.

Brown, Murray and Rosenberg, Nathan (1961) 'Patents, Research and Technology in the Machine Tool Industry.' *The Patent, Trademark and Copyright Journal of Research and Education* 5 (Spring).

Burawoy, Michael (1976) 'The Organization of Consent: Changing Patterns of Conflict on the Shop Floor, 1945–1975.' Unpublished doctoral dissertation, University of Chicago.

Christiansen, Erik (1968) *Automation and the Workers*. London: Labour Research Development Publications, Ltd.

Diebold, John (1952) *Automation*. New York: Van Nostrand.

Drucker, Peter F. (1967) 'Technology and Society in the Twentieth Century.' In *Technology in Western Civilization*, edited by Kranzberg and Pursell. New York: Oxford University Press.

Leaver, E. W. and Brown, J. J. (1946) 'Machines Without Men.' *Fortune* (November).

Lundgren, Earl (1969) 'Effects of N/C on Organizational Structure.' *Automation* 16 (January).

Lynn, F., Roseberry, T. and Babich, V. (1966) 'A History of Recent Technological Innovations.' In National Commission on Technology, Automation and Economic Progress, *The Employment Impact of Technological Change*, Appendix, vol. II. *Technology and the American Economy*. Washington, D.C.: Government Printing Office.

Mathewson, Stanley B. (1969, originally 1931) *Restriction of Output Among Unorganized Workers*. Carbondale, Ill.: Southern Illinois University Press.

Melman, Seymour (1959) 'Report on the Productivity of Operations in the Machine Tool Industry in Western Europe.' European Productivity Agency Project No. 420.

Montgomery, David. 1976b. 'Workers' Control of Machine Production in the Nineteenth Century.' *Labor History* 17 (Fall): 486–509.

Piore, Michael (1968) 'The Impact of the Labor Market Upon the Design and Selection of Productive Techniques Within the Manufacturing Plant.' *Quarterly Journal of Economics* 82.

Rosenberg, Nathan (1963) 'Technical Change in the Machine Tool Industry, 1840–1910.' *Journal of Economic History* 23: 414–43.

Ross, Douglas (1978) 'Origins of APT Language for Automatically Programmed Tools.' Softech, Inc.

Stephanz, Kenneth (1971) 'Statement of Kenneth Stephanz.' In *Introduction to Numerical Control and Its Impact on Small Business.* Hearing before the Subcommittee on Science and Technology of the Select Committee on Small Business, U.S. Senate, 92nd Congress, 1st session (June 24, 1971).

Vonnegut, Kurt (1952) *Player Piano*. New York. Delacorte Press.

Wagoner, Harless (1968) *The United States Machine Tool Industry from 1900 to 1950.* Cambridge, Mass.: MIT Press.

15 ▶ The material of male power

Cynthia Cockburn

A skilled craftsman may be no more than a worker in relation to capital, but seen from within the working class he has been a king among men and lord of his household. As a high earner he preferred to see himself as the sole breadwinner, supporter of wife and children. As artisan he defined the unskilled workman as someone of inferior status, and would 'scarcely count him a brother and certainly not an equal' (Berg 1979: 121). For any socialist movement concerned with unity in the working class, the skilled craftsman is therefore a problem. For anyone concerned with the relationship of class and gender, and with the foundations of male power, skilled men provide a fertile field for study.

Compositors in the printing trade are an artisan group that have long defeated the attempts of capital to weaken the tight grip on the labour process from which their strength derives. Now their occupation is undergoing a dramatic technological change initiated by employers. Introduction of the new computerized technology of photocomposition represents an attack on what remains of their control over their occupation and wipes out many of the aspects of the work which have served as criteria by which 'hot metal' composition for printing has been defined as a manual skill and a man's craft.[1]

In this paper I look in some detail at the compositors' crisis, what has given rise to it and what it may lead to in future. Trying to understand it has led me to ask questions in the context of socialist-feminist theory. These I discuss first, as preface to an account of key moments in the compositors' craft history. I then isolate the themes of *skill* and *technology* for further analysis, and conclude with the suggestion that there may be more to male power than 'patriarchal' relations.

► ## PRODUCING CLASS AND GENDER

The first difficulty I have encountered in socialist-feminist theory is one that is widely recognized: the problem of bringing into a single focus our experience of both class and gender. Our attempts to ally the Marxist theory of capitalism with the feminist theory of 'patriarchy' have till now been unsatisfactory to us (Hartmann 1979a).

One of the impediments I believe lies in our tendency to try to mesh together two static structures, two hierarchical systems. In studying compositors I found I was paying attention instead to *processes*, the detail of historical events and changes, and in this way it was easier to detect the connexions between the power systems of class and gender. What we are seeing is *struggle that contributes to the formation of people within both their class and gender simultaneously.*

One class can only exist in relation to another. E. P. Thompson wrote 'we cannot have two distinct classes each with an independent being and then bring them into relationship with each other. We cannot have love without lovers, nor deference without squires and labourers.' Likewise, it is clear we cannot have masculinity without femininity: genders presuppose each other, they are relative. Again, classes are made in historical processes. 'The working class did not rise like the sun at an appointed time. It was present at its own making . . . Class is defined by men [sic] as they live their own history' (Thompson 1963). The mutual production of gender should be seen as a historical process too.

So in this paper I set out to explore aspects of the process of mutual definition in which men and women are locked, and those (equally processes of mutual creation) in which the working class and the capitalist class are historically engaged. Capital and labour, through a struggle over the design and manipulation of technology that the one owns and the other sets in motion, contribute to forming each other in their class characters. Powerfully-organized workers forge their class identity vis-a-vis both capital and the less organized and less skilled in part through this same process. And men and women too are to some extent mutually defined as genders through their relation to the same technology and labour process. In neither case is it a balanced process. By owning the means of production the capitalist class has the initiative. By securing privileged access to capability and technology the man has the initiative. Each gains the power to define 'another' as inferior.[2] I will try to draw these occurrences out of the story of the compositors as I tell it.

► ## COMPONENTS OF POWER

The second theoretical need which an examination of skilled workers has led me to feel is the need for a fuller conception of the material basis of male power, one which does not lose sight of its physical and socio-political ramifications in concentrating upon the economic.

As feminism developed its account of women's subordination, one problem that we met was that of shifting out of a predominantly ideological mode and narrating also the concrete practices through which women are disadvantaged. Early literature relied on 'sexist attitudes' and 'male chauvinism' to account for women's position. Socialist feminists, seeking a more material explanation for women's disadvantage, used the implement of Marxist theory, unfortunately not purpose-designed for the job but the best that was to hand. The result was an account of the economic advantages to capital of women as a distinct category of labour and their uses as an industrial reserve army. The processes of capitalism seemed to be producing an economic advantage to men which could be seen uniting with their control over women's domestic labour to form the basis of their power.[3]

Many feminists, however, were dissatisfied with what seemed a narrow 'economism' arising from Marx (or through misinterpretation of Marx according to the point of view). The ideological vein has been more recently worked with far more sophistication than before, in different ways, by Juliet Mitchell on the one hand and Rosalind Coward on the other (Mitchell 1975; Coward 1978). But, as Michèle Barrett has pointed out, while 'ideology is an extremely important site for the construction and reproduction of women's oppression . . . this ideological level cannot be dissociated from economic relations' (Barrett 1980).

There is thus a kind of to-ing and fro-ing between 'the ideological' and 'the economic', neither of which gives an adequate account of male supremacy or female subordination. The difficulty lies, I believe, in a confusion of terms. The proper complement of *ideology* is not the *economic*, it is the *material*.[4] And there is more to the material than the economic. It comprises also the *socio-political* and the *physical*, and these are often neglected in Marxist-feminist work.

An instance of the problems that arise through this oversight is Christine Delphy's work, where a search for a 'materialist' account of women's subordination leads her to see marriage in purely economic terms and domestic life as a mode of production, an interpretation which cannot deal with a large area of women's circumstances (Delphy 1977).

It is only by thinking with the additional concepts of the socio-political and the physical that we can begin to look for material instances of male domination beyond men's greater earning power and property advantage. The socio-political opens up questions about male organization and solidarity, the part played by institutions such as church, societies, unions and clubs for instance.[5] And the physical opens up questions of bodily physique and its extension in technology, of buildings and clothes, space and movement. It allows things that are part of our practice ('reclaiming the night', teaching each other manual skills) a fuller place in our theory.

In this account I want to allow 'the economic' to retire into the background, not to deny its significance but in order to spotlight these other material instances of male power. The socio-political will emerge in the shape of the printing trade unions and their interests and strategies. The physical will also receive special attention because it is that which I have found most difficult to understand in the existing framework of Marxist-feminist thought. It finds expression in the compositor's capability, his dexterity and strength and in his tools and technology.

▶   ## PHYSICAL EFFECTIVITY IS ACQUIRED

One further prefatory note is needed. In 1970, when Kate Millett and Shulamith Firestone, in their different ways, pinned down and analysed the system of male domination they spoke to the anger that many women felt (Millett 1971; Firestone 1971). But many feminists were uneasy with the essentialism inherent in their view, and especially the biological determinism of Firestone and its disastrous practical implications.

Marxist-feminist theory has consequently tended to set on one side the concept of the superior physical effectivity of men, to adopt a kind of agnosticism to the idea, on account of a very reasonable fear of that biologism and essentialism which would nullify our struggle. I suggest however that we cannot do without a politics of physical power and that it need not immobilize us. In this article I use the term physical power to mean both corporal effectivity (relative bodily strength and capability) and technical effectivity (relative familiarity with and control over machinery and tools).

To say that most men can undertake feats of physical strength that most women cannot is to tell only the truth. Likewise it is true to say that the majority of men are more in their element with machinery than the majority of women. These statements are neither biologistic nor essentialist. Physical efficiency and technical capability do not belong to men primarily by birth, though DNA may offer the first step on the ladder. In the main they are appropriated by males through childhood, youth and maturity. Men's socio-political and economic power enables them to do this. In turn, their physical presence reinforces their authority and their physical skills enhance their earning power.

Ann Oakley, among others, has made the fruitful distinction between biologically-given sex (and that not always unambiguous) and culturally constituted gender, which need have little correlation with sex but in our society takes the form of a dramatic and hierarchical separation (Oakley 1972).

The part of education and that of child-rearing in constituting us as masculine and feminine *in ideology* is the subject of an extensive literature (e.g. Wolpe 1978; Belotti 1975). But there is evidence to show that *bodily difference* is also largely a social product. With time and work women athletes can acquire a physique which eclipses the innate differences between males and females (Ferris 1978). Height and weight are correlated with class, produced by different standards of living, as well as with gender.[6]

Boys are conditioned from childhood in numberless ways to be more physically effective than girls. They are trained in activities that develop muscle, they are taught to place their weight firmly on both feet, to move freely, to use their bodies with authority. With regard to females they are socialized to seize or shelter them and led to expect them in turn to yield or submit.

While so much of the imbalance of bodily effectiveness between males and females is produced through social practices it is misguided to prioritize that component of the difference that may prove in the last resort to be inborn.[7] More important is to study the way in which a small physical *difference* in size, strength and reproductive function is developed into an increasing relative physical *advantage* to men and vastly multiplied by differential access to technology. The process, as I will show, involves several converging practices: accumulation of bodily capabilities, the definition of tasks to match them and the selective design of tools and machines. The male physical advantage of course interacts with male economic and socio-political advantage in mutual enhancement.

The appropriation of muscle, capability, tools and machinery by men is an important source of women's subordination, indeed it is part of the process by which females are constituted as women. It is a process that is in some ways an analogue of the appropriation of the means of production by a capitalist class, which thereby constituted its complementary working class. In certain situations and instances, as in the history of printers, the process of physical appropriation (along with its ideological practices) has a part in constituting people within their class and gender simultaneously.

 ## THE HAND COMPOSITOR: APPROPRIATION OF TECHNIQUE

Letterpress printing comprises two distinct technological processes, composing and printing. Before the mechanization of typesetting in the last decade of the nineteenth century compositors set the type by hand, organizing metal pieces in a 'stick', and proceeded to assemble it into a unified printing surface, the 'forme', ready for the printer to position on the press, coat with ink and impress upon paper.

The hand compositor, then, had to be literate, to be able to read type upside down and back to front, with a sharp eye for detail. He had to possess manual dexterity and have an easy familiarity with the position of letters in the 'case'. He had to calculate with the printers' 'point' system of measurement. Furthermore he had to have a sense of design and spacing to enable him to create a graphic whole of the printed page, which he secured through the manipulation of the assembled type, illustrative blocks and lead spacing pieces. The whole he then locked up in a forme weighing 50 lbs or more. This he would lift and move to the proofing press or bring back to the stone for the distribution of used type. He thus required a degree of strength and stamina, a strong wrist, and, for standing long hours at the case, a sturdy spine and good legs.

The compositor used his craft to secure for himself a well-paid living, with sometimes greater and sometimes less success depending on conditions of trade. Through their trade societies (later unions) compositors energetically sought to limit the right of access to the composing process and its equipment to members of the society in a given town or region, blacking 'unfair houses' that employed non-society men.

Comps deployed all the material and ideological tactics they could muster in resistance to the initiatives of capital in a context of the gradual, though late, industrialization of printing. Capitalists continually aimed for lower labour costs, more productive labour processes, the 'real subordination' of labour. Their two weapons were the mobilization of cheap labour and the introduction of machinery. They repeatedly assaulted the defences of the comps' trade societies. The organized, skilled men saw their best protection against capital to lie in sharply differentiating themselves from the all-but-limitless population of potential rivals for their jobs, the remainder of the working class.

They sought to control the numbers entering the trade and so to elevate their wage-bargaining position by a system of formal apprenticeship. They tried to limit the number of apprentices through an agreed ratio of boys to journeymen and to keep the period of apprenticeship as long as possible. The introduction of unapprenticed lads, 'the many-headed monster', the 'demon of cheap boy labour' was always a source of fear to compositors. Comps' jobs were kept within the class fraction by the custom of limiting openings wherever possible to members of existing printer families.

Thus the struggles over physical and mental capability and the right of access to composing equipment was one of the processes in which fractions of classes were formed in relation to each other.

How did women enter this story? The answer is, with difficulty. Women and children were drawn into industrial production in many industries in the first half of the nineteenth century but in printing their entry was almost entirely limited to the bookbinding and other low-paid finishing operations held to require no skill. Girls were not considered suitable for apprenticeship. Physical and moral factors (girls were not strong enough, lead was harmful to pregnancy, the social environment might be corrupting) were deployed ideologically in such a way that few girls would see themselves as suitable candidates for apprenticeship. A second line of defence against an influx of women was of course the same socio-political controls used to keep large numbers of boys of the unskilled working class from flooding the trade.

Women who, in spite of these barriers, obtained work as non-society compositors were bitterly resisted and their product 'blacked' by the society men, i.e. work typeset by women could not be printed. Their number remained few therefore (Child 1967). After 1859 a few small print shops were organized by philanthropic feminists to provide openings for women. It is worth noting that these enterprises did prove that women were in fact physically capable, given training and practice, of typesetting and imposition, though they did not work night shifts and male assistants were engaged to do the heavy lifting and carrying. These projects were dismissed by the men as 'wild schemes of social reformers and cranks'.[8]

The process of appropriation of the physical and mental properties and technical hardware required for composing by a group of men, therefore, was not only a capitalist process of class formation, as noted above, but also a significant influence in the process of gender construction in which men took the initiative in constituting themselves and women in a relation of complementarity and hierarchy.

▶ THE MECHANIZATION OF TYPESETTING: APPROPRIATION OF THE MACHINE

The compositors' employers had for years sought to invent a machine that could bypass the labour-intensive process of hand typesetting. They hoped in so doing not only to speed up the process but to evade the trade societies' grip on the craft, introduce women and boys and thus bring down the adult male wage. The design of such a machine proved an intractable problem. Though various prototypes and one or two production models were essayed in the years following 1840, none were commercially successful. It was only when high speed rotary press technology developed in the 1880s that typesetting became an intolerable bottleneck to printing and more serious technological experiment was undertaken. Among the various typesetting machines that then developed, the overwhelmingly successful model was the Linotype. It continued in use almost unchanged for sixty or seventy years.

The Linotype was not allowed to replace the hand typesetter without a struggle.[9] The men believed the Iron Comp would mean mass unemployment of society members. They did not (as an organized group) reject the machine out of hand, however. Their demand was the absolute, exclusive right of hand comps to the machine and to improved earnings. Their weapons were disruption, blacking and deliberate restriction of keyboard speeds. The outcome of the struggle was seen finally by both print employers and compositors as a moderate victory for both sides. There was unemployment of hand compositors for a few years in the mid-nineties, but with the upturn of business at the end of the Great Depression the demand for print grew fast and the demand for typesetters with it. Indeed, the first agreement between the London Society of Compositors (LSC) and the employers on the adaptation of the London Scale of Prices to Linotype production was a disastrous error for the capitalists, who had underestimated the productive capacity of their new force of production and over-estimated the strength of the organized comps. The bosses only began to share fully in the profits from their invention when the agreement was revised in 1896. A lasting cost to the comps was an increasing division of labour between the two halves of their occupation: typesetting and the

Figure 15.1 The Young–Delcambre composing machine. The girl on the left is justifying the lines

subsequent composing process. They did succeed however in continuing to encompass both jobs within the unitary craft and its apprenticeship as defined by their societies.

Those who really lost in the battle scarcely even engaged in it. They were the mass of labour, men and women who had no indentured occupation and who, if organized at all, were grouped in the new general unions of the unskilled. Jonathan Zeitlin firmly ascribes the success of the compositors (in contrast to engineers) in routing the employers' attempt to break their control of their craft in the technological thrust of the late nineteenth century to the former's success in ensuring that during the preceding decades no unskilled or semi-skilled categories of worker had been allowed to enter the composing room to fill subordinate roles (Zeitlin 1981). And the incipient threat from women had been largely averted by the time the Linotype was invented. An exception was a pocket of female compositors in Edinburgh who had entered the trade at the time of a strike by the men in 1872 and had proved impossible to uproot.

A more sustained attempt was made by employers ten years later to introduce women to work on another typesetting process that was widely applied in the book trade: the Monotype. The Monotype Corporation, designers of the machine, in contrast to the Linotype Company Ltd., opened the way to a possible outflanking of the skilled men by splitting the tasks of keyboarding and casting into two different machines. Men retained unshaken control of the caster, but an attempt was made by employers to introduce women onto the keyboards, which had the normal typewriter lay.

In 1909–10 the compositors' societies organized a campaign, focusing on Edinburgh, to eliminate women from the trade once and for all. They succeeded in achieving a ban on female apprentices and an agreement for natural wastage of women comps and operators. This male victory was partly due to an alliance between the craft compositors and the newly organized unions of the unskilled men in the printing industry (Zeitlin 1981).

That there were large numbers of women, literate, in need of work and eminently capable of machine typesetting at this time is evidenced by the rapid feminization of clerical work that accompanied the introduction of the office typewriter, in a situation where the male incumbent of the office was less well organized to defend himself than was the compositor (Davies 1979). Men's socio-political power, however, enabled them to extend their physical capabilities in manual typesetting to control of the machine that replaced it. (The gender-bias of typesetting technology is discussed further below.) The effect has been that women's participation in composing work, the prestigious and better-paid aspect of printing, was kept to a minimum until the present day, not excluding the period of the two World Wars. The composing room was, and in most cases still is, an all-male preserve with a sense of camaraderie, pin-ups on the wall and a pleasure taken in the manly licence to use 'bad' (i.e. woman-objectifying) language.

 ELECTRONIC COMPOSITION: THE DISRUPTION OF CLASS AND GENDER PATTERNS

In the half-century between 1910 and 1960 the printing industry saw relatively little technical change. Then, in the nineteen-sixties two big new possibilities opened up for capital in the printing industry as, emerging from post-war restrictions, it looked optimistically to expanding print markets. The first was web offset printing, with its potential flexibility and quality combined with high running speeds. The logical corollary was to

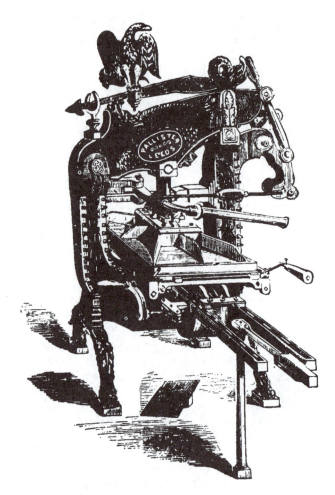

Figure 15.2 A nineteenth-century iron hand-press

abandon the machine-setting of metal type and to take up the second component of 'the new technology': letter assembly on film or photographic paper by the techniques of computer-aided photocomposition. The new process began to make inroads in to the British printing industry in the late sixties and swept through the provincial press and general printing in the seventies. The last serious redoubt of hot metal typesetting and letterpress printing is now the national press in Fleet Street.

Photocomposition itself has gone through several phases of development. At first, the operation comprised a keyboarding process whereby the operator tapped a typewriter-style keyboard producing a punched paper tape. The operator worked 'blind', that is to say he saw no hard copy of his work as he produced it. The 'idiot tape' was fed into a computer which read it, made the subtle line-end decisions formerly the responsibility of the operator and output clean tape. This second tape drove a photosetter, each impulse producing a timed flash of light through a photographic image on a master disc or drum. The result was a succession of characters laid down on film or bromide paper. The columns of text were taken by the compositor, cut up, sorted and pasted in position on a prepared card, later to be photographed as a whole and reproduced on a printing plate.

In the latest electronic composing technology there is no such photo-matrix of characters. The computer itself holds instructions that enable it to generate characters, in an almost limitless range of type faces and sizes and at enormously rapid speeds, on the face of a cathode ray tube. The inputting operation is performed with a keyboard associated with a video display unit on which the operator can assist computer decisions and 'massage' the copy into a desired order before committing it to the computer memory. The matter is transmitted direct from computer to photosetter and may now be produced in complete sections as large as a full newspaper page, making paste-up unnecessary.

The process is clearly seen by capital as a means of smashing the costly craft control of the compositor. The system is greatly more productive and requires less manpower. It would require less still if operated in the manner for which it is designed, i.e. avoiding two keyboarding processes by having typists, journalists, editors and authors key matter direct onto the computer disc for editing on screen and thence to direct output.

The work is much lighter, more sedentary. The abilities called upon are less esoteric, more generally available in the working population outside print. Inputting requires little more than good typing ability on the QWERTY board, something possessed by many more women than men. The implications for compositors of this twist in their craft history are dramatic. Combined with a recession it is causing unemployment in the trade, something unknown since the thirties. The individual tasks in the overall process have become trivialized and the men feel the danger of increased sub-division, routinization and substitution of unskilled workers.

The union response has not been to reject the new technology. Instead it has fought an energetic battle to retain the right to the new equipment as it did to the old. It resists 'direct input' by outsiders, asserts exclusive right to the photosetting keystroke (if necessary to a redundant second typing), to paste-up and the control of the photosetters, and where possible the computers. It is demanding increased pay and reduced hours in exchange for agreement to operate the new technology. And it is insisting (in principle at least) that all composing personnel get the chance to retrain for all aspects of the *whole* photocomposing job . . . an uphill struggle for reintegration of the now transformed craft.

▶ **SKILL AND ITS USES**

An extensive literature has demonstrated the effect of craft organization on the structure of the working class. 'The artisan creed with regard to the labourers is that the latter are an inferior class and that they should be made to know and kept in their place' (Hobsbawm 1964). The loss of demands on manual skill brought about by electronic photocomposition does not necessarily mean the job has become more 'mental'. On the contrary, present-day compositors feel their new work could be done by relatively unskilled workers. Many members feel they have lost status and some resent the strategic necessity to seek amalgamation of the National Graphical Association (NGA) with the unions representing the less skilled.

Our account shows, however, that the purposeful differentiation between skilled and unskilled workers was also a step in the construction of *gender*. This is a more recent conception. Heidi Hartmann has suggested that 'the roots of women's present social status' lie in job segregation by sex and demonstrates the role of men and their unions in maintaining women's inferiority in the labour market by deployment of skill (Hartmann 1979b). The fact that females in the closed-shop NGA (which embodies a large proportion of the better paid workers in the printing industry) until recently amounted to no more than 2% of its membership is directly connected with the fact that women's average earnings have always been lower relative to men's in printing than in manufacturing occupations as a whole. Through the mechanisms of craft definition women have been constructed as relatively lacking in competence, and relatively low in earning power. Women's work came to be seen as inferior. Now that the new composing process resembles 'women's work' stereotypes it is felt as emasculating. The skill crisis is a crisis of both gender and class for comps.

Anne Phillips and Barbara Taylor propose that skill is a direct correlate of sexual power. 'Skill has increasingly been defined against women . . . far from being an objective economic fact, skill is often an ideological category imposed on certain types of work by virtue of the sex and power of the workers who perform it' (Phillips and Taylor 1980).

It is important to recognize this ideological factor. It has become increasingly important in printing with the advance of technology. The compositor sitting at a keyboard setting type is represented as doing skilled work. A girl typist at a desk typing a letter is not – though the practical difference today is slight. Nonetheless, the formulation here again, posing the ideological as foil to the economic, leads to an under-emphasis on the material realities (albeit socially acquired) of physical power and with them the tangible factors in skill which it is my purpose to reassert.

Phillips and Taylor cite several instances of job definition where the distinction between male and female jobs as skilled and unskilled is clearly no more than ideological. But in printing, and perhaps in many other occupations too, unless we recognize what measure of reality lies behind the male customary over-estimate of his skill we have no way of evaluating the impact of electronic photocomposition, the leeching out of the tangible factors of skill from some tasks and their relocation in others, out of the compositor's reach.

What was the hot metal compositor's skill? He would say: I can read and calculate in a specialized manner; I can understand the process and make decisions about the job; I have aesthetic sense; I know what the tools are for and how to use them; I know the sequence of tasks in the labour process; how to operate, clean and maintain the machinery; I am dexterous and can work fast and accurately under pressure, can lift heavy weights and stand for hours without tiring. No one but an apprenticed compositor can do ALL these things.

There are thus what we might call tangible factors in skill – things that cannot be acquired overnight. They are both intellectual and physical and among the physical are knack, strength and intimacy with a technology. They are all in large measure learned or acquired through practice, though some apprentices will never make good craftsmen. The relative importance of the factors shifts over time with changing technology. Skill is a changing constellation of practical abilities of which no single one is either necessary or sufficient. Cut away the need for one or two of them and the skill may still be capable of adaptation to remain intact, marketable and capable of defence by socio-political organization.

The tangible factors in skill may be over-stated for purposes of self-defence and are variably deployed in socio-political struggle. Thus, against the unskilled male, defined as corporally superior to the skilled, hot metal comps have defended their craft in terms of (a) its intellectual and (b) its dexterity requirements. Against women, with their supposed superior dexterity, the skilled men on the contrary used to invoke (a) the heavy bodily demands of the work and (b) the intellectual standards it was supposed to require.[10] (Among comps today it is sometimes done to keep a list of the 'howlers' they detect in the typescripts coming to them from the 'illiterate' typists upstairs.)

The bodily strength component of the compositor's craft may be isolated to illustrate the politics involved. Men, having been reared to a bodily advantage, are able to make political and economic use of it by defining into their occupation certain tasks that require the muscle they alone possess, thereby barricading it against women who might be used against them as low-cost alternative workers (and whom for other reasons they may prefer to remain in the home). In composing, the lifting and carrying of the forme is a case in point. Nonetheless, many compositors found this aspect of the work heavy and it was felt to be beyond the strength of older men. They were always torn between wishing for unskilled muscular assistants and fearing that these, once ensconced in part of the job, might lay claim to the whole.

The size and weight of the forme is arbitrary. Printing presses and the printed sheet too could have been smaller. And heavy as it is, the mechanization exists which could ease the task. It is, in printing, purely a question of custom at what weight the use of hoists and trolleys to transport the forme is introduced.

Units of work (hay bales, cement sacks) are political in their design. Capitalists with work-study in mind and men with an interest in the male right to the job both have a live concern in the bargain struck over a standard weight or size. But the political power to design work processes would be useless to men without a significant average superiority in strength or other bodily capability. Thus the appropriation of bodily effectivity on the one hand and the design of machinery and processes on the other have often converged in such a way as to constitute men as capable and women as inadequate. Like other physical differences, gender difference in average bodily strength is not illusory, it is real. It does not necessarily matter, but it can be made to matter. Its manipulation is socio-political power play.

Above everything, a skill embodies the idea of wholeness in the job and in the person's abilities, and what this 'whole' comprises is the subject of a three-way struggle between capital, craftsman and the unskilled. The struggle is over the division of labour, the building of some capabilities into machines (the computer, the robot), the hiving off of some less taxing parts of the job to cheaper workmen, or to women. Craft organization responds to capitalist development by continually redefining its area of competence, taking in and teaching its members new abilities. Wholeness has become of key significance to the compositors' union as electronic technology has trivialized and shifted the pattern of the individual tasks. Socio-political organization and power have become of paramount import-ance as the old tangible physical and intellectual factors have been scrapped along with the old hardware.

▶ **CONTROL OF TECHNOLOGY**

Capitalists as capitalists and men as men both take initiatives over tech-nology. The capitalist class designs new technology, in the sense that it commissions and finances machinery and sets it to work to reduce the capitalist's dependency on certain categories of labour, to divide, disorgan-ize and cheapen labour. Sometimes machinery displaces knack and know-how, sometimes strength. Yet it is often the knowledge of the workers gained on an earlier phase of technology that produces the improvements and innovations that eventually supersede it. For instance, in a radical working men's paper in 1833, claiming rights over the bosses' machines, the men say: '*Question*: Who are the inventors of machinery? *Answer*: Almost universally the working man' (Berg 1979: 90).

In either case, it is overwhelmingly males who design technological processes and productive machinery. Many women have observed that mechanical equipment is manufactured and assembled in ways that make it just too big or too heavy for the 'average' woman to use. This need not be conspiracy, it is merely the outcome of a pre-existing pattern of power. It is a complex point. Women vary in bodily strength and size; they also vary in orientation, some having learned more confidence and more capability than others. Many processes could be carried out with machines designed to suit smaller or less muscular operators or reorganized so as to come within reach of the 'average' woman.

There are many mechanized production processes in which women are employed. But there is a sense in which women who operate machinery, from the nineteenth-century cotton spindles to the modern typewriter, are only 'lent' it by men, as men are only 'lent' it by capital. Working-class men are threatened by the machines with which capital seeks to replace them. But as and when the machines prevail it is men's hands that control them. Comps now have twice adopted new technology, albeit with bad grace, on the strict condition that it remain under their own control. They necessarily engage in a class gamble (how many jobs will be lost? will wages fall?) but their sexual standing is not jeopardized.

The history of mechanized typesetting offers an instance of clear sex-bias within the design of equipment. The Linotype manufacturing company has twice now, in contrast to its competitors, adopted a policy that is curiously beneficial to men. A nineteenth century rival to the Linotype was the Hattersley typesetter. It had a separate mechanism for distributing type, designed for use by girls. The separation of the setting (skilled) from dissing (unskilled) was devised as a means of reducing overall labour costs. A representative of the Hattersley company wrote 'it would be a prostitution of the object for which the machine was invented and a proceeding against which we would protest at all times' to employ *men* on the disser (Typographical Association 1893).

The Linotype machine on the other hand did not represent the destruction but merely the mechanization of the comp's setting skills as a *whole*. In fact, the LSC congratulated the Linotype Company Ltd. 'The Linotype answers to one of the essential conditions of trade unionism, in that it does not depend for its success on the employment of boy or girl labour; but on the contrary, appears to offer the opportunity for establishing an arrangement whereby it may be fairly and honestly worked to the advantage of employer, inventor and workman' (Typographical Association 1893). While Linotype were not above using male scab trainees when driven to it by the comps' ca'canny, they never tried to put women on the machines and indeed curried favour with the LSC by encouraging employers who purchased the machine to shed female typesetters and replace them with union men.

Ninety years on, Linotype (now Linotype Paul) are leading designers and marketers of electronic composing systems. Most present-day manufacturers, with an eye to the hundreds of thousands of low-paid female typists their clients may profit from installing at the new keyboards, have designed them with the typewriter QWERTY lay, thus reducing Lino operators at a single blow to fumbling incompetence. Linotype Paul is one of the few firms offering an optional alternative keyboard, the 90-key lay familiar to union comps. Once more, they seem to be wooing the organized comp as man and in doing so are playing, perhaps, an ambivalent part in the class struggle being acted out between print employers, craftsmen and unskilled labour (since employers would profit more by the complete abolition of the 90-key board).

Now, electronic photocomposition is an almost motionless labour process. The greatest physical exertion is the press of a key. The equipment is more or less a 'black box'. The intelligence lies between the designers, maintenance engineers and programmers and the computer and its

Figure 15.3 The Victoria Press in Great Coram Street, London, for the employment of women compositors, 1861

peripherals. Only the simplest routine processes and minimal decisions are left to the operator.

Two factors emerge. It is significant that the great majority of the electronic technical stratum are male (as history would lead us to expect). Male power deriving from prestigious jobs has shifted up-process leaving the compositor somewhat high and dry, vulnerable to the unskilled and particularly to women. In so far as he operates this machinery he has a 'female' relationship to it: he is 'lent' it by men who know more about its technicalities than he does.

The NGA, faced with a severe threat to composing as a craft, has been forced into innovatory manoeuvres in order to survive as a union. It is widening its scope, radically re-designing and generalizing its apprenticeship requirements, turning a blind eye to the fact that some of the new style comps it recruits 'on the job' come in without apprenticeships (and a handful of these are now women who have graduated from typing to simple composers). It is seeking to recruit office workers, a proportion of whom will be female typists who are seen as a weapon employers may try to use against comps. They are to be organized in a separate division within the union and thus will be under supervision by the union but not permitted to invade the area of existing comps' work.

▶ ## CONCLUSION: MEN'S POWER AND PATRIARCHY

This study has been of the workplace. Marxist theory proposed the workplace as the primary locus of capitalist exploitation, while women's disadvantage was seen as having its site in the property relations of the family. The corollary of this view was the belief (disproved by the passing of time) that women would evade their subordination to men when they came out into waged work (Engels 1972). Feminists have shown on the contrary that the family, as the throne of 'patriarchy', has its own malevolent effectivity within capitalism and capitalist relations, it pursues women out into waged work (Kuhn 1978; Bland 1978).

Many women, however, are relatively detached from conjugal or paternal relationships. Many are single, childless, widowed, live independently, collectively, without husbands, free from fathers. Can 'the family' satisfactorily account today for the fact that they hesitate to go to the cinema alone, have to call on a man to change a car wheel, or feel put out of countenance by walking into a pub or across a composing room floor? Our theories of sexual division of labour at work have tended to be an immaculate conception unsullied by these physical intrusions. They read: women fill certain inferior places provided by capitalism, but do so in a way for which they are destined by the shackles of family life. The free-standing woman, the physical reality of men, their muscle or initiative, the way they wield a spanner or the spanner they wield, these things have been diminished in our account.

The story of compositors, for me, throws doubt on the adequacy of the explanation that the sexual relations of work can be fully accounted for as a shadow cast by the sex-relations of the family. It seems to me that the construction of gender difference and hierarchy is created at work as well as at home – and that the effect on women (less physical and technical capability, lack of confidence, lower pay) may well cast a shadow on the sex-relations of domestic life.

In socialist-feminist thought there has been a clear divide between production (privileged site of class domination) and the family (privileged site of sexual domination). The patriarchal family is recognized as adapted to the interests of capital and the capitalist division of labour as being imprinted with the patterns of domestic life. They are conceded to be mutually effective, but are nonetheless still largely conceived as two separate spheres, capitalism holding sway in one, patriarchy in the other.

Yet the compositors' story reveals a definable area of sex-gender relations that cannot be fully subsumed into 'the family', an area which has tended to be a blind spot for socialist-feminist theory. It is the same as that spot within the class relations of wage labour and capitalist production, invisible to Marxist theory, in which male power is deployed in the interests of men – capital apart.

In our analysis we can accommodate men as 'patriarchs', as fathers or husbands, and we can accommodate capitalists and workers who are frequently men. But where is the man as male, the man who fills those spaces in capitalist production that he has defined as not ours, who designs the machines and thereby decides who will use them? Where is the man who decorates the walls of his workplace with pin-ups of naked women and whose presence on the street is a factor in a woman's decision whether to work the night shift?

It was an incalculable breakthrough in the late sixties when the sexual relations of private life came to be more generally recognized as political. But somehow those sexual relations have remained ghettoized within the family. Only slowly are we demolishing the second wall, to reveal in theory what we know in practice, that the gender relations of work and public life, of the factory and the street, are sexual politics too.[11]

It is in this sense that the prevailing use of the concept of 'patriarchy' seems to me a problem. Some feminists have argued, I think rightly, that it is too specific an expression to describe the very diffuse and changing forms of male domination that we experience, and that it should be reserved for specific situations where society is organized through the authority of fathers and husbands over wives and offspring and of older men over younger men (Young and Harris 1976).

Such a 'patriarchy' would usefully enable us, for instance, to characterize certain historical relations in the printing industry: the archaic paternalism of journeyman-apprentice relations, the handing of job from father to son, the role of the 'father' of chapel in the union etc. But these practices are changing in printing – just as Jane Barker and Hazel Downing have shown that patriarchal relations of control in the office are being rendered obsolete by the new capitalist office technology (Barker and Downing 1980).

Do we then assume that male supremacy is on the wane in the workplace? I think not. The gap between women's earnings and men's in printing has widened in the last few years. What we are seeing in the struggle over the electronic office and printing technology is a series of transformations within gender relations and their articulation with class relations. The class relations are those of capitalism. The gender relations are those of a wider, more pervasive and more long-lived male dominance system than patriarchy. They are those of a sex-gender system[12] in which men dominate women inside and outside family relations, inside and outside economic production, by means which are both material and ideological, exercising their authority through both individual and organizational development. It is more nearly andrarchy[13] than patriarchy.

Finally, in what practical sense do these questions matter to women? Seeing bodily strength and capability as being socially constructed and politically deployed helps us as an organized group in that we can fight for the right to strengths and skills that we feel to be useful. On the other hand, where we do not see this kind of power as socially beneficial, our struggle can seek to devalue it by socio-political means in the interests of a gentler world (or to prevent our being disadvantaged by what may turn out to be our few remaining innate differences).

Identifying the gendered character of technology enables us to overcome our feelings of inferiority about technical matters and realize that our disqualification is the result not of our own inadequacy, nor of chance, but

Figure 15.4 A made up forme

of power-play. Understanding technology as an implement in capital's struggle to break down workers' residual control of the labour process helps us to avoid feeling 'anti-progress' if and when we need to resist it. Understanding it as male enables us to make a critique of the exploitation of technology for purposes of power by men – both over women and over each other, in competition, aggression, militarism.

Unless we recognize what capital is taking away from some men as workers, we cannot predict the strategies by which they may seek to protect their position as men. As one technology fails them will they seek to establish a power base in another? Will they eventually abandon the de-skilled manual work to women, recreating the job segregation that serves male dominance? Or will the intrinsic interdependency of keyboard and computer force a re-gendering of 'typing' so that it is no longer portrayed as female? As men's physical pre-eminence in some kinds of work is diminished will they seek to reassert it heavily in private life? Or is the importance of physical effectivity genuinely diminishing in the power relations of gender? Can the unions, so long a socio-political tool of men, be made to serve women? We need to understand all the processes that form us as workers and as women if we are to exert our will within them.

Thanks for helpful criticism of this paper in draft to Marianne Craig, Jane Foot, Nicola Murray, Anne Phillips, Eileen Phillips, Caroline Poland, Mary Slater, Judy Wajcman, Kate Young and members of the *Feminist Review* collective.

NOTES

1 The article is based on a project in progress, 'Skilled printing workers and technological change', funded by the Social Science Research Council and carried out at The City University, London. The paper was first given at the annual conference of the British Sociological Association in 1981.

2 The fact that a mode of production and a sex-gender system are two fundamental and parallel features of the organization of human societies should not lead us to expect to find any exact comparability between them, whether the duo is capitalism/'patriarchy' or any other. In the case of a sex-gender system there is a biological factor that is strongly, though not absolutely, predisposing. This is not the case in a class system. The historical timescale of modes of production appears to be shorter than that of sex-gender systems. And the socio-political and economic institutions of class seem to be more formal and visible than those of gender – though one can imagine societies where this might not be the case.

3 Michèle Barrett's recent book reviews in detail the progress of this endeavour (Barrett 1980). An important contribution to the 'appropriation of patriarchy by materialism' has been Kuhn and Wolpe (1978).

4 I adopt here Michèle Barrett's useful re-assertion of the distinction between ideology and 'the material', in place of a simplistic fusion 'ideology is material'. She cites Terry Eagleton, 'there is no possible sense in which meanings and values can be said to be "material", other than in the most sloppily metaphorical use of the term . . . If meanings *are* material, then the term "materialism" naturally ceases to be intelligible' (Barrett 1980: 89–90).

5 Heidi Hartmann's definition of patriarchy is novel in including 'hierarchical relations between men and *solidarity among them*' (Hartmann 1979b).

6 For instance, children whose families' low income entitles them to free school milk are shorter than the average child (demonstrated in an article in *The Lancet* 1979). More information relating class and stature should be available from Department of Health and Social Security 'Heights and Weights Survey' (1982).

7 Griffiths and Saraga (1979) have argued the same of sex difference in cognitive ability.

8 A fuller account exists in Cynthia Cockburn (1980) 'The losing battle: women's attempts to enter composing work 1850–1914', Working Note No. 11, unpublished.

9 I have traced the course of this technological development in 'The Iron Comp: the mechanization of composing', Working Note, No. 10, 1980, unpublished.

10 For an interesting discussion of 'dexterity' versus 'skill' in relation to gender see Ramsay Macdonald (1904).

11 A sign of change in this direction was Farley (1980), concerning sexual harassment of women at work.

12 Gayle Rubin's term (Rubin, 1975).

13 Rule by *men* as opposed to rule by fathers or male heads of household or tribe, cf. androgynous, polyandry, andro-centrism.

▶ **REFERENCES**

Barker, Jane and Downing, Hazel (1980) 'Word processing and the transformation of patriarchal relations of control in the office' *Capital & Class* No. 10.

Barrett, Michèle (1980) *Women's Oppression Today*. London: Verso.

Belotti, Elena (1975) *Little Girls*. London: Virago.

Berg, Maxine (1979) (ed.) *Technology and Toil in Nineteenth Century Britain*. London: CSE Books.

Bland, Lucy *et al.* (1978) 'Women "inside and outside" the relations of production' in Women's Studies Group, Centre for Contemporary Cultural Studies (1978).

Child, John (1967) *Industrial Relations in the British Printing Industry*. London: Allen & Unwin.

Coward, Rosalind (1978) 'Rethinking Marxism' *m/f* No. 2.

Davies, Margery (1979) 'Woman's place is at the typewriter' in Eisenstein (1979).

Delphy, Christine (1977) *The Main Enemy*. London: Women's Research and Resources Centre.

Eisenstein, Zillah (1979) (ed.) *Capitalist Patriarchy and the Case for Socialist Feminism*. New York and London: Monthly Review Press.

Engels, Frederick (1972) *The Origin of the Family, Private Property and the State*. London: Pathfinder Press.

Farley, Lin (1980) *Sexual Shakedown*. USA: Warner Paperback.

Ferris, Elizabeth (1978) 'Sportswomen and medicine, the myths surrounding women's participation in sport and exercise' in Report of the 1st International Conference on Women and Sport, Central Council of Physical Recreation, London.

Firestone, Shulamith (1971) *The Dialectic of Sex*. London: Jonathan Cape.

Griffiths, Dorothy and Saraga, Esther (1979) 'Sex differences and cognitive abilities: a sterile field of enquiry' in Hartnett (1979).

Hartmann, Heidi (1979a) 'The unhappy marriage of Marxism and feminism: towards a more progressive union' *Capital & Class* No. 8.

Hartmann, Heidi (1979b) 'Capitalism, patriarchy and job segregation' in Eisenstein (1979).

Hartnett, O. *et al.* (1979) *Sex-role Stereotyping*. London: Tavistock Publications Ltd.

Hobsbawm, E. J. (1964) *Labouring Men*. London: Weidenfeld and Nicolson.

Kuhn, Annette (1978) 'Structures of patriarchy and capital in the family' in Kuhn and Wolpe (1978).

Kuhn, Annette and Wolpe, AnnMarie (1978) (eds) *Feminism and Materialism*. London: Routledge & Kegan Paul.

Millett, Kate (1971) *Sexual Politics*. London: Rupert Hart-Davis.

Mitchell, Juliet (1975) *Psychoanalysis and Feminism*. London: Penguin.

Oakley, Ann (1972) *Sex, Gender and Society*. London: Temple Smith.

Phillips, Anne and Taylor, Barbara (1980) ' Sex and skill: notes towards a feminist economics' *Feminist Review* No. 6.

Ramsay MacDonald, J. (1904) (ed.) *Women in the Printing Trades, a Sociological Study*. London: P. S. King and Son.

Reiter, Rayna R. (1975) (ed.) *Toward an Anthropology of Women*. New York: Monthly Review Press.

Rubin, Gayle (1975) 'The traffic in women: notes on the political economy of sex' in Reiter (1975).

Thompson, Edward P . (1963) *The Making of the English Working Class*. London: Victor Gollancz.

Typographical Association, Report of the Delegate Meeting in Sheffield, December 4, 1893.

Wolpe, AnnMarie (1978) 'Education and the sexual division of labour' in Kuhn and Wolpe (1978).

Women's Studies Group, Centre for Contemporary Cultural Studies (1978) *Women Take Issue*. London: Hutchinson.

Young, Kate and Harris, Olivia (1976) 'The subordination of women in cross cultural perspective' in *Papers on Patriarchy*. London: PDC and Women's Publishing Collective.

Zeitlin, Jonathan (1981) 'Craft regulation and the division of labour: engineers and compositors in Britain 1890–1914' PhD Thesis, Warwick University.

► What machines can't do:
politics and technology in the
industrial enterprise

Robert J. Thomas

 CASE 1: THE FLEXIBLE MACHINING SYSTEM

Automation has been of great interest to engineers, managers, and academics for well over three decades (Bright 1958; Piore and Sabel 1984). For most of that time, 'fixed' automation, dedicated to the production of enormous volumes of identical parts, was the dominant approach. However, over the past decade new approaches have greatly reduced the limits of fixed automation; these more 'flexible' machining systems (FMS) allow a wider variety of small volume parts to be made at a much lower cost (Noble 1984; Jaikumar 1986; Shaiken 1985). In 1980, when this case begins, FMS technology was in its infancy; it represented a major departure from a large, US-based commercial aircraft company's prior experience and, arguably, put it at or near the leading edge of its industry.

This change took nearly four years to complete. When put into operation in 1984, the FMS cell departed in significant ways from the process that preceded it (the cell is sketched in Figure 16.1). Formerly, each of the three machines at the heart of the cell was operated independently by a skilled machinist. Now the three machines form a system controlled by a computer and linked by a conveyor. The computer not only directs the operation of each machine but also orchestrates the movement of parts from one machine to the next.

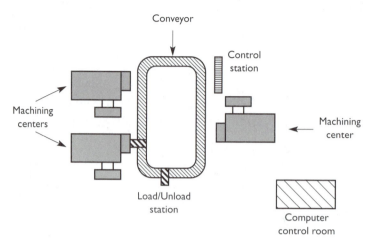

Figure 16.1 Flexible machining system

Choosing between technologies

My introduction to the FMS came in the form of a stack of documents pulled from company archives. The centerpiece was a lengthy proposal asking for several million dollars to be allocated for research and development of what was characterized as a 'dramatic new approach to precision machining.' Along with the proposal, however, I also received a sheaf of memoranda, letters, and notes (some of which were hand-written) from the manufacturing R&D group that had drafted the formal request for funds. My review of the proposal and documents surrounding the FMS left me puzzled. On the one hand, it was not clear that the authors of the FMS proposal had much confidence in either the technology or their ability to develop it. Internal memos from the R&D group expressed concern about the failure of more experienced companies with similar efforts. Other reports suggested that the level of R&D's expertise was not sufficient to the task: engineers would have to be sent to special seminars, and other skills would have to be acquired through 'additional hires.' On the other hand, the proposal submitted to corporate executives showcased very explicit and attractive return-on-investment (ROI) calculations. Great confidence was expressed in the ability of the R&D organization to develop the FMS – to the extent that all the major work should be done in-house instead of going to a more experienced equipment manufacturer. Moreover, no reference was made in the calculations to the expense of acquiring new skills and new people or the potential costs of delay. Thus, even before I had a chance to ask why this technology was chosen, I was puzzled by the difference between R&D's confident portrayal of the FMS to upper management and its private tentativeness, uncertainty, and possible underestimation of the system's true costs.

Although these contradictory impressions were irksome, I held them aside to focus my first interviews on contextual factors, especially on the role that organizational objectives played in the choice of problems to be solved by means of new technology. If the company had chosen to pursue an expensive example of advanced manufacturing technology, I expected it to reflect some broader strategic goals. In a lengthy interview with the divisional operations manager in charge of manufacturing and manufacturing technology, I asked whether corporate management specified the direction of production technology development. He responded by suggesting that corporate management's guidance was indirect at best:

> Corporate doesn't exactly dictate which direction we ought to go. They're more influential when it comes to how much money we have to spend. If you're given a budget of so many dollars, it's up to you to come up with a scheme for making the best use of that money.

Less expensive changes in manufacturing technology are made at the discretion of the divisions, but large expenditures or 'big ticket items' like the FMS have to undergo scrutiny by corporate executives and financial staff to make sure they are in line with corporate strategic objectives. Even then, he said, corporate objectives were not particularly explicit; they consisted in three simple rules: 'curb costs, increase productivity, and lose heads' (i.e., reduce the number of production workers).

If corporate executives did not give guidance but still retained their right to review divisional spending, perhaps divisional management played that steering role. When asked about the budget cycle in which the FMS was initiated, the operations manager said he'd offered 'hints' to his line managers and to the R&D department about what he felt were some major production problems. But, he added, his hints were largely 'echoes of the message we got from corporate.' In passing on the message to his subordinates, he said 'I sort of threw down the gauntlet to my people to see who would respond.'

Symptoms versus causes In this case, the gauntlet took the form of a memo in June 1980 requesting proposals for investments in new technology. By October a dozen proposals had been submitted. The majority requested replacement of worn-out machines or extension of ongoing development projects. However, two stood out – in terms of both their projected expense and their focus on technologies that were unprecedented in the division and the company. One, expressed in a two-page memo from a pair of production superintendents, called for a computer-controlled system with robot carriers to automate the storage and retrieval of parts in several shops. According to the proposal, the system would enable the shops to track their inventory more closely and to reduce the time it took individual parts to get through the machining process. It offered little in the way of expectations for the cost of the system or the savings it might generate.

The second proposal came in from manufacturing research and development, neatly bound and complete with transparencies and preliminary cost justifications. It urged consideration of a flexible machining system. According to the proposal, the division's largest machine shop was in serious trouble: 'Of the total time a part spends in the shop, only 5 per cent is spent on a machine. For the remainder of the time, the part is either waiting for processing or is in transit from one station to the next.' The proposal identified two major causes for the shop's problem. First, it argued, the current process for scheduling the movement of parts through the maze of machines and machining steps was 'archaic' and 'out of control.' It could be saved only by a *'modern material handling system'* (emphasis in original). Second, the proposal went on, the existing process was overly labor intensive. The lack of adequate control over scheduling and parts movement allowed 'outdated staffing practices to persist, especially compared to modern (Japanese) methods.' In other words, there were not only too many workers, but too many workers standing idle for long periods of time. The proposal went on to point out that several major Japanese companies were already running their FMS systems unmanned in 'lights-out' factories.

A side-by-side comparison of the two proposals revealed several significant similarities and differences. For example, both proposals targeted the movement of parts through the shop as a major cause of idle machines and workers. The shop superintendents' proposal implied that the system for routing, tracking, and storing parts was the culprit. The R&D proposal essentially concurred, but in its use of phrases such as 'out of control' it hinted that human systems were as much to blame as paper systems. The point was underscored by reference to 'outdated staffing practices' that, by contrast to 'modern (Japanese)' methods, were 'archaic.' The evocative phrasing and the polished form of the R&D proposal highlighted another important difference: new technology was their business; they were prepared to launch it; they were, by comparison to the shop, purveyors of modern production practices.

This textual analysis was corroborated in interviews with the authors of the FMS proposal. An R&D manager who had supervised the writing recalled his explicit efforts to use 'the same words and even the same phrases' he'd heard the operations manager use to describe an 'antiquated factory' run by a competitor company. He told me proudly:

> I really think he bought the idea because of the way we presented it. We pushed the right buttons . . . cost reduction, losing heads. I think we succeeded because I understood him. I understood how to sell this guy something.

Familiar imagery would enhance the proposal's chances, but equally important was the distinction to be drawn between R&D and the shop: R&D was in touch with new technology; the shop was not.

However, those interviews also revealed that R&D's advocacy of the FMS system was hardly coincidental. The proposal came close on the heels of a major machine tool show in Chicago that had featured the first FMS created by a major equipment manufacturer. Several engineers had attended the exhibition; at least one had gone with the explicit intent of viewing the FMS. Moreover, their wording of the FMS proposal borrowed liberally from brochures that extolled the virtues of the system on display. Inspired by the sales brochures and videotapes of Japanese factories, the FMS advocates had found a solution. Hints from the operations manager provided them with a problem. Thus, a proposal was conceived.

When it came to judging the relative merits of the proposals he had received, the operations manager gave credence to the R&D group's assessment of the importance of imagery in decision making. He preferred the FMS because 'harder-working machines' and fewer people were something he thought his superiors could understand. The automated storage system might have been worthwhile, but the proposal was not nearly as compelling as the FMS and it did not promise to eliminate any workers. According to the operations manager, the automated storage system was a 'band-aid':

> It didn't get to the heart of the action: getting control over the process by getting the parts in and out of the machines. You don't get that unless you have a way to keep bottlenecks like people and paper from getting in the way.

A closer look at the 'heart of the action' brings to the surface an objective familiar to labor process theory – control – but it also reveals that the problem to which the operations manager referred was not limited to workers. On the one hand, the FMS would significantly alter the allocation of workers to different jobs in the machine shop. In theory, at least, the use of conveyors, sensors, and computer controls would eliminate the existing practice of allocating one operator per machine. It could also do away with many of the stock handlers who moved parts between machines and the set-up workers who changed the tools and fixtures between machining operations. In other words, by substituting machines for workers, control over the process would be hard-wired. Moreover, by exercising its contractual right to introduce new production technology, management could avoid a direct confrontation with the union over work pace and production standards.

On the other hand, when I pressed the operations manager about the issue, he admitted that idle workers were as much a symptom as they were a cause of what he perceived to be the shop's inefficiencies. He pointed to shop management as contributors to the situation:

> It's hard for them to see the real problems sometimes. They don't always see that they're part of the problem. . . . So, they set their own rates about how much work should get done. They allow things to slack off. There's no shortage of excuses, of course. And maybe some are valid. A shop like that can be a tough place, especially with the kind of union we have. And a lot of foremen came out of the hourly ranks, so they have commitments to this guy or that way of doing things. But my job is to get the work out. To keep things under control.

He then tempered that remark by arguing that he could understand the reasons behind the shop management's 'myopia':

> Shop supervision had the same concern about job security that the guys operating the machines did. In other words, if you don't need as many

machine operators, you don't need as many supervisors to supervise them.

In other words, a real attraction of the FMS was its promise to alter dramatically the social organization of the shop and its historical accumulation of customs and informal norms – seemingly independent of whatever tangible economic benefits the technology might provide. The operations manager confirmed this assessment when he added that it would 'let us clean house and start all over again. Wipe the slate clean.'

These comments, as well as the slippery distinction between symptoms and underlying causes, led me to search out the superintendents who had authored the competing proposal. One had retired, but the other provided several important points of contrast with what I had been told by the operations manager and the R&D engineers. When I queried him as to the justification for an automated storage and tracking system, he smiled, held up three fingers, and launched into a monologue:

> First, you have to understand that we are not in charge of our own world here. We don't schedule what comes in here; a logistics department does. They get orders from engineering [design] about what they want and when they want it, and then they tell us what to do. So we don't have the chance, really, to control the flow of work in the shop. Second, the parts tracking system is old and a mess. Every part has a number, but we still use paper to track everything. So if paper is lost or a part gets sidetracked, we probably don't know where it is. We thought that a new system could solve that. Maybe we could use bar codes and light wands. We couldn't convince the guys in logistics that they should do it, so we thought we'd try to do it ourselves. And, third . . . we wanted to build something that would help us cope with engineering. You see, engineering is the boss in this company. . . . They need a part today, and so they tell you to drop everything else and to make their part. Problem is, there are all kinds of 'theys.' There's the they who's got this big order in and he's expecting you'll do it on schedule. Then there's the they who's got these parts he needs made today and he didn't think about sending it in until yesterday. And then there's the they who sent in his order two days ago and wanted it done today but now he's decided he wants to make a change and still have it today. That's what we were up against. Still are, to be honest with you.

The shop, it seemed, wanted control over itself. The automated system would bring discipline to the logistics department responsible for scheduling and provide a means for the shop to cope with the unpredictable and largely uncontrollable demands from engineering.

New technology in the form of an automated system was not, however, intended to replace workers with machines. From the superintendent's perspective, there was 'always room for improvement in productivity' by finding ways to increase the attentiveness of machine operators; but, he added:

> If the machines are idle 95 per cent of the time like you said, you can't blame the guys on the floor. Sure, they could work harder. But if they're waiting because our logistics and scheduling are all fouled up, who's to blame?

This comment explained the absence of calculations for the number of heads that would be lost in the superintendent's proposal. It was not at all clear that heads would be lost or, for that matter, whether any needed to be.

Bringing interests to the surface   At this point in the research, I found it hard to see this decision-making process as an example of strategic choice (Child 1972). Although all participants in the contest struggled to justify their choices in the language of corporate objectives, the actual decisions seemed to be influenced as much by the availability of solutions as they were by the importance of problems. Indeed, the FMS appeared to be a clear instance of a solution searching for a problem (see Cohen, March, and Olsen 1972).

There were, however, two further dimensions to the story that argued for the incorporation of a political perspective on the choice of technology. First, as I questioned the operations manager about his reasons for choosing the expensive FMS and forgoing other requests (including the automated storage system), I found that the FMS also coincided with a distinctly personal objective: 'I wanted to be the guy who did it first.' Being the first operations manager in the company with the technology could have significant career implications, suggesting that the fact of innovation may be as important as its usefulness. It would also be very much in character with this particular manager: his aggressive pursuit of resources to support modernization of manufacturing processes had won him fierce loyalty among manufacturing engineers.

In order to do it first, he argued, he had to be sure that it was 'done right, done cheaply, and done quietly.' Thus, after reviewing the initial proposal, he authorized R&D to begin by investigating how other companies had set up an FMS; learning from others' mistakes could help it be done right and cheaply. He also cautioned R&D to proceed without fanfare. As he told me:

> Those guys [from a larger division] were looking over my shoulder all the time. If they had caught on to what I was thinking about, you can be sure they would have said they deserve to be the first ones with an FMS.

Second, the FMS held a major attraction for R&D engineers. It gave them the opportunity to distinguish themselves – to do what their department's formal mandate espoused but what many said they rarely had a chance to do. For example, one engineer was inspired by what he had seen at the machine show and learned from peers in other companies. He listed the opportunities submerged in the proposal:

> This gives us the chance to be real engineers. To do what we do best, to do what we're trained to do, and to do what this company needs. We need to be at the front end of technological change.

When I asked what he meant by 'real engineering,' he thought for a moment and then replied:

> It's hard to say exactly. In my mind it has to do with creating something, . . . seeing a problem and using all your skills to solve it. Something that's real, important, and creative. Something you can see through to the end. I'll tell you what it's not: it's not the firefighting we usually do: the fixes to somebody else's mistakes, the leaky hose, the faulty programming, the band-aid that squeezes two more months out of a machine that should have been scrapped ages ago.

In a more personal vein, a colleague in the project added:

> Projects like this don't come around that often. You have more of a chance to bring some positive attention to yourself when you're onto something new. It's not exactly a career maker, but it sure doesn't hurt either.

The opportunity to be 'real' engineers was important in both relative and absolute terms. In relative terms, engineers in the manufacturing R&D organization felt themselves to be, at best, subordinate in status to product designers. Many with whom I spoke chafed at the fact that they were not even referred to as engineers. One younger machine-control specialist expressed an anger shared by many of his colleagues:

> You know, I have a good mechanical engineering degree from a very good school. I just happen to like this end of things . . . the manufacturing end of the company. I like it better than design. I graduated from a better place with a higher GPA than at least a dozen guys I know over there [pointing to an adjacent building housing a design group].

Differences in pay scales and promotional opportunities exacerbated the manufacturing engineers' resentment.[1] In absolute terms, the FMS held out the possibility that the manufacturing R&D engineers could accomplish by themselves a smaller version of the much larger and more expensive FMS they had seen on display. In other words, on top of introducing an innovation to the company, they could engage in something much closer to invention.

Thus, early in the history of this change process – before any money was spent – important choices had been made and significant parameters for later choices had already been set. Of the alternatives generated by the request for proposals, the FMS was deemed the more appealing because it aligned corporate objectives with the personal and professional interests of the divisional manager and R&D engineers. By defining the problem as one of control, however, managers and engineers made choices that virtually guaranteed that neither workers nor shop-level supervisors would have a direct role to play in the choice process. Indeed, the change proposal initiated by shop management was dismissed as myopic and the FMS proposal was supported because it promised to forcibly alter the web of obligations and traditions that characterized social relations in production.

At this stage, the contradictory impressions I had received from my earlier review of the documents became more pronounced. Return-on-investment calculations had initially appeared to serve as a filter for screening out unprofitable investments. However, nothing I heard in the early interviews suggested that profitability really mattered in the choice among alternatives. Moreover, the lack of in-house expertise neither deterred development of the proposal nor created excessive concern about the ultimate cost of the undertaking. Instead, what seemed most important was the identification of an attractive concept: one that evoked powerful images (heads lost, a shop under control) and opportunities (career advancement, 'real engineering' work).

However, because of its anticipated cost, the FMS required corporate review. If strategic objectives had practical force over the politics of divisional decision making, then profitability might take back center stage.

Choosing within the technology

Corporate approval depended heavily on the ability of R&D and divisional management to justify the FMS financially. Direct labor costs figure prominently in ROI calculations, and, as I was told, they are critical in a proposal's funding chances in competition with other investment alternatives. A cross-country tour of FMS vendors and users showed the R&D project group that labor savings were not guaranteed. One engineer suggested that 'even when you held constant the size of the cell and what they were making, the staffing solutions were all over the map,' that is, from less than one machine operator for every three machines to one person per machine.

Finessing the ROI   The high level of uncertainty about the true cost of the FMS combined with the lack of experience among R&D engineers do create a great deal of tension. According to the previous engineer:

> I had to come up with a payback . . . and the best I could come up with was four years. Around here, the corporate rule, the ground rule, is if the thing can't be paid back in less than two years, preferably one year, your request has a chance of a snowball in hell of flying. Well now, that either makes you damned smart or a good liar.

The manager of R&D sent this engineer off to work with staff from the industrial engineering department to come up with more acceptable figures. Industrial engineering staff apparently helped; however, when I sought to find out how those figures were arrived at, I was told: 'I'm not going to tell you how we generated the ROI because it was really silly. We had a number to hit and we hit it.'

The R&D engineers, with new calculations in hand, went back to divisional management. However, the numbers were still not acceptable, particularly to those wary of the discerning eyes of corporate 'bean counters.' The operations manager resolved the staffing problem by making a supervisor the system operator. In the words of one participant, this arbitrary change enabled the assembled group to 'finesse the ROI.'

Given the amount of work by the operations manager and the R&D project team, it might seem reasonable to conclude that the story would end at this point. However, many elements of the *content* of the FMS still remained to be determined, and these elements could not be fixed outside the context of its use. The proposal reflected the perceptions and the encapsulated interests of only a subset of the organization. For example, one of the principal engineers came back from the cross-country tour enthusiastic about the technical possibilities and challenges but worried about organizational and procedural consequences of the new technology. As he explained to me later:

> A user contemplating the system has a lot of homework to do. You have got to consider combining project requirements. In that very statement there – about combining project requirements – you run into organization restructuring. Because we had [listing of current programs] and each one of these programs has a program manager, and under that program manager is a manufacturing manager, and so on. . . . What that means is that when you start saying all your parts are going to go through a common machining cell, politics start entering.

In other words, he feared that a single program or contract alone would not be enough to support the project.

Pitching the technology   As a result, the idea and its proponents would have to move from a center in R&D in three different organizational directions: outward to embrace the functional groups it would affect; downward to the shop where it would be housed; and upward to corporate executives and staff on whose approval the proposal ultimately rested. The different tactics employed to organize support – or placate resistance – not only varied among these three groups but demonstrated the importance of adroit political strategy for accomplishing technological change.

Outward   Project team members were dispatched as emissaries to allied functions (e.g., quality control, scheduling, maintenance). Each of these departments had its own set of interests and its own chain of command. A close examination of their prepared scripts made it clear that, despite higher-level backing, the project representatives were willing to bargain with allied groups over the ultimate configuration of the FMS. For example, a dispute arose over the method for linking the machines: one group wanted robot carriers to shuttle parts between the machines, but the group that would have ultimate responsibility for maintaining the system detested the idea and held out for a conveyor line. Because the conveyor would reduce the overall cost of the project and stifle resistance, the conveyor was chosen.

As important as the bargaining and trading, however, the presentations were designed to provide allied groups with a way to pitch the projects to their own bosses. A software engineer described the process:

It wasn't anything new, really. You always wind up briefing a guy who doesn't really know what the hell he's talking about to send him up to another guy who doesn't know what the hell he's talking about, but may have heard enough pitches that he can ask some good questions and have some kind of expectations of an answer.

Thus, a coalition was formed to back the proposal.

Downward   Because shop-level supervisors and workers had already been identified as part of the problem to be 'solved' by the FMS, they were given little opportunity to affect the content of the proposal. This did not mean, however, that they could be ignored entirely. The operations manager, it will be recalled, did not want to telegraph his plans to other divisions. Thus, the project team set out to co-opt lower-level management. Specifically, the operations manager instructed the project team to assemble a 'user group' with a shop-level supervisor as its nominal leader. Engineers acknowledged the appropriateness of the tactic but still complained that the exercise was futile since, as one put it bluntly, 'Most of those guys have to be dragged kicking and screaming into the twentieth century.'

The general supervisor put in charge of the user group admitted to being opposed to the FMS project when he first heard about it. He anticipated a repetition of past experiences, with R&D developing a new system or machine in a curtained-off area, unveiling it to supervisors, and 'leaving the shop to pick up the pieces.' Over time his opposition turned to cautious acceptance: 'They may not have listened to me, but at least they heard me out.'

Neither the union nor workers in the area were included as part of the user group. In fact, the union learned of the FMS long after the machinery was bolted into the ground and on its way to operation – three and a half years after the project was initiated.

Upward   The operations manager's presentation to his boss and then to corporate management drew heavily on his earlier work with R&D managers to find the right numbers and the right language with which to pitch the project. When asked how he described the FMS to his bosses, he was very clear: 'I'd be kidding you if I didn't say we leave out some of the negative things when we're trying to sell this stuff upstairs. So we forget about some of those things.' And when I asked how technical he gets in making the pitch, he responded:

Oh, I don't try to get technical. I go on emotion as much as anything. It's got to be something that turns them on. They can visualize in their own mind that this is going to be a good idea, it's going to solve the problem – at least in the way we've set up the problem – and it'll get good savings.

The process was not complete without supportive economic figures. However, financial considerations had been well covered with the 'finessed' ROI calculations – so well covered, in fact, that the final proposal promised to eliminate direct labor entirely. This meant that there would be no direct cost associated with the operation of the FMS. In the end, funding was approved, and the FMS project was launched into its final stages.

Thus, profitability took back center stage in the closing moments of the second act. But it played a curious role. On the one hand, corporate review required ROI figures in support of the proposal; this requirement, in a sense, encouraged FMS proponents to play games with the numbers. On the other hand, divisional management and R&D took the corporate review seriously enough to make bold claims despite their fragile numbers. The net result was that technological choices were made largely on the basis of personal and professional interests and what was perceived to be an archaic social context in the shop that could not be altered without the external pressure of technological change. In other words, two years before the FMS went into operation, R&D had committed itself to a course of action that dramatically reduced the degrees of freedom available for modification of the system once it was in place.

Implementing the technology

The installation and construction of the FMS occurred under guarded conditions. The machines were taken over by engineers, and the project team draped a large canvas curtain around the area. They spent the next twenty months in relative seclusion. Supervisors and workers were discouraged from eavesdropping on the work in progress.

As the work of implementation proceeded, the system began to prove far more complex and sensitive to bottlenecks than anticipated. For example, debris from certain metals jammed up the conveyor, slowing the overall process and rendering it far less 'flexible.' Software engineers experienced great difficulty integrating the system; eventually, an outside company was hired to write an entirely new program. A full-time technician had to be added to oversee the computer, and a programmer was assigned part-time to respond to intermittent glitches.

Symbolic sensors   Problems with the sensors proved most intractable, but they also demonstrated the significance of the entrenched interests that had guided technological choices made early on in the process. Sensors, like those that open and close supermarket doors, use light beams and electronic eyes to identify the position of a metal part and to signal the beginning and ending of machining activities. They are intended to replace the eyes and ears of the individual machinist. When sensors malfunction, damage can be done to the metal parts or, more commonly, to the machines doing the cutting and grinding. Despite continuous setbacks the engineers doggedly continued in their efforts. In separate interviews two engineers provided virtually identical explanations for their persistence: (1) the sensors were one of the 'real' engineering challenges associated with the FMS, and in the words of one, 'We had a lot of pride invested in solving that one'; (2) without the sensors, someone would have to monitor the individual machines, meaning that some heads would not be lost; and (3) the sensors were a central part of the tracking system that had promised to 'get the shop under control.' Thus, failure to fully integrate the sensors would not only shoot down the expected return on investment and undermine the effort to reorganize the shop but also threaten the interests and careers of the 'real' engineers.

The symbolic importance of the sensors was so great that when a production superintendent suggested that workers be brought back into the system, the engineers nearly revolted. As evidence of his claims, the superintendent circulated a memo written by an R&D engineer when the initial FMS proposal was drafted. The memo warned that the sensors would be a 'linchpin' in the whole undertaking. He argued that replacing the eyes and ears of the skilled machinist would be difficult and cautioned against excessive reliance on the electronic alternative. The engineer's warnings had been ignored, he told me in a later interview, because they 'were just the ravings of a guy from the old school.' Under pressure from the operations manager, who was becoming increasingly nervous about excessive delays, the project team reluctantly scaled back the use of sensors and a former machinist was assigned to monitor the machines.

At the end of the nearly two years set aside for construction and implementation, the FMS was officially turned over to production management. According to the formal 'close-out' report in the files, the project was completed on time and only slightly over budget. A comparison between the proposal and the actual system revealed, however, that the FMS was not exactly complete. The scaled-back use of sensors had curtailed many of the more sophisticated feedback and control functions originally intended for the machine. The system required not one but three full-time hourly employees to operate it. Additional software programming support was necessary, and even more was anticipated. And problems with cooling and debris removal persisted, restricting the kinds of parts that could be run through the system. The project team's success in meeting formal schedule and budget criteria were thus artifacts of these relatively invisible amendments to the original design.

After a year of operation, the FMS demonstrated modest improvements in the flow time for parts in the shop, but it did not come close to the time and cost savings predicted in the proposal. Technical problems continued to plague the system, and the higher level of staffing (direct and indirect) rendered the FMS much more costly to run than had been anticipated.

Despite the vigorous attention that ROI figures had received early on, the company never did review the economic performance of the machine. According to my interviews, the absence of an audit was not unusual, even for an expenditure of this magnitude. What mattered, I was told, was that the close-out report did not draw attention to any economic and performance shortcomings. But those shortcomings did not prevent the FMS from garnering 'managed' attention. As one participant recalled, shortly before the system was turned over to the shop, the operations manager ordered a crew to 'paint and polish' the machines and set up a demonstration run that would allow him to 'put on a show for corporate brass.' A symbol had been sold and a symbolic demonstration was in order. Shortly after the start-up of the system, the operations manager was promoted to a higher-level post in the company.

Conclusion

New technology in the form of a flexible machining system 'impacted' this organization. The number of hourly workers used in the area was reduced from seven to four.[2] Machine operating time increased by about 20 per cent. Control over the shop appeared to have been centralized; at least the installation of a central computer allowed the work of the three machines to be more closely monitored by a supervisor.

However, it would be extremely difficult to understand why those impacts occurred – or, more specifically, why that technology appeared – without attending directly to the process through which it was given meaning and purpose. This case study has shown that the choice of the FMS was not a simple matter of rational economic calculation and that the evolution of the system was not simply a technical affair. Both choice and development were directly affected by the perceptions and interests of managers and engineers and by their efforts to shape the context within which it would be used. Those perceptions and interests were largely opaque to the corporate decision makers who sat in judgment on the FMS proposal.

Organizational objectives were influential in the choice of both the problem and the solution. However, the *interpretation* of organizational objectives seemed most paradoxical. As I suggested in the first chapter, it should not be particularly surprising to find that the distribution of expertise and resources influences the range of problems and solutions an organization identifies and from which it selects. But it is surprising that organizational objectives should be deemed so important and then be so clearly ignored. The admonition to 'get rid of heads' told operations management and R&D what they would have to promise in order to get the resources – and the opportunity to gain attention and to do 'real engineering' work – but in practice the ROI calculations and all the preliminaries appeared to be little more than a charade. The rules of the game required that all participants present themselves as rational, calculating utility maximizers, but beneath the official rules an entirely different game was being played. . . .

▶ THE LIMITS OF SOCIAL CHOICE

Although findings from the case studies point up serious limits on the applicability of technological determinist arguments, they do not necessarily suggest that the opposing view – from social choice – ought to be embraced without reservation. Indeed, they suggest rather strongly that social choice arguments – especially the notion of strategic choice advanced by Child (1972) and the assertion of a managerial 'control imperative' in labor process theory (e.g., Braverman 1974; Edwards 1979) – provide an equally partial depiction of the process through which technological choices are made and implemented.

Whose strategy? Whose choice?

The concept of strategic choice as laid out by Child (1972) and applied to the study of technological change by Buchanan and Boddy (1983), Child (1985), and Clark *et al.* (1988) draws explicit attention to the active role that top managers and executives play in influencing, if not directly determining, organizational structure. It portrays strategic choice, and strategy formation in general, as an *interpretive process*: organizational leaders may gather and assess information from a wide variety of sources (external as well as internal to the organization), but the strategies they formulate and the choices they make are interpretations of the world that surrounds them. As Child argued in his seminal article (1972), the world – whether it is defined as the environment, history, or the organization itself – is no more real or immediate to those at the top of an organization than it is to anyone else. What distinguishes those at the top, however, is the power they possess to *enact* their interpretations as plans and directives that structure the organization. In this sense Child makes an important but often overlooked point: the behavior of organizational leaders is as likely to be affected by broadly normative, ideological, and political considerations as is that of any other group. In this formulation he builds into the concept of strategic choice a conception of politics that goes well beyond simple self-interest to include distinctive perspectives or beliefs about how people and organizations *should* work.

Given this conceptual base, we might expect analyses rooted in a strategic choice perspective to be far more sensitive to the mediating role of organizational process and politics in relation to technological choice. Yet we find instead that those studies have, for the most part, taken *a* strategic choice – or, more accurately, a formal decision to adopt a given technology – as the *starting point* for analysis. As a result, all that remains to be explained is the effect of the people and groups who occupy the layers between the top and the bottom of the organization on the manner in which 'strategic choices' are *implemented*. We may be alerted to the existence of substrategies and to their ability to attenuate strategic intent, but the process of change is rendered analytically static: that is, the only interaction is that which is occasioned by the technology as it passes from one layer to the next. Politics is once again reduced to narrow self-interest, and substrategies are engaged only in reaction to hierarchical orders.

However, the case studies suggests that although this conception of strategic choice may apply in some situations, the process of change is on the whole far more dynamic and interactive than prior work would lead us to believe. Moreover, they show that politics is far more central to the choice of technology than has been recognized to this point. Three points from the case studies are relevant here.

First, choices between and within technologies – and by extension, choices between and within structures associated with those technologies – are not limited to the higher echelons of management. They may, in fact, originate at some distance removed (in time and space) from the top of the organization. Such choices and the activities that go into the framing of alternatives *prior* to the formal decision represent a critical part of the process of technological change that is all but invisible when one begins, as the strategic choice perspective does, with the formal decision to proceed with a change.

For example, the adoption of the FMS and surface mount technologies [SMT] in the aircraft and computer companies, respectively, could be presented as outcomes of top-level strategic choice. After all, top managers' approval set each change in motion. But characterizing these changes as the outcome of strategic choice would seriously understate the significance of activities that preceded formal decisions. The decision to pursue the development of the FMS was made *in substance* in the middle of the organization, where the divisional operations manager resided. The choice was *ratified* or approved by top management, but little of the content of the change or the full range of its purposes or intended impacts was revealed in the process. Arguably, the choice was linked to organizational strategy, but given that, as the divisional manager explained, strategy consisted largely of the admonition to 'cut costs, improve productivity, and lose heads,' it would be difficult to argue that strategy was an explicit guide to choice. Rather, the FMS option was chosen on the basis of a particular *interpretation* of that strategy, the organization, and its environment and on the way each fit with that manager's world views. Thus, the divisional manager and his staff may have lacked the formal power with which to enact their choice, but they did exercise considerable influence in the *framing of the decision* to be made by higher-level executives. They exercised this influence by carefully constructing a story about the FMS that would, to all appearances, speak directly to broader organizational strategy – even if it required finessing the ROI.

The computer case makes the same point from a different angle. On the surface the 'official story' about SMT seems to be a classic example of strategic choice. Yet detailed examination revealed that the official account told at best half the story. Missing was an explanation of how (or why) SMT came to be sufficiently strategic to warrant corporate action. Overlooked entirely were the arduous and in many ways quixotic efforts of manufacturing engineers to *make* SMT a strategic choice. Once again the point is not that top managers were uninvolved or that the organization lacked a competitive strategy; rather, it is that the choice alternatives *presented to* top management were the outcome of a complex set of events that took place long before and far away from the moment of strategic choice.

Second, the 'substrategies' employed by different actors in the choice process are not just passive filters, nor can they be easily reduced to simple self-interest or structurally defined objectives. Instead, the case studies suggest that organizational actors – both in formulating proposals for change and in choosing how to respond to changes initiated by others – engage in no less complex a process of interpretation than do those to whom a strategic choice perspective pays greatest attention (i.e., top decision makers). In other words, subordinate position in an organizational hierarchy does not preclude the possibility that people harbor or even desire to enact objectives they feel to be important; nor does it mean that those objectives will be limited in scope to immediate self-interest or to the specific domain in which people work.

In fact, as the case studies demonstrate repeatedly, much of the overt contention that surrounds the process of technological change derives from differences in the interpretive frameworks – the worldviews – of the actors involved. Although often derided as 'just politics,' contention and conflict emanated from differences in what people believed to be important or necessary not only for themselves or for the positions they occupied but *for the organization as a whole*. Thus, even in what seemed to be the least ambitious or expansive example of change, the CNC case in chapter 2, shop managers saw in the new technology a means by which to achieve what they believed to be vital organizational objectives, as well as what might be viewed as specific positional objectives. Similarly, the actions of the divisional manager in the aircraft FMS case, the manufacturing engineers in the computer company, and the plant and union leaders in the auto company could very easily be dismissed as extensions of narrow career or economic interests; but such an explanation would overlook the way their actions also reflected visions of *the way things should work*. In the absence of formal power or the authority with which to impose their interpretations on the rest of their organization, each sought – albeit in different ways – to influence the premises on which decisions were made. Failing that, they sought to imprint on a decision already made their unique interpretation of the way it ought to be implemented.

The third and in many respects most significant limit on the utility of a strategic choice perspective involves the role of strategy – or *lack thereof* – in the process of technological choice. In short, in none of these organizations, and certainly in none of the cases, was there anything that could be identified as a technology or a manufacturing strategy. This is not to say that these companies did not routinely create elaborate strategic plans or forecasts; indeed they did. However, those plans and forecasts pertained primarily (if not exclusively) to *products* and only secondarily to *processes* for making their products.[3] With the possible exception of the aluminum company case, products, product technologies, and product needs were designated as the *drivers* of process technology and of manufacturing more generally. Although not unusual in assembled goods industries – especially in the United States (see Hayes, Wheelwright, and Clark 1988; Clark and Fujimoto 1991; Dertouzos, Lester, and Solow 1989) – the hegemony of the product has critical implications for the process of technological choice that, to date at least, have been largely ignored in the strategic choice literature.

Most important, the subordination of process to product and of manu-facturing to product design left managers and engineers in manufacturing without a strategy to guide their choices other than what they could infer from careful scrutiny of new product developments. In this regard the case studies offer a measure of support to critics of the strategic choice perspec-tive (e.g., Rose and Jones 1987; Barley 1986) who contend that broad statements of strategic intent offer little insight as to the outcomes of technological change. However, the absence of an explicit strategy did not prevent managers and engineers from constructing strategies to guide their choices; instead, it meant that the strategies they constructed were likely to be implicit, rather than explicit, and therefore invisible except through detailed observation of the process of technological choice.

The case studies are especially useful for what they reveal about the managers' and engineers' behavior in the absence of an explicit strategy to guide their choice of technology. Most directly, they suggest that the subordination of manufacturing to design reduced the incentive – or raised the perceived risks – of engaging in technological innovation. Departures from precedent were rare, and opportunities to innovate with major new technologies were both coveted and feared – as we saw in virtually every case. When departures were contemplated, considerable care was taken to create fallback positions, to align powerful allies, and to shield development activities from the view of 'outsiders,' including other departments as well as lower-level managers and workers. Even when the opportunity to innovate was actively sought out, history and power rela-tions played an important role in shaping the choice and configuration of new process technology. In the computer company case, for example, it took an unconventional effort – a virtual social movement among manu-facturing engineers – to garner the knowledge, resources, and visibility necessary to make a case for surface mount technology. Still, once the technology proved feasible, the insurgency was rapidly co-opted by higher-level management and the design labs, leaving the status of manu-facturing engineers and the structure of the organisation largely un-changed.

Moreover, the subordination of process to product helps explain why manufacturing managers and engineers would adhere to traditional return-on-investment metrics. In most cases those metrics provided a functional substitute for an explicit manufacturing strategy. That is, the problem was *not* that other metrics were unavailable or that suitable ones could not be devised with some imagination; rather, in the absence of a manufacturing strategy, those other metrics *lacked meaning and therefore influence*. Thus, when manufacturing managers and engineers were faced with product design organizations hostile to changes in manufacturing processes that did not meet with their prior approval and, more generally, when they were faced with higher-level decision makers for whom alternative measures made no sense, the default option was also the safest: to restrict the search for both problems and solutions that fit with traditional measures – even when doing so might produce deleterious consequences (e.g., increases in the volume of indirect labor). On those few occasions when managers and engineers chose to go out on a limb, they ardently resisted arguing for alternative measures because 'having the numbers ' – even numbers they may have ridiculed in private – enabled them to argue that their choices were legitimate. The numbers were legitimate because they had been screened through a *procedure* that was deemed to be legitimate and defensible.

In sum, the strategic choice perspective remains useful for the attention it draws to the *capacity* of people in organizations to choose technology and, through technology, to affect organizational structures and practices in ways that are consistent with particular interpretations of the organization and its environment. However, the case studies suggest that the strategic choice perspective has limited applicability because (1) it presumes that choices made at the top of the organization are strategic by definition, when in fact they may only be ratifications of decisions made elsewhere; (2) it characterizes the substrategies of the groups that occupy the intervening layers between the top and the bottom as unidimensional when they are instead multidimensional and, more important, proactive as well as reactive; and (3) it presents organizational strategy as an explicit and comprehensive guide to all an organization's major activities when, by contrast, strategy may pertain only to activities that are deemed strategic: that is, others may be left to operate within constraints over which they can exercise little direct control and within which, if they are to innovate at all, they must do so by nonsanctioned means.

The case study findings thus lead to an ironic conclusion: although the concept of strategic choice was formulated as a response to the failure of functionalist theories (e.g., technological determinism) 'to give due attention to the agency of choice by whoever have the power to direct the organization' (Child 1972: 2), the strategic choice perspective restricts our attention to the behavior of top-level managers and to the implementation of their decisions. Like the functionalist theories it critiques, it leads us to conclude that history – as embedded in prior decisions about technology and structure – is largely irrelevant to understanding the range of new technologies or new social systems an organization is capable of or willing to adopt. And, finally, it leaves us to wonder how the very strategic choices that result in major technological changes *actually come about*.

Class action?

Labor process theory – the other and in many ways more emphatic variant on social choice – also deserves scrutiny on the basis of the case study findings. Although Marx's admonition to enter the 'hidden abode' of production spurred prodigious efforts to analyze the labor process under capitalism, the principal advocates of labor process theory in the United States largely ignored organizational processes and relations that did not seem to have a direct influence on the structuring of skills or the distribution of control over work. Braverman (1974), for example, argued that capitalists, aided by the managers and engineers they hired, used physical technology as a tool with which to divide, deskill, and ultimately dominate workers. Yet he gave only passing mention to the organizational forms and processes in and through which those technologies were designed and deployed. In other cases (e.g., Edwards 1979; Gordon, Edwards and Reich 1982) differences in organizational size, scale, and market power were used to distinguish between 'monopoly' and 'competitive' or 'core' and 'peripheral' enterprises; but again, organizational and intraorganizational levels of analysis were treated as virtually transparent. Even some of the most insightful field studies, such as Burawoy's (1979) study of a machine shop, resort to conjecture and theoretical assertion instead of empirical observation when trying to bridge the gap between the social organization of the shop floor and the macrodynamics of capitalist enterprises and economies.

Whether the researchers were content to assume that organizational processes and relations were inconsequential, were hampered by the difficulties of trying to assimilate organizations into a theoretical framework that only had room for classes, or were unable to gain access to data beyond the shop floor, the result was the same: researchers analyzed work processes in virtual isolation from the rest of the organizations in which they were located. Moreover, in treating technology as a tool in the hands of capitalists and managers, they gave it the appearance of being infinitely elastic and mutable, governed almost exclusively by a logic of class domination. Technological choice, in other words, would be little more than an extension of managerial objectives.

From the perspective of the case studies, labor process theory comes up short in three distinct areas. First, labor process theory assumes not only that capitalists and their agents share a common consciousness but also that they work in concert to enact common objectives through the design of both the technical and the social systems of production. Although *control* was frequently referred to in the cases as a concern or an objective associated with the introduction of new technology, the object of control was as likely to be *other* managers or departments as it was to be workers. Proponents of the aircraft company FMS declared the entire machining factory 'out of control'; shop managers in that factory wanted an automated storage and retrieval system to get control over their parts flow. Shop managers in the CNC case wanted control over parts ordering and programming in order to serve their customers. Manufacturing engineers in the computer company wanted control over prototyping, board assembly, and even board design. Top-level auto company executives saw the design of the FMS cell as part of a strategy for enhancing control over the plants, and so on. At one level, each change proponent attached a different meaning to the term *control*; however, the breadth of concern and contention over control suggests that no group or objective could be identified as the *singular* focus of these many meanings of control.

If anything, what emerges from the case studies is a singular concern with creating spheres of activity within which *any group* (be it executives, managers, engineers, production supervisors, or workers) could enact their particular view of *the way things should work*. Indeed, as I argue in the next section, that concern represents one of the most important themes to come out of the case studies. It is also one of the most important underpinnings for a power-process analysis of technological choice and change in organizations.

Second, labor process theory substantially overlooks the role of engineers in the design of new technology. Theorists have been content to assume not only that engineers are an undifferentiated group but that they unconditionally and unreflectively obey the commands of their organizational superiors. Even Noble (1984) portrays engineers as largely devoid of aspirations or worldviews apart from a kind of presocial or antisocial consciousness in which machines are more trustworthy than human beings. Yet the case studies present a picture that differs in important ways. In some instances the engineers who design physical processes share their superiors' suspicion and disdain for workers *and* for lower-level managers. But far more often they harbor conceptions of work systems that could just as easily upgrade as downgrade worker skills. What they tend to lack is the capacity, the language, and, most important, the power with which to make those visions real. They are, as I pointed out in the last section, no less constrained by history and power relations in these organizations than are the managers for whom they work or the plant supervisors and workers who are ultimately 'impacted' by the technologies they design.

This point is most evident when, under admittedly unusual circumstances, opportunities arise that allow (or force) engineers to pursue conceptions of the labor process that depart from precedent. For example, in the auto and aluminum company cases, engineers were engaged in a way that changed the parameters of their assignments and enabled them to work collaboratively with managers, supervisors, and workers to solve problems of *common* interest rather than narrowly defined problems of departmental, professional, or hierarchical interest.

Third, labor process theory has a great deal of difficulty explaining situations in which workers themselves play a central and collaborative role in the design and introduction of new technology. The only real allowance made for this occurrence is a variant on 'false consciousness': workers are enticed into a form of self-exploitation by what only appear to be opportunities to gain greater control over the labor process. Certainly this has been at the core of labor process theorists' critiques of worker participation (cf. Parker 1985; Thomas 1988; Kelly 1987; Grenier 1988), Japanese management techniques (Parker and Slaughter 1988; Burawoy 1985; Yamamoto 1981), and new approaches to work organization such as lean production and flexible specialization (Hyman 1989; Bergren, Bjorkman, and Hollander 1991). Yet, although there is empirical evidence to support many of these criticisms, the case studies suggest that at least some of the success of new technologies like the rod caster and the effort to jointly design the auto company FMS must be attributed to the efforts of workers to realize in practice their desires to use new technology in ways that gave them a stake in both the process and the products of manufacturing. In other words, the ardent efforts of those workers to realize their own worldviews about the way things should work may reflect a false consciousness, but those same efforts also represented a source of enormous pride, too.

To draw attention to the limits of labor process theory is, once again, not to deny it completely as a source of insight on the social context within which technologies are chosen and used. There is no reason to quibble with the historical significance of class conflict as a 'driver' of managerial efforts to enhance their control over production. However, it is to argue that class conflict is not the only axis of contention in the design and deployment of new technology. Other lines of cleavage are not inconsequential, and, more important, they are not simple derivations of a primordial struggle between labor and capital. Only in the absence of detailed study of the *process* of technological change is it possible to arrive at the conclusion that class conflict alone is the driver of technological choice.

In sum, the case studies do not call for a complete disregard for the insights to be drawn from the social choice perspective – any more than they warrant the total rejection of technological determinism. The value of a social choice perspective resides in its insistence on the inclusion of social and organizational context as mainstays in the analysis of technological change. Yet as I've tried to show, research based on the social choice perspective has not been attentive *enough* to the multiple, the complex, and the historically situated objectives called into play in the process of change.

 NOTES

1 At the time of the study, the company was working to equalize titles, pay scales, and promotional tracks.
2 The other three 'heads' were transferred to jobs in an adjacent area of the shop. No one was laid off as a result of transfer.
3 In every company I requested and received access to information about market conditions, strategic plans, and forecasts. In some cases this information took the form of planning documents and analyses; in others I was briefed by representatives from corporate or divisional strategic planning offices. I had signed non-disclosure agreements with all four companies and was given no reason to believe that strategic objectives were withheld or obscured for fear that I might leak them. By no means did these companies lack for research and planning. However, what's at issue is the extent to which those studies and plans *actually* informed decision making around new technology; my contention is that their effect was indirect at best.

 REFERENCES

Barley, Stephen (1986) 'Technology as an Occasion for Structuring.' *Administrative Science Quarterly* 31: 78–108.
Bergren, Christian, Torsten Bjorkman and Ernst Hollander (1991) 'Are They Unbeatable?' Working paper, Royal Institute of Technology, Stockholm, Sweden.
Braverman, Harry (1974) *Labor and Monopoly Capital.* New York: Monthly Review Press.

Bright, J. R. (1958) 'Does Automation Raise Skill Requirements?' *Harvard Business Review* 36 (4): 85–98.

Buchanan, David and David Boddy (1983) *Organizations in the Computer Age*. Aldershot, Eng.: Gower.

Burawoy, Michael (1979) *Manufacturing Consent*. Chicago: University of Chicago Press.

Burawoy, Michael (1985) *The Politics of Production: Factory Regimes Under Capitalism and Socialism*. London: Verso.

Child, John (1972) 'Organization Structure, Environment, and Performance: The Role of Strategic Choice.' *Sociology* 6: 1–22.

Child, John (1985) 'Managerial Strategies, New Technology and the Labour Process.' In D. Knights, H. Willmott and D. Collinson, eds, *Job Redesign: Critical Perspectives on the Labour Process*, 107–41. London: Gower.

Clark, Jon, Ian McLoughlin, Howard Rose and Robin King (1988) *The Process of Technological Change*. Cambridge: Cambridge University Press.

Clark, Kim and Takahiro Fujimoto (1991) *Product Development Performance*. Boston: Harvard Business School Press.

Cohen, M., James March and J. Olsen (1972) 'A Garbage Can Model of Organizational Choice.' *Administrative Science Quarterly* 17: 1–25.

Dertouzos, Michael, Richard Lester and Robert Solow (1989) *Made in America*. Cambridge: MIT Press.

Edwards, Richard (1979) *Contested Terrain*. New York: Basic Books.

Gordon, David, Richard Edwards and Michael Reich (1982) *Segmented Work, Divided Workers*. New York: Cambridge University Press.

Grenier, Guillermo (1988) *Inhuman Relations*. Philadelphia: Temple University Press.

Hayes, Robert, Steven Wheelwright and Kim Clark (1988) *Dynamic Manufacturing*. New York: Free Press.

Hyman, Richard (1989) 'Flexible Specialization: Miracle or Myth?' In *New Technology and Industrial Relations*, edited by R. Hyman and W. Streeck, 48–60. London: Basil Blackwell.

Jaikumar, Ramchandran (1986) 'Postindustrial Manufacturing.' *Harvard Business Review* 64 (6): 69–76.

Kelly, John (1987) 'Management's Redesign of Work.' In D. Knights, H. Willmott and D. Collinson, eds, *Job Redesign*, 30–51. Aldershot, Eng.: Gower.

Noble, David (1984) *Forces of Production: A Social History of Industrial Automation*. New York: Knopf.

Parker, Mike (1985) *Inside the Circle: A Union Guide to QWL*. Boston: South End Press.

Parker, Mike and Jane Slaughter (1988) *Choosing Sides: Unions and the Team Concept*. Boston: South End Press.

Piore, Michael and Charles Sabel (1984) *The Second Industrial Divide*. New York: Basic Books.

Rose, M. and B. Jones (1987) 'Managerial Strategy and Trade Union Response in Work Reorganization Schemes at the Establishment Level.' In D. Knights, H. Willmott and D. Collinson, eds, *Job Redesign*, 81–106. Aldershot, Eng.: Gower.

Shaiken, Harley (1985) *Work Transformed*. New York: Holt, Rinehart, Winston.

Thomas, Robert J. (1988) 'Participation and Control: A Shopfloor Perspective on Employee Participation.' In S. Bachrach and R. Magjuka, eds, *Research in the Sociology of Organizations*, 1988. Greenwich, Conn.: JAI.

Yamamoto, Kyoshi (1981) 'Labor-Management Relations at Nissan Motor Co., Ltd.' *Annals of the Institute of Social Science* (University of Tokyo) 21: 24–44.

17 ▶ Writers, texts and writing acts: gendered user images in word processing software

Jeanette Hofmann

The purpose of this chapter is to make word processing programs 'legible' as reified interpretations and definitions of the gendered process of text production in the office world. A useful point of departure for deciphering the conceptions underlying word processing software is to explore what is called the user interface, that part of software which is visible and accessible to the writer. Among other things, the different design strategies at the interface between human being, program, and digital machine – from menus to function keys and interactive graphic objects – reflect the degree of competence and control that the writer is granted or required to have. The user interface, in the words of Akrich, 'the outside of an object', expresses 'a line of demarcation traced, within a geography of delegation, between what is assumed by the technical object and the competencies of other actants' (Akrich 1992: 206). Such technically manifested distributions of competence and their correspondent conceptions of users do not arise randomly; as I want to show, they are distinctly related to practices of text production associated with the technological forerunners of computer-based writing, the typewriter.

By comparing three kinds of word processing systems, a few aspects of such digital conceptions of writing will be discussed: notions about the gender and skills of writers, the type of texts, and the ways in which the texts are produced – all as reflected in the user interfaces of dedicated systems (Displaywriter, WangWriter), early Microcomputer software (WordStar, WordPerfect) and the first graphical interface (Xerox Star computer).

THE USER INTERFACE DESIGN: WHO COMMANDS WHAT AND IN WHICH WAY?

'The design of the user interfaces of computer software is a fairly flat space. It can be done hundreds of thousands of different ways' (interview with Belleville[1]). If there are, as Belleville states, unlimited ways to lay out the user interface, which considerations guide its design? Before I look at concrete user images inscribed in word processing programs, I will make some brief remarks about the relationship between computer and user and the role of user interface therein.

To have a computer perform a task, it is not necessary to know how a computer works; it is only necessary to find out the semantic and syntactic expressions with which to translate one's intentions. The symbolic manipulation of digital machines has become established in the metaphor of 'command'. As in the military, the writer gives commands in the expectation that the computer will carry them out.

The term command suggests a constellation between human and machine in which the control lies with the writer and the computer serves as an instrument that carries out instructions. However, in view of the conditions on which the operator's role is based, the appropriateness of this metaphor appears to be more than doubtful. Before commands can be given to a computer at all, it must first be assured that the machine understands the instructions. To put it bluntly, the operator has no means to sanction the computer to impose her will; on the contrary, it is the operator who must change her procedure if the computer fails to respond (see Hartmann 1992: 215–16).

The obvious limits of the operator's power point to a peculiarity in the relation between human and computer that is at odds with the widespread associations attached to the metaphor of command. In contrast to what one would expect from command-obey relations, based on the giving and receiving of commands, it is the one issuing the orders who must acquire the language and the rationality of the obeyer. In order to use the computer for writing, the act of writing must be presented in a way compatible with the binary logic of the digital machine. That is, it must be broken down into unambiguous operational units and translated into the order and symbols that conform to the structure of the specific program. In other words, it is the particular program's set of instructions and its way of presentation that defines what and how the writers are permitted to command. As Zoeppritz (1988: 112) puts it, 'communication with a machine is always asymmetric'.

The price to be paid for the role as the issuer of commands thus consists in subordinating writing practices to a formal algorithmic order. This submission takes place as the acquisition of a specific manner of expression, a kind of 'language' by means of which the permitted acts of writing are transformed into conventions of giving commands. To understand the meaning of user interface design in program development, it is important to realize that using computers for writing requires learning a kind of foreign language, even though this gradually becomes subconscious and seemingly natural. The design of the user interface can be seen as a translation of the interaction language's lexemes into manual actions to be carried out by the user: menu items to be selected, various buttons on the keyboard to be pressed all of which have to take place in a prescribed syntactical order.

As understood by software developers, the user interface serves to offer the user a kind of conceptual guide to, or 'model' of, the way in which the program operates. As stated by the designers of the Xerox Star computer:

> A user's conceptual model is the set of concepts a person gradually acquires to explain the behavior of a system, whether it be a computer system, a physical system, or a hypothetical system . . . The first task for a system designer is to decide what model is preferable for users of the systems.
>
> (Smith *et al.* 1982b: 248)

The user interface is supposed to present the program's interaction language in a compatible way to what system designers assume to be the average user's experiences and competencies. A striking example of this is given by a user interface designer who described his conceptual considerations regarding the development of a word processing program for a publishing company: 'My model for this was a lady in her late fifties who had been in publishing all her life and still used a Royal typewriter' (Smith and Alexander 1988: 110).

The choice of the 'Royal typewriter lady' as a model had several important consequences for the program design. Among them were the exclusion of other users (the company's editors, for example) and such program features (as 'tools for organizing thought') which were assumed to be useless for Royal typewriter ladies. Instead, 'a fairly simple program for word processing and page layout' (Smith and Alexander 1988: 110) was regarded as being the appropriate solution to the secretaries' typical tasks.

Thus, the user interface of a word processing program mirrors the conceptions which the program developers have about the writers, including conceptions of the conceptions that the writers themselves may have about the program. One deals, as it were, with multiple reflections of imagined realities. Computer programs embody a generalized idea, a 'script' of actions that they digitalize. The user interface can be regarded as the staging of that script, a presentation intended to help operators find their way into the realities that it simulates.

The general frame of reference for the development of user images was found at first in the conventional forms of office organization. The traditional division of labour within writing, which separates the process of composing text from that of typing, became the model for the design of word processing software. Accordingly, digital writing systems of the early 1980s were aimed at a special target group. As made clear by photographs printed in the relevant office and computer journals, this target group consisted of secretaries, that is, mostly women. To be sure, there is no question about the gendered character of user images inscribed in word processing software of that time. But, as I want to show in the following sections, the explicit targeting of women does not say anything about the specific design of the program interface. A glance at the range of then-current word processing programs suggests the conclusion that utterly contrary assumptions about the characteristics of female secretaries must have been underlying the focus on that clientele. This will be demonstrated by contrasting two types of digital writing that reflect entirely different contexts of development: dedicated systems and word processing programs which emerged from text editors. Subsequently, a third type of program will be introduced which paved the way for a new kind of use and users: the first commercial program with a graphical user interface intended to meet the needs of male writers.

TWO ANSWERS TO THE QUESTION OF WHAT WOMEN ARE ABLE TO DO: THE WRITER IN THE PASSENGER SEAT

Dedicated systems, which can be thought of as a kind of cross between a typewriter and computer, came from the area of office automation and were one of the important antecedents of today's word processing software. The machine and the program were a hard-wired unit, which is why they were also called 'stand-alone computers'. The major firms selling this equipment included office-machine manufacturers such as IBM (Displaywriter, 1980; DisplayWrite in a software version, 1984) and Wang (Wang WPS, 1976; WangWriter, 1981).

Common to both WangWriter and IBM's Displaywriter was that they simulated writing as a series of selection procedures. Metaphorically speaking, the characteristic of 'menu-operated' software is that the writers are not required to learn actively to speak the program code, but merely to remember the various terms, which must be chosen from lists.

The very act of turning on the Displaywriter unavoidably made the first menu appear on the screen asking the operator to classify her intentions (see Seybold 1980: 3):

a: Typing tasks
b: Work diskette tasks
c: Program diskette tasks
d: Spelling tasks

The only way to get rid of the menu was to follow the selection procedure. After typing in the desired alphanumeric character, a submenu appeared on the screen calling for further specifications of the intended task. Wang-Writer insisted on a similar step-by-step 'reception': 'Whatever option the operator chooses leads to either prompts or a second menu with further options. In other words, there is not much that an operator can do wrong' (Seybold 1982: 3).

The menu-guided program interface of dedicated systems imposed a rigid sequential order on the operator. As pointed out by Brennan (1990: 398) menus can be described as 'well-trodden paths' which enforce their own hierarchy on the user's goal. Dedicated systems, therefore, had the reputation of being especially suitable for beginners. Because the given range of action was so limited only few chances for secretaries to do anything wrong were left, a feature repeatedly stressed in journal reviews.

Writing and editing, too, were organized as a hierarchical selection of commands supplemented by a number of function keys on the keyboard. In keeping with the 'verb-noun approach' then prevailing in the office industry, the secretary first had to define and select from the menu the operation to be performed (verb) on the text and then select the proper body of text (noun).

Actions to be required or permitted were communicated from the system to the typists by means of questions and instructions:

> To delete text, the operator places the cursor under the first character to be deleted and presses the *delete* key. The system prompts: 'Delete what?' at the bottom of the screen, and highlights the character above the cursor line. The user can respond either by typing *enter* to delete the highlighted character, or he can define more text to be deleted by moving the cursor through the ensuing text, or by typing any character . . . Once he has identified the amount of material to be deleted, the operator presses *enter* to execute the function.
>
> (Seybold 1980: 5)

The verb-noun convention (also called 'prefix') truncates text editing into at least three steps, not counting interim questions: (1) select action, (2) select object, (3) execute. It is thus necessarily one step longer than the shortest path of the reverse procedure. (Today's dominant 'noun-verb' sequence first has the operator identify the desired unit of text so that the subsequent specification of the editing instruction can coincide with its execution.) The tedious verb-noun procedure thus reflects a certain distrust of the writer's instructions by taking an additional step to assure their correctness.

A conspicuous feature of the user interface design of dedicated systems was the endeavor to keep the writer under ongoing supervision and to provide her with comprehensive lectures about the behaviour compatible with the system's current status. No writing act went uncommented. All actions were guided to preclude errors as far as possible. Accordingly, these selection and command procedures could not be circumvented in many text systems, including IBM's Displaywriter. No matter how experienced and competent secretaries became in working with a dedicated system, they always had to follow the same 'traditional question-and-answer computer-assisted instructions' (Shneiderman 1983: 63).

> This style of guided dialogue, while useful for novices, is often frustrating and annoyingly time consuming for experts. Furthermore, it enforces sequential specification of multiple operands, without providing the ability to edit them. Worst, of course, is that the user is locked into

the dialogue, and cannot leave to browse or to collect information with which to complete the command.

(Meyrowitz and van Dam 1982: 403)

By denying even the experienced user a way to bypass such solicitude, the user interfaces of dedicated systems tended to treat writers as perennial beginners. As stated by the head system designer of a Wang dedicated system, 'We wanted to make something for the really dummy user who doesn't have any idea of the technology' (interview with Moran). One of the most striking characteristics of the typist as inscribed in the user inter-face of dedicated systems was apparently that she not only is a 'dummy', but also remains one.

The WangWriter and several other dedicated systems did not only restrict the writer's scope of action but also the access to their product. Secretaries could prepare and edit texts but were not allowed to copy, rename, or even delete them. Such file-related operations could only be executed outside the text editor – at the system level, to which access could be blocked. In the digital world, producing text can be separated from physical disposition over the outcome (for other means of supervision and control by dedicated systems, see Barker and Downing 1980: 92).

It is not only from today's perspective that the user interfaces of dedicated systems seem more (the IBM version) or less (the Wang version) inelegant and unwieldy. The multiple, superimposed menus made typing slow and cumbersome. The program's selection procedures structured not only the sequence but also the pace of text production. For their low computational capacity, dedicated systems bogged writers down with frequent pauses during which they had to wait for additional menus, questions, and instructions. Commenting on Displaywriter, Seybold (1980: 3) wrote, 'Apart from the minor tedium of wading through menus . . . experienced users will find the system less than optimally responsive. It seems to take at least four seconds for the system to retrieve a page of information from diskette, and it can be a lot longer'.

However, the sluggishness of dedicated systems cannot be properly ap-preciated without taking into account that they actually aimed to increase writing productivity. The introduction of dedicated systems in the late 1960s and the following decade was motivated by the hope of thoroughly streamlining clerical work (see Hofmann 1994). The obvious divergence between stating and achieving this general purpose suggests the dilemma that burdened the design of user interfaces and, furthermore, indicates to whose benefit it was resolved. Faced with the choice between optimizing the efficiency of text production and protecting the dedicated system and texts as much as possible from potential blunders by the supposedly dummy operators, the system developers took the latter option.

It is these goal conflicts and compromises in software design that expose how female users of word processing systems are perceived. The imagined addressees of the dedicated system were regarded as eternal beginners whose technical competence was so modest that costs incurred by unproductive text production were estimated to be lower than those that would be incurred by mistakes in operating the devices. While it is certainly not astounding that such conceptions of women's technical aptitude are accepted as a representative model for technology development, the simultaneous existence of word processing programs which epitomized contrary images of female users may indeed be regarded as a surprise.

▶ ## THE WRITER IN THE DRIVER'S SEAT

An entirely different conception of women and secretaries was expressed by a second type of computer-assisted writing. WordStar and WordPerfect belonged to the first commercially successful word processing programs developed for microcomputers. Unlike such systems as WangWriter and Displaywriter, these programs did not come from the field of office machines but harked back instead to program editors for mainframe computers and, hence, to a quite different clientele – programmers. MicroPro, the manufacturer of WordStar (founded in 1978) and WordPerfect (founded in 1979) were among the companies that formed leeward of the hardware-oriented computer industry in order to profit from the growing demand for microcomputer software.

The first version of WordStar for the CP/M operating system appeared on the market in 1979; the first version of WordPerfect (at that time still for a shared-logic system under the name of 'SSIWP') one year later. It is hardly known that WordStar and WordPerfect were originally developed for a target group very similar to that for dedicated systems. The first version of WordPerfect was a kind of commissioned production for the city administration of Orem, Utah. Like many writing programs at that time, Word-Perfect began as a program editor ('P-edit'), which was gradually expanded to include more and more functions (and, characteristically, was deprived of an elementary one, namely, its programmability). WordStar was addressed expressly to touch typists. One of the stated design objectives, which the two programs shared with dedicated systems, was 'high typing productivity'.

Interestingly enough, however, this goal was met by WordStar and WordPerfect in very different ways from those of dedicated systems. With WordStar, writing on the computer was organized modally by means of the letters on the keyboard. Modes means that the act of writing fell at least into two different program statuses: actual typing of text, and editing. In combination with the control key, the user of WordStar moved into the control or command mode, which changed the meaning of the keys on the keyboard. As the developers of WordStar saw it, the typical forms of text production suggested this kind of user interface. Writing via letters and control keys would enable touch typists to keep their hands on the keyboard, thereby interrupting the rhythm of writing as little as possible. That is why WordStar developers at first avoided exploiting function keys (of which WordPerfect made full use): 'I can't touch [the function keys] without looking down. I can type a combination of keys that are right near my fingers more easily than I can type a single key that you have to look at with your eyes to find' (interview with Barneby).

Used in combination with the control key (in the command mode), certain letters on the keyboard served to activate menus; others symbolized machine operations. In contrast to the practice followed by such systems as IBM's Displaywriter, however, all menus could be circumvented. Indeed, experienced users were virtually invited to do just that.

Coded as combinations of three keys, WordStar offered an impressively broad repertoire of actions and expressions for producing text that was never again achieved by one of the subsequent commercial word processing programs. For example, the typist had many options to choose from as far as marking text passages was concerned. As well as defining text elements by putting markers at the beginning and the end, WordStar included special expressions for tagging paragraphs, sentences, parts of sentences (to the left or to the right of the cursor), and letters. The disadvantage of such manifold possibilities of expression lay in the relatively high expectations of the secretary's willingness to learn: 'WordStar had a lot of two-key sequences. There were about four times as many commands as there were keys on the keyboards . . . I expected people to learn the keys . . . The ones you use most commonly you would learn and then the menu wouldn't slow you down' (interview with Barneby).

Compared to menu-structured programs, text production by means of letter and control keys is more direct and faster. The price for these advantages, however, was that the writers had to translate what they did into many different combinations of characters that had to be memorized and actively used. These learning processes were aggravated by the obviously erratic relation between the act of writing and the symbols used to codify it. WordStar's semantic arrangement of letters did not make use of acronyms such as 'D' for delete or 'M' for move; instead the interactions language was designed as a precaution taken against the errors that writers might commit. As a 'foolproof convention' the user interface anticipated weaknesses of the users:

> Considerations were: the things you used commonly and weren't dangerous should be easy to type. The things that did deletions, for instance, that would be harmful for your document if you did them by accident were supposed to be farther from where your fingers are resting. 'Control y' and 'control t': 'control y' deletes a line, and 'control t' deletes a word. Both of them are away from you and you have to move your finger over them . . . So, deleting a line was a fairly dangerous command. I didn't want the delete key to be where you would hit it by accident if you let your hand go down while you were thinking or something.
>
> (Interview with Barneby)

As became apparent from the considerations of the program designer quoted above, the imagined user of WordStar also had shortcomings against which precautions had to be built in. 'Idiot-proof' is the term used to designate the corresponding antidotes in user interface design. As Hansen (1971: 527) describes the problem:

> the designer must remember that human users share two common traits: they forget and they make mistakes. With any interactive system problems will arise – whether the user is a high school girl entering orders or a company president asking for a sales break-down. . . . Good system design must consider such foibles and try to limit their consequences.

Among such precautions are well-known, excessive question-and-answer dialogues ('Are you sure you want to delete this file?'); the undo key, which enables the operator to retract a varying number of actions and the auto-matically stored backup copy of a document (a sort of simulated carbon copy of a typed page).

A comparison between word processing programs shows that the type and implementation of such measures against the follies of users gives clues about the characteristics attributed to them by developers. For example, the weaknesses which WordStar tried to compensate differ recognizably from those against which Displaywriter and WangWriter offered protection. Tellingly enough, the preventive measures built into WordStar's keyboard arrangement are directed against accidents that could happen to any sort of user as a result of, say, rapid typing.

The design of WordPerfect's user interface reflected a perception of the target group's needs which was very similar to that of WordStar. The primary goal was to facilitate a rapid and direct simulation of the writing act. Mnemonic aids such as on-screen menus were therefore dispensed with.

The comparatively high independence and competence that Word-Perfect attributed to users was symbolized by the almost completely empty screen, which resembled a blank page. Six characters on the screen merely informed writers of the cursor's current location. WordPerfect's user interface clearly stood in the tradition of text editors for time-sharing systems, for which the slowness of character representation was a major bottleneck and cost factor.[2] Time economy and aesthetic criteria formed an integrated whole, declaring the empty screen, with its sparing use of the computer's capacity, the ideal base for producing text. It was 'perfectly simple'.[3]

In WordPerfect, editing and formatting operations were carried out by means of 12 function keys, each of which had four different meanings depending on how they were used in combination with the shift, alt, or control keys. The memory aid that WordPerfect provided to help users recall the function keys' semantics clearly aimed at avoiding taking up limited computer capacity. Instead of menus, dialogues or other digital means, WordPerfect provided a simple cardboard template:

> Having used the computer for several years, we knew what we wanted to do, and we wanted to be able to do it quickly rather than having

people memorize dot commands. We gave them a push-button approach where we put a little template above the function keys. Then we said 'All you have to do is type. It is just like a piece of paper and you just type and . . . if you want to bold, you hit the bold key and type and then you hit the bold key to turn it off.'

(Interview with Ashton)

In the eyes of the developers, using function keys to communicate instructions to the computer had the advantage of sparing writers the constant switching between different program modes. With WordPerfect, only the function keys changed their meaning. Nevertheless, secretaries working with WordPerfect found themselves confronted by the same expectation as with WordStar: fluid work on the computer required active mastery of the program's specific language. Both WordPerfect and WordStar thus had the same reputation of being programs that were difficult to learn:

We had specifically designed the product for productivity. We didn't have a lot of menus leading you through. And we were criticized for being very hard to learn because we didn't take you by the hand and say, answer this menu, answer this menu, answer this menu, answer this menu and finally you could then begin to type.

(Interview with Ashton)

Basically, the development of WordStar and WordPerfect had to struggle with the same dilemma as dedicated systems. At a time when the computing and storage capacity of microcomputers was very small, every character that was shown on the screen competed with the goal of organizing the writing process in real time. The more information that appeared on the screen, the slower the computer's reactions were, the more its meagre working storage space was taken up, and the less room remained for the text to be produced. Interestingly, WordStar and WordPerfect represent different solutions to this dilemma than dedicated systems. Here, the conflict between speed and 'solicitous' attention to the user was decided in favour of the former while ease of learning and use was treated as a subordinate goal.

The reason why the developers of WordStar and WordPerfect regarded the amount of effort it took to learn their programs as acceptable although it was generally deplored relates to the conception they had of their target group. It was assumed that professional typists would first go through a training course and then use the writing program regularly and often: 'It was meant for somebody who used it often enough so you weren't thinking about the words [commands] to type them' (interview with Barneby). Because typing was one of the main activities of secretaries, the developers figured that secretaries could be expected to learn even difficult and memory-intensive program expressions quickly: 'After a month, you learn whatever it is, and it doesn't matter an awful lot' (interview with Barneby). 'Muscle memory' is the term used for this kind of learning process (Hansen 1971: 530).

As perceived by the authors of WordStar and WordPerfect, typists were not technically hopeless beings from which computers and texts had to be protected. Instead they were seen as professionals believed capable of mastering any program syntax precisely because typing was their profession and main task: 'Once they knew [the program syntax], they became very productive and could produce letters and documents very quickly' (interview with Ashton).

While the designers of the Displaywriter's insuppressible menus conceptualized secretaries as neophytes, WordStar and WordPerfect made it a prerequisite that secretaries go through a learning phase at the end of which they would have memorized the program's cryptic codification and would be able to work competently and smoothly. The same discrepancy surfaced with regard to the setting up of the various writing systems. The very lack of standardization among CP/M computers made it necessary for would-be users of WordStar first to adjust all the relevant parameters to their specific computer configuration. The users were thereby forced to acquire the necessary knowledge about the internal program structure. Whereas the WordStar user had broad freedom to control the program (much greater freedom, incidentally, than that granted by current word processing programs that can be 'customized'), a dedicated system usually came with a training and service contract.

A comparison between the user interfaces of different word processing programs shows that technical artifacts can attribute different, even opposite, characteristics to the same target group. As a consequence of such heterogeneous user images, secretaries were therefore confronted with quite divergent expectations of their qualifications. The secretary who used WordStar or even one of the precedent text editors was expected to acquire a much wider range of competence than a secretary who worked on stand-alone computers. As van Dam (in interview) recounts:

> In the old days, we trained our secretaries to use these command line interfaces . . . For people who type, it's much faster than having to pull down menus. Once you memorize the key combinations – people on 'VI' or 'Emacs' are good at that – it can go much faster than experienced [operators working on] word processors.

Conversely, the comparison between these types of program shows that different user interfaces can indeed be accounted for taking the same target group into consideration. The traits and potential that secretaries have in the eyes of IBM or Wang are quite different from those seen by the developers of WordPerfect and WordStar. Such images of the user become real and effective to the extent that they actually succeed not only in structuring the development of technology but also the operator's daily actions. This notion recalls the thesis of Madeleine Akrich (1992: 22), according to which, 'technical objects and people are brought into being in a process of reciprocal definition in which objects are defined by subjects and subjects are defined by objects'. Word processing software may be a good example of such reciprocal definition, provided one bears in mind that the power to define the other is distributed unevenly between the parties involved, namely developers, artifacts, and users. According to

Schachtner (1993: 114), software developers create models of order by means of which reality is formed in order to let these formal models become reality itself. User images can be regarded as part of such ordering models. Their ascriptions participate in the process of defining gender.

Yet if technical objects and acting subjects define each other, then the category of gender cannot be taken as an external, independent variable that would permit self-evident conclusions about the quality of technology. As becomes clearer in the following section, simply categorizing technologies as distinctly 'male' or 'female' runs the risk of duplicating existing social gender constructs instead of exposing the conditions giving rise to them. Another way to look at the relationship between word processing software and men is to explore what the characteristics and competencies of men look like in terms of software developed in the late 1970s. This question is cleared up with a glance at the only word processing program I know of ever to have been developed for men.

► THE RISE OF THE DILETTANTE AS THE LEADING FIGURE IN SOFTWARE DEVELOPMENT

The Xerox Star computer, brought onto the market by Xerox in 1981, enjoys the dubious claim to fame of having laid the commercially unsuccessful cornerstone for what went on to become a model of commercial and conceptual success to this day: the famous Apple-Windows graphical user interface. Xerox Star was the first marketable computer to organize text production by means of pictorial symbols, or 'icons', a kind of pictographic language.

The 'vocabulary' of Xerox Star was a combination of graphic objects, a few menus, and more than 20 function keys. The range of expression and the actions that could be simulated were intentionally confined to a narrow spectrum of so-called 'generic' or 'universal' commands: move, copy, delete, show properties, copy properties, again, undo, and help. These functions were available across all hard-wired applications and programs. Conspicuously absent among those functions, however, were operations like 'mail merge' and 'search and replace', which were regarded as part of the standard repertoire of word processors (see Seybold 1981). This lack was no accident. Xerox Star's developers considered such features, which were seen to be essential for secretarial services, to be only peripheral for the intended target group. As David Smith, one of the designers of Xerox Star's user interface, explains: 'The first version of Star had fewer text editing features than many dedicated word processors at the time . . . We couldn't build all known functions all at once, nor did we want to. We decided to concentrate on the core functionality in each area' (1994, quoted from e-mail communication).

Apparently, the secretaries' digital tools did not count as what was seen to be 'core functionality'. Xerox Star was the first commercial computer to make use of the input device called the 'mouse', which had been developed more than ten years before by Doug Engelbart. Clicking with the mouse on graphic objects is somewhat euphemistically known as 'direct manipulation' (Shneiderman 1983) because it conveys more convincingly than other procedures the illusion of direct user control of the text. The mouse-operated user interface permits the selection and editing of text passages with a few precise hand movements instead of through syntacticosemantic translation. The degree of instruction required in order to specify the operator's intended action, as well as switching back and forth between writing and editing modes, could thereby be lessened significantly. In the words of the designers:

> Commands in Star take the form of noun-verb pairs. You specify the object of interest (the noun) and then invoke a command to manipulate it (the verb). Specifying an object is called 'making a selection' . . . Place the cursor over the object on the screen you want to select and click the first (SELECT) mouse button . . . Once selected, documents may be moved, copied, or deleted by pushing the MOVE, COPY, or DELETE key on the keyboard.
>
> (Smith *et al.* 1982a 5, 11)

As with stand-alone computers, the operator wrote or delegated by means of selection. All computer operations could be activated by a click of the mouse or single command buttons. An important dissimilarity between the Xerox Star computer and the dedicated-system model, however, was the avoidance of hierarchical, linearly linked sequences of commands. It differed from WordPerfect in that each function key was assigned one operation rather than several. Almost all available operations were visible either as icons on the screen or as a label on a key – an arrangement that kept the demands on the user's memory deliberately low:

> We decided to create electronic counterparts to the physical objects in an office: paper, folders, file cabinets, mail boxes, and so on – an electronic metaphor for the office. We hoped this would make the electronic 'world' seem more familiar, less alien, and require less training.
>
> (Smith *et al.* 1982b: 252)

Introducing graphic objects to imitate the office environment can be understood as the software designers' effort to oblige what they presumed to be the weak memories of the people to whom they were addressing their product. The pictographic interface was also supposed to reduce a sense of alienation from the digital world: '[the design] was to be simple and easy to use even for the type of person who feels threatened by anything which is in the least "computerish"' (Seybold 1981: 8). What sensitivity the designers of Xerox Star showed to their target group's frame of mind!

Yet such accommodation in interface design carried a price, both literally and figuratively. In the early 1980s the simulation of text production as a pressing of keys and a clicking on images entailed quite a number of problems. The first one was the sales price. The 'bit-mapped screen' required by a graphical interface was comparatively expensive to manufacture (so expensive, incidentally, that Microsoft decided to design its operating system, 'DOS', and its writing program, 'Word', only for character-oriented screen display). Xerox Star's costs thus lay above the usual range for dedicated systems. A second problem was posed by the computer's low processing capacity. What was true of display menus was even truer of graphic objects, for the fancy user interface required the users' considerable patience. A third problem with the digital simulation of text production still exists: the lack of complexity of the interaction language. Compared to programs like WordStar, Xerox Star's word processing program provided a very restricted repertoire of expressions.

As summed up by Brennan (1990: 399), '[direct manipulation] is fine for very small worlds, if what you want to do can be represented by dragging around icons'. Clicking with the mouse on graphic symbols is a form of expression that, as Erickson (1990: 12–13) states, is essentially comparable to 'pidgin language', by means of which people from different language cultures communicate with each other:

> First and foremost, pidgins are easy to learn. Ease of learning is the raison d'être of a pidgin; its users are typically those who lack the time or inclination to learn a language – they simply want to get on with business. The other characteristics shared by pidgins are clearly related to their ease of learning. Pidgins employ only simple sentences, with very regular, if awkward, syntax . . . Pidgins have a very limited vocabulary, which may be buttressed by the use of pointing and other gestures. Finally, pidgins typically allow their speakers to deal with a very narrow segment of the culture.

Whereas fumbling around, gesticulating, and pointing within the context set by a supply of shared, rudimentary meanings may suffice for the most necessary messages, it is not possible on this basis to communicate nuances, detailed instructions, and discerning insights into other people's worlds. The restrictions of pidgin languages pertain as much to transmitting information as to understanding it. Graphical user interfaces share all these obstacles of pidgin languages, as Erickson (1990: 13) concludes in reference to the Macintosh computer, a descendent of Xerox Star: 'You can get your basic tasks done, but that's about it'.

The system developers regarded all such weaknesses as defensible – again because of the characteristics of the target group for whom Xerox Star had been designed: managers and knowledge workers who in future would neither have secretaries nor the time or desire to study the idiosyncrasies of complicated operating systems and word processing programs. Knowledge workers were said to constitute an especially difficult target group because of the broad variety of their activities. For these addressees, it was presumed that writing would always be only one of many tasks – and a secondary one at that. The Xerox Star designers thus expected that anything the knowledge workers learned about using the computer would be quickly forgotten because they would not be using the technology regularly:

Star's designers assumed that the target users were interested in getting their work done and not at all interested in computers. Another important assumption was that Star's users would be casual, occasional users rather than people who spent most of their time at the machine. This assumption led to the goal of having Star be easy to learn and remember.

(Johnson 1989: 11)

In other words, the intention and challenge of Xerox Star's development was to come up with a user interface for the abiding and self-assured dilettante, that is, for the user who could afford to have no interest in computers. All features and functions were therefore supposed to be understandable, manageable, and, hence, even simpler and less demanding on the user's technical skills than the features and functions of word processors. The user was to be able to master the system without a training course and breaking-in period. Accordingly, the design criteria of the user interface called for an interaction language that would be comprehensible to laypersons and as self-explanatory as possible. It meant enabling the user to point to and pull a manageable number of objects that represented familiar objects of the office world in the least technical way: 'To file a document, you move it to a picture of a file drawer, just as you take a piece of paper to a physical filing cabinet' (Smith *et al.* 1982a: 3). The expectations of the 'office professional's' ability and willingness to learn were so low that falling below the level of variety and speed customary for text production at that time was consciously accepted. To avoid asking too much of these predominantly male clients, the designers neither assumed them to have nor seriously called upon them to show notable competence in the field of text production. It was this new target group of part-time writers that made 'usability' both a criterion of quality and a subject of scientific research in the sphere of program development.

Hence, the primary objective of the Xerox Star project was not to enhance the writer's productivity but to establish a fundamentally new conception of computer-assisted writing. This innovative intention, in turn, was directly related to the new kind of target group. The ideal-type product inscribed in Xerox Star's design has as little to do with a conventional typewriter text as Xerox Star's imaginary text producer has to do with a secretary.

This aspect brings up a second, as yet unmentioned, dimension of modelling reality expressed by word processing programs. Not only do program interfaces reflect certain ideas about users, but these images of users prove to be part of more comprehensive conceptions related to the production of texts. Xerox Star's design highlights the point that dedicated systems and word processing programs such as WordPerfect, for all the differences in the way secretaries are perceived in them, start from similar notions of the type of texts to be written and the way they are to be produced. First of all, and along with traditionally discrete mechanical solutions, typewriting had been conceptualized as something apart from other office activities. By virtue of their very existence, word processing programs treat writing as a separate procedure and thereby perpetuate the traditional vertical and horizontal division of labor in the sphere of administration. Following the mental heritage of the typewriter, conventional word processing programs basically simulated typing as a copying of drafts or manuscripts that had been composed already – usually by somebody else.

The general model for word processing systems was not the creative process of composing texts but the typing, editing, and formatting of prepared manuscripts. This difference also becomes apparent from design objectives of the user interface discussed above. High text production speed, the goal of developers of dedicated systems and word processing programs alike, can be achieved primarily when the time-intensive element in producing a text – the mental work – has already been completed.

By contrast, Xerox Star's program interface represented office work in a then unknown way. One of the early protagonists in the development of text editors has described the diverging perspectives on the future users and usages of office computers as follows:

> The big difference is that our vision in the mid until late 1960s was to get the office on-line. That was not the vision of the people who produced word processors. People who produced word processors just wrote their programs for secretaries, who would transcribe marked up manuscripts that their employers would give them. We said, 'all good and well, but that's not using the medium'.
>
> (Interview with van Dam)

Software intended for knowledge workers was supposed to represent office work more holistically and thereby increase the managers' creativity. Such a scenario had as little room for special word processing programs as for a division of labor between conceptualizing, thinking, and typing. On the contrary, the whole point was to improve the intellectual quality of work by 'using the medium' for all creative tasks.

Influenced by Doug Engelbart's life project, the development of Xerox Star was one of the first commercial attempts to develop an office computer tailored more to creative than routine activities. Engelbart's vision and professional commitment centred on a computer and text processing system named 'Augment', which was 'designed for augmenting human intellectual capabilities. It was targeted particularly toward the core work of professionals engaged in "tough knowledge work" – e.g. planning, analyzing, and designing in complex problem domains' (Engelbart 1984: 1). Engelbart imagined the computer as a constant 'working companion' that would support all possible forms of intellectual production and communication (1973: 221). As he emphasizes: 'Our lab was concerned with augmentation, not automation' (1988: 246). Engelbart's project paved the way for a new understanding of the kind of texts to be produced as well as for the means needed to produce them:

> The idea was that professionals in a business or organization would have workstations on their desks and would use them to produce, retrieve, distribute, and organize documentation, presentations, memos,

and reports . . . The applications included in the system were those that office professionals would supposedly need: documents, business graphics, tables, personal databases, and electronic mail. The set was fixed, always loaded, and automatically associated with data files, eliminating the need to obtain, install, and start the right application for a given task or data file.

(Johnson 1989: 11)

It is no coincidence that Xerox Star's word processing program was not given a name, and neither were its other available applications. The intention, symbolized by the metaphor of the 'desktop', was to provide users with the whole spectrum of actions to be executed by the computer simultaneously.

The user interface was to revolve around the blank page, from which the writer had direct access to all available operations of the digital machine: 'In a desktop metaphor system, users deal mainly with data files, oblivious to the existence of programs. They do not invoke a "text editor", they "open a document"' (Johnson 1989: 13). Such a concept meant deliberately abandoning the model of the classical typewritten page, which generally contained nothing but text consisting of many letters and few numerals. Instead, documents to be prepared by knowledge workers were to consist of any desired combination of texts, tables, and graphics. In this sense, 'using the medium' was meant to exploit the computer as a universal tool rather than modelling it after traditional 'one purpose machines'.

In many respects Xerox Star pioneered today's application software, exploring, for example, program 'suites' and graphic user interfaces. Fundamental design elements such as the metaphorical window, scroll bars in the window frame, the cursor's function as transmitter of information about the program's status (e.g. the hourglass), as well as many icons, originated with Xerox Star, which, in turn, was inspirited by Engelbart's project 'Augment'.

Surprisingly then, it turns out that the graphic design of program interfaces which has meanwhile become dominant in word processing harks back to the orientation to male knowledge workers as a target group. The characteristics, needs, and skills attributed to men have become reinterpreted, taking gender into consideration, and generalized into universal standards. As Webster (1993: 120) put it, 'Word processing, the activity and the technology, is shifting in its domain, and the boundaries of the old gendered division of labour in the office are being redrawn'. The popular notion of icon-based, user-friendly interface design correlates with the image of a universal user who is no longer either female or male.

Characteristically enough, the 'ungendering' of the user image corresponds to a re-evaluation of the relation between composing and typing of text. The recently observable ambition of user interface designers to develop intuitively comprehensible program interfaces can be interpreted as a sign that the typing of text is gradually losing recognition as a separate occupation which requires prior training. Originally thought of as digitalized instruments for women to type other people's texts, word processing programs are being more and more reconceptualized as devices that aid in the thinking and designing process. Yet the more the 'holistic' creation of texts emerges as the general purpose of word processing software, the more the typing part of the activity appears to lose importance, even seeming to take place by itself. If, as Shneiderman writes, the focus of user interface design becomes the task to be done and the need of the user, 'then the interface will work, not because it doesn't exist, but because it blends so smoothly with the task to be done that it disappears from consciousness' (Norman 1990: 218–19). With the attention on knowledge workers as occasional users, digital writing as it is generally understood has evidently been turning into a natural process akin to, 'say, moving furniture around in a house' (Kay 1990: 201).

The meaning of gender and the meaning of technical artifacts are phenomena so intertwined that any change of one element necessarily affects the others. This involves not only the re-evaluation of skills, tasks, and qualifications, and thereby gendered divisions of labour, but also the very distinction between women and men. After roughly 20 years of word processing systems, who would dare to tell the difference between male and female users?

PRODUCTIVE ARTIFACTS, CHANGEABLE GENDER CONCEPTS: INSIGHTS ABOUT AN UNSTABLE RELATIONSHIP

This chapter arose from two intentions. First, I wanted to show that decisions concerning software design are related to specific, gendered ideas about use and users of the program in question. Second and more importantly, I wanted to demonstrate that this relationship is by no means static. The comparison of different word processing systems served as an example of what obviously cannot be mapped: the scope of (re-)interpretation and change inherent to the relationship between gender and technology.

At first glance, the question of how users of word processing programs are perceived by software designers seemed to bring forth rather trivial results. Until the early 1980s, virtually all word processing programs addressed specific user groups, most of them female, excepting one slightly exotic and consequently commercially unsuccessful program that targeted men.

At second glance, however, gender categories only played an indirect role in the concerns of software developers. The principal aspects of user interface design were the users' occupations and tasks, which were taken to give information pertaining to average technical competencies and needs. Thus, women were perceived as typists or secretaries while men were addressed as knowledge workers. And what being a secretary or knowledge worker meant revealed itself as imputed (writing) skills, tasks, interests, or time frames. In other words, gender did not appear as an independent category but as a consequence of specific design considerations.

An examination of the interaction language of word processing software illuminated some of the questions by means of which user interface designers approached and 'configured' (Woolgar 1991) their clientele: How many expressions or commands are users able to memorize? How much are they willing to learn? Which precautions are necessary to protect the system against the users' foolishness? What kind of texts are they going to produce and how often? As has been illustrated, the users' assumed characteristics became part of the program's constitution. The lexemes and syntax of the program's interaction language – the arrangement of buttons, letters, words, and pictures – came to symbolize different degrees of technical competence. Letters and words, originally typed or selected by female users were regarded as being too difficult for men who were offered a picture-based command language to be 'spoken' by clicking upon something called a mouse. The alphabet for women, pictures for men – this strange taxonomy was very short-lived.

The answers given to the questions about the prospective users manifest themselves as sets of rules and symbols intended to prescribe the users' subsequent actions. In this sense, technical artifacts are productive. They actively contribute to forming social behaviour, competencies, and hierarchical distinctions among them in a prestructuring and ordering manner. Pictographic interaction languages which have come to be celebrated as user-friendly mean *de facto* a loss of the user's control over digital machines. Although, as Akrich (1992) and Latour (1992) point out, the artifacts' power to constitute actors and actions shouldn't be misunderstood as unilateral. Technical 'presumptions' regarding users and uses may indeed be answered by subsequent reinterpretations.

The specific scenarios within which user images of software were developed relate to what is considered to be 'normal' office life. The concepts of writing underlying all early word processing systems clearly reflected procedures of text production which where established with the rise of the mechanical typewriter. Among other aspects, this concerns the gendered division of labour between composing and typing texts and, consequently, the reduction of typing to copying manuscripts. The example of word processing programs thus indicates that software development – which, at least in principle, is free of physical constraints inherent to its mechanical predecessors – shows a clear tendency to reproduce traditional sociotechnical arrangements. As Webster (1993: 114) notes, word processing 'brought electronic technologies to the typewriting task, rather than taking text production technologies to the computing activity'.

However, as the comparison of different approaches to digital word processing shows, each program offered different conclusions as to how text production in an office environment should be digitally simulated. This is especially striking for the distribution of competencies and control between women, men and digital machines. Whereas dedicated word processing systems treated female users as eternal beginners, programs like WordStar and WordPerfect epitomized the opposite view according to which secretaries could be expected to learn even complicated interaction languages. Amazingly enough, the least competence regarding all kinds of application software was attributed to men.

Neither of these gendered ascriptions was compelling nor completely implausible. Furthermore, neither proved stable in the long run. Although they left traces of their history, gendered approaches to user interface design disappeared in favor of universal programs for everybody. And even though technology development bears a tendency to reproduce and maintain long-established social gendered arrangements and institutions, this endeavour never really succeeds.

There are two reasons for this. For one thing, technology development refers to actions to be automated by interpreting and thereby gradually redefining them. Word processing software transformed the mechanical heritage of text production. Among other things, the typing of manuscripts as a special occupation is going to disappear. For another, the meaning of every artifact draws on the manner in which it is used. The history of digital word processing software provides several examples of how users diverge from the course of action assumed by system designers. The most recent case for such user-based reinterpretations is provided by new applications of the Internet. The World Wide Web consists of multimedia documents based on hypertext which will pose a challenge to the linear, print-based notion of text which forms a constituent part of the design of current word processing software.

From the perspective of word processing systems, the relationship between gender and technology appears as a context-dependent matter consisting of mutual ascriptions to prospective users and programs. The meaning of gender – portrayed as imagined skills, competencies, and tasks – varies depending on concrete practical circumstances of software development and use. From a research point of view, the discovery of divergent gender images inscribed into software design supports Knapp's (1994: 270) observation of 'the enormous flexibility and, simultaneously, the firmness of apparent naturalness and inevitability of gender constructions which support gendered division of labor and gender hierarchies'. It is this hidden flexibility and, particularly, the conditions of taking advantage of it which needs to be explored in more detail. 'How questions', which illuminate the relationship between gender and technology from a procedural rather than from a structural point of view, may serve as a suitable tool for this project.

▶ **NOTES**

1 The interview with Belleville and other interviews cited in this article were conducted by the author in 1993.
2 'On the early teletypes, you wanted to have as little feedback as possible when you typed in because it took so long to print the characters out . . . All the editing was done on paper . . . We really looked at having a clean screen for very productive type work' (interview with Ashton).
3 In the meantime, the opposite ideal has come to predominate user interface design. As Donald A. Norman (1988: 178) puts it: 'Ever sit down to a typical computer? If so, you have encountered "the tyranny of the blank screen". The person sits in front of the computer screen, ready to begin. Begin what? How? The

screen is either completely blank or contains noninformative symbols or words that give no hint of what is expected . . . It is almost as frightening as being taken to a party filled with strange people, being led to the center of the room and let go.'

▶ **REFERENCES**

Akrich, M. (1992) The de-scription of technical objects, in W. E. Bijker and J. Law (eds) *Shaping Technology/Building Society: Studies in Sociotechnical Change*, pp. 205–24. Cambridge: The MIT Press.

Barker, J. and Downing, H. (1980) Word processing and the transformation of the patriarchal relations of control in the office. *Capital & Class*, 10: 64–99.

Brennan, S. E. (1990) Conversation as direct manipulation: an iconoclastic view, in B. Laurel (ed.) *The Art of Human-Computer Interface Design*, pp. 393–404. Reading: Addison-Wesley.

Engelbart, D. C. (1973) Design considerations for knowledge workshop terminals. *AFIPS Conference Proceedings*, 42: 221–7.

Engelbart, D. C. (1984) Authorship provisions in Augment. Reprint of *COMPCON '84 Digest, Proceedings of the 1984 COMPCON Conference*, San Francisco: 1–19.

Engelbart, D. C. (1988) Working together. *Byte*, December: 245–52.

Erickson, T. D. (1990) Working with interface metaphors, in B. Laurel (ed.) *The Art of Human-Computer Interface Design*, pp. 1–16. Reading, MA: Addison-Wesley.

Hansen, W. J. (1971) User engineering principles for interactive systems. *AFIPS, Fall Joint Computer Conference*: 523–32.

Hartmann, C. (1992) *Technische Interaktionskontexte. Aspekte einer sozialwissenschaftlichen Theorie der MenschComputer-Interaktion*. Wiesbaden: DUV.

Hofmann, J. (1994) Two versions of the same: the text editor and the automatic letter writer and contrasting conceptions of digital writing, in A. Adam, J. Emms, E. Green and J. Owen (eds) *Women, Work and Computerization*, pp. 129–42. Amsterdam: Elsevier.

Johnson, J. (1989) The Xerox Star: a retrospective. *IEEE Computer*, 22 (9): 11–28.

Kay, A, (1990) User interface: a personal view, in B. Laurel (ed.) *The Art of Human-Computer Interface Design*, pp. 191–207. Reading: Addison-Wesley.

Knapp, G-A. (1994) Politik der Unterscheidung, in Institut für Sozialforschung Frankfurt (ed.) *Geschlechterverhältnisse und Politik*, pp. 262–87. Frankfurt: Suhrkamp.

Latour, B. (1992) Where are the Missing Masses? The Sociology of a Few Mundane Artifacts, in W. E. Bijker and J. Law (eds) *Shaping Technology/Building Society: Studies in Sociotechnical Change*, pp. 223–56. Cambridge, MA: The MIT Press.

Meyrowitz, N. and van Dam, A. (1982) Interactive editing systems: Part I & II. *Computing Surveys*, 14 (3): 321–415.

Norman, D. A. (1988) *The Psychology of Everyday Things*. New York: Basic Books.

Norman, D. A. (1990) Why interfaces don't work, in B. Laurel (ed.) *The Art of Human-Computer Interface Design*, pp. 209–19. Reading: Addison-Wesley.

Schachtner, C. (1993) *Geistmaschine. Faszination und Provokation am Computer*. Frankfurt: Suhrkamp.

Seybold (1980) IBM's Displaywriter: building block for the '80s? *The Seybold Report on Word Processing*, 3 (11): 1–13.

Seybold (1981) The Xerox Star: a professional workstation. *The Seybold Report on Word Processing*, 4 (5): 1–19.

Seybold (1982) The WangWriter: stand-alone that missed the mark? *The Seybold Report on Office Systems*, 5 (5): 1–10.

Shneiderman, B. (1983) Direct manipulation: a step beyond programming languages. *IEEE Computer*, August: 57–69.

Smith, D. C., Harslem, E., Irby, C. and Kimball, R. (1982a) The Star user interface: an overview. Proc. of National Computer Conference; Reprint: Xerox: *Office Systems Technology*, 1984, pp. 1–14.

Smith, D. C., Irby, C., Kimball, R., Verplank, B. and Harslem, E. (1982b) Designing the Star user interface. *Byte*, 7 (4): 242–82, Reprint: Xerox: *Office Systems Technology*, 1984, pp. 15–28.

Smith, D. K. and Alexander, R. C. (1988) *Fumbling the Future. How Xerox Invented, then Ignored, the First Personal Computer*. New York: Morrow.

Webster, J. (1993) From the word processor to the micro: gender issues in the development of information technology in the office, in E. Green, J. Owen and D. Pain (eds) *Gendered By Design? Information Technology and Office Systems*, pp. 111–23. London: Taylor & Francis.

Woolgar, S. (1991) Configuring the user: the case of usability trials, in J. Law (ed.) *A Sociology of Monsters: Essays on Power, Technology and Domination*. London: Routledge.

Zoeppritz, M. (1988) 'Kommunikation' mit der Maschine, in R. Weingarten and R. Fiehler (eds) *Technisierte Kommunikation*, pp. 109–21. Opladen: Westdeutscher Verlag.

18 ▶ Learning by trying: the implementation of configurational technology

James Fleck

 IMPLEMENTATION

Some technologies are easily implemented; a television set, for instance, requires plugging into a suitable power source, several adjustments to tuning, and then it is immediately ready for use. Large-scale company-wide information systems, such as CAPM [Computer-aided production management (CAPM) refers to the use of computers in managing and controlling the materials, machinery and manpower employed in manu-facturing industry], however, lie at the opposite extreme. It takes great effort, over substantial periods of time, to bring such large-scale technolo-gies to the point where they can be used effectively in day to day operation.

The term, 'implementation', is particularly appropriate here. According to the Shorter Oxford English Dictionary, the verb 'to implement' means 'to complete, perform, to fulfil . . . to supplement'; or 'to provide with imple-ments'. As a noun, 'implement' refers to 'things that serve as equipment . . . the apparatus, instruments employed in any trade or in executing a piece of work . . . a tool, instrument'. There is also a substantive use (in the context of Scots law) meaning 'full performance'.

These connotations are all very germane to the implementation of com-plex technology, where there is a sense of carrying out all the disparate activities necessary to complete a technological set-up so that it becomes an effective tool, capable of achieving 'full performance'. This latter charac-teristic clearly distinguishes implementation from the demonstration of technical feasibility which often comprises an early stage of the overall process.

Implementation is the process through which technical, organizational and financial resources are configured together to provide an efficiently operating system. This, it should be noted, is much broader in scope than the use of the term in common parlance and in software engineering often denotes. In the context of the system design/development life-cycle, implementation is often used to refer only to the final stages of putting a system into productive operation, i.e. more specifically to installation (see, for example, [8]). Other authors, however, do recognize the wider scope of the broader overall process of implementation: Voss, for instance, observes that 'the process of implementation has its roots in the firm's background and history, and includes both pre-installation and post-installation factors' [13]; and Swanson, with reference to information systems, comments that 'implementation is best thought of as a bridge between design and utilization' and that it must be understood in the context of the development process as a whole, over the full life course of the system [12].[1]

I shall argue in this paper that implementation in this broad sense is the key process through which CAPM is evolving. Rather than the technology being unproblematically available from technical suppliers outside the user firms, new developments are being forged during implementation within user organizations, albeit in close cooperation with generic technology suppliers. This puts greater emphasis on the crucial role of people inside the user firm and hence greater emphasis on the organizational structures within which and through which they operate. This helps to explain some of the particular features of the user-supplier relationships that may be observed, and also points up the strategic scope and creative opportunity that is in fact available to user organizations in this type of technology development. The resulting picture is more challenging, but much more enlightening and promising than the conventional view. That view tends to see the situation in terms of the mere routine application of information technology to production management, and thus plays down the contribution of the user firm.

The structure of the implementation process

One starting point to obtain a clear grasp of the structure of the implementation process is via the software systems development life-cycle. Different authors present more or less similar analysis of this cycle. Typical examples are the 'waterfall' model (see, for example, [8]), and various 'stages' models (for example, [6]). Perhaps the most sustained attempt to date to apply such a systems analysis approach to the implementation of manufacturing technology has been carried out with respect to contracts for the US Air Force [5].

But we are not concerned in this paper with detailed discussion of the various steps and tasks involved in such models. It is sufficient here to observe that the implementation process does indeed involve a number of more or less distinct stages, each of which requires consideration at certain times in the overall development cycle and involves a large amount of iterative interaction. Moreover, many different partitionings are possible, not all of which are equally effective [2]. Thus the process is, most emphatically, not straightforwardly linear.[2]

The non-linearity of development is of central importance. It is associated with the problem of clearly defining requirements. Requirements are notorious for changing in the light of the provisional solutions entertained, and 'prototyping' approaches are frequently advocated for helping to overcome this major difficulty. This instability of requirements represents the central bugbear of software engineering, but it also offers an important opportunity for innovation, and is therefore of considerable strategic significance. Where requirements are uncertain, their clarification and identification may well lead to an important advance in terms of what the technology can be used to do.

The time distribution and mutual dependencies between the different tasks and stages of implementation are now usually approached using project management techniques. However, it is difficult to identify a single general pattern across different projects: the situation is one in which each particular case has to be dealt with in terms of the contingencies of its own context (cf. [8, chapter 17]). Thus, rather than an over-simplified linear depiction such as suggested by some models, the situation far more resembles a jigsaw puzzle, as suggested by Swanson [12]. Commonalities can be identified in terms of the 'pieces' of the puzzle that have to be put together in the light of the particular organizational contingencies, even though no strong generalizations can be made about sequencing or the overall pattern of dependencies. Learning how to solve the puzzle for a particular technology and organization constitutes the implementation process, and requires a variety of different sorts of expertise.

The fundamental implementation equation

In terms of the essential inputs of expertise to the implementation process, we can usefully summarize the situation in a simple statement, the fundamental implementation equation:

successful implementation *requires*
  generic technology knowledge
    + local practical knowledge

The local practical knowledge component is, of course, highly contingent to the particular firm concerned. It deals with the specific 'knowledge base' [4], built up over many years of experience in carrying out the firm's business. Moreover, this local component is usually distributed among the many operatives involved in day-to-day activities. In many cases, this knowledge is tacitly embodied in skills and practices which have been gradually formed over a period, and are not available in any other centralized or formalized form. In such cases, therefore, users constitute the only available repository of the local knowledge component which might be essential for achieving successful implementation.

Large-scale, company-wide and even sector-wide configurations of technology are highly specific to particular operating requirements and environmental contingencies. In these cases, the right-hand term of the
equation, the local knowledge element, becomes a crucial source of innovation. Over a number of successful instances of implementation, local
knowledge becomes gradually absorbed into generic knowledge; i.e. firm
specific knowledge becomes appropriated by suppliers. At the same time,
existing generic technology knowledge becomes implemented in a form
suitable for exploitation by the user.[3] As suppliers' experience accumulates
over time and across a particular sector, the immanent possibilities for
sector specific technology developments become thoroughly explored and
progressively developed (indeed almost literally crystallized) into actual
artefacts.

As this happens, the suppliers' understanding of specific requirements
possibilities grows. Eventually, in some cases, we move away from a
situation in which more or less unique configurations reflecting local
contingencies are important, to a situation where requirements are so clear
and stable that standard or generic systems emerge. In these latter cases,
development becomes increasingly technically oriented, often in terms of
defined efficiency or performance measures, and increasingly in the hands
of the suppliers. However, this is not an inevitable outcome. The variety
afforded by configurations which continue to incorporate organizationally
specific requirements and bodies of knowledge and practice may always be a
source of competitive positioning and differentiation.

Assembly line production technology is a case in point. While at one time
it appeared that a stable system – Ford's single product-dedicated, high-
volume task-fragmented line – had emerged, there were limitations intrinsic to the way in which less technical components, such as models of work
organization and quality control, were incorporated. This led to declining
quality and productivity and increasing labour turnover. As a response, and
precisely because of the configurational possibilities of the situation, various quite different configurations were explored: the Volvo Kalmar approach with autonomous working groups; the Toyota 'jidoka' system; and
Just-In-Time layouts such as U-shaped and parallel lines.

More recently, a particular combination of practices, including elements
of JIT, simultaneous engineering, total quality management, continuous
improvement policies, worker empowerment, and marketing-production
integration, appear to be culminating in what the directors of the MIT
International Automobile Study have dubbed 'lean production', a radically
different paradigm for manufacturing [16].

Just as with Schumpeter's famous dictum about the fundamental discontinuity between railroads and stagecoaches, there is no way that the
lean production paradigm could be attained by merely adding in more
high-tech automation or optimization algorithms, as GM's enormously
expensive and apparently fruitless efforts in the 1980s have singularly
demonstrated. Lean production is a radically different configuration of
components, some common to the previous mass-production paradigm,
and some peculiarly new.

▶ CASE STUDY: COMPANY B

Company B was a large diversified engineering company, with historical roots in Scotland and a range of activities around the world. The part of Company B with which we were concerned covered two different sites. One site, employing 500 people, manufactured two major product lines, initially as two separate business divisions, though latterly moving towards integration. The first involved the assembly of large heavy electrical transformers with some minor fabrication work (for the outer casings). These products were all one-offs, made to contract, although the generic structure of each transformer (core, windings, insulating oil, outer casing, connection gear) was essentially standard and really quite simple. The precise rating and characteristics of each transformer, however, had to be individually calculated for each customer's requirements: for example, different lengths of copper had to be wound on to steel laminate core material, precisely cut to dimension.

The second product division on the first site manufactured and assembled two ranges of heavy electrical motors (for compressors etc.). These were also made as one-offs to customer order. Again the generic product structure was relatively simple. Only some 20 basic parts were involved, along with many small bits and pieces bolted on. The products were customized to provide specific operational and torque characteristics.

The second site, employing about 300, manufactured a range of lighter duty transformers. These were similar in structure to the big transformers made at the other site, and were manufactured using much the same process, but tended to be made in small batches, rather than as one-offs.

In all cases, the production processes throughout were almost entirely manual. Many semi-skilled/unskilled women and men were employed, the latter doing some of the heavier lifting tasks. A few specialist machines were used for numerically controlled mitre cutting of steel laminates, and ovens were used for eliminating moisture from the windings. The low-volume, high-variety production profile essentially precluded a high level of manufacturing automation.

In the early 1980s, as part of a planned 'grand system', the company decided to introduce a large MRP II system [manufacturing resource planning – a computer-based inventory and materials control system, which also provides facilities for capacity planning; i.e. for scheduling the use of manufacturing resources more generally, including machinery and manpower, as well as controlling material flows [11]. This system, comprising the modules illustrated in Figure 18.1, was supplied by a large American computing concern (which we shall call 'Supplier C'). It is recognized in the textbook literature as a standard (out of the 200 or so systems available). Moreover, this particular system (let us call it SCMRP, short for Supplier C's MRP) was explicitly modelled on the MRP guru 'Ollie' Wight's recommendations. Wight has claimed trenchantly that it is a misconception that:

Each company requires a unique 'system' designed for them to solve their unique problems. In practice the problems of scheduling a factory, scheduling the vendors, and coordinating the activities of marketing, engineering, manufacturing and financing are not particularly unique company to company. There is a standard logic for MRP, and we have yet to see a company that had to reinvent this logic, or for that matter, one that tried to reinvent it and made it work.

(Newsletter from Oliver Wight, 1977, quoted in [10]).

This is a strong claim, and one which makes our case particularly interesting. According to the above, and in light of the standard nature of the MRP system adopted, it should provide a test case (similar to a critical experiment in science) for the straightforward standard installation of such systems, rather than anything approaching the innofusion model outlined above. And as with the critical experiment in science, only one countervailing instance is sufficient to throw serious doubt on the claim.[4] We do not require battalions of examples to challenge the standard installation hypothesis, though as we briefly note later, this case is not a peculiar exception.

What do we find?

The process of implementation of the SCMRP system was a sorry tale of almost unmitigated disaster.

First, as was clear from the various and sometimes contradictory views expressed in our interviews, there was evident confusion and uncertainty about the scope and intent of the original CAPM system. For instance, top management (the Financial Director) initially envisaged a very large-scale system, integrated across both sites, and all divisions. But in actuality, development took place at individual site level. Moreover, this confusion appeared to have resulted in too big and sophisticated a system being purchased; one suitable for the turnover of the company as a whole, rather than the turnover of the separate sites, the actual level at which implementation took place.

The confusion was further compounded by the usual sorts of local company politics:

the Computer Services Manager went out the door. Unfortunately he went out the door with most of the knowledge of the system because he hadn't written anything down. He left no records. He instructed his people to destroy everything, so I came into a department with no written knowledge.

(Interview, Computer Operations Manager, Company B, 1989)

Second, each site in fact evolved its own distinct approach to implementation (perhaps further reflecting tensions between managers on the different sites). In particular, as indicated in Figure 18.1, each site made use of a quite different selection of the available modules of the overall system. This different selection of components clearly reveals the configurational nature of the technology.

Third, originally a grand integrated CAPM system combining separate

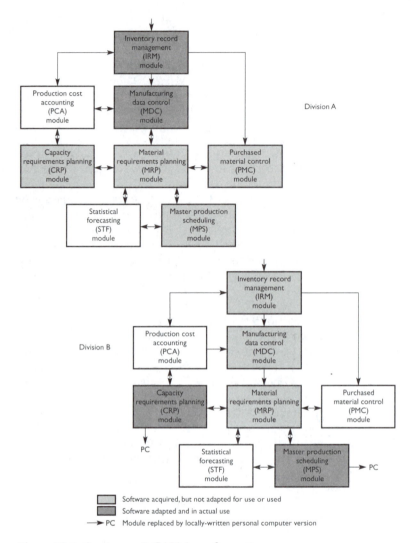

Figure 18.1 Company B CAPM configurations

packages for computer-aided design (CAD), MRP II, and a Financial System was envisaged (again, it would appear, by the very top management). But, even though these packages were all supplied by the one company (who also supplied the computer mainframe, another essential component of the configuration), it turned out they could not communicate with one another. The suppliers claimed that they 'knew where all the hooks were' to link these various packages together, but it had not been done before and therefore further development was necessary. Indeed at a very early stage, apparently (it was very difficult to establish conclusively precisely what happened because of the prevailing uncertainty), the supplier was going to develop a new drawing office management system specifically to mediate between the CAD and the MRP II packages. This was to be designed specially for Company B, using them as the 'guinea pig', in consideration of which they would get the work done at discounted rates. However, 'some of the people in Supplier C moved on to other places, time passed etc.', and this particular system never materialized.

In the event, some elements of the CAD package were in fact implemented, but to differing extents in the drawing offices for the motors and for the transformers (though some respondents in Company B claimed that they already had better in-house programs). Only a few of the MRP II modules were ever used, and different ones on each site, as already noted above, and the accounting package was abandoned:

> efforts were being made to implement [the accounting package], the Ledgers system. It was just foundering. So one of the first decisions I made, it really stopped, there was nothing being done, so I decided because this system was horrifically expensive, and maintenance was more than if we bought a new accounting system every year and threw it away. The maintenance was colossal and we were spending all this money on maintenance on a system we couldn't use. So I took [the package] off the system altogether and I cancelled the maintenance on it. I decided it was just out-with our capabilities. We would look for a system that was manageable, even maybe a PC-based system.
>
> (Interview, Computer Operations Manager, Company B, 1989)

Fourth, several major problems were encountered in trying to get the system into operation.

(a) The package as designed could cater for integer units in the Bill of Materials (essentially the structure of related parts making up the product), but could not easily cope with measured parameters: so many metres of winding, such and such a volume of oil, etc. Yet these were crucial for the customized manufacturing operation in which the company was engaged. The specific products designed for particular customers differed not in their generic structure, but in the amounts and distribution of materials used, which provided the distinctive operating characteristics and power ratings, etc., required.

(b) The package as offered ostensibly contained a 'contract traceability' function. This was essential given the make-to-contract nature of the business carried out by Company B. In particular, materials traceability was required by the Ministry of Defence. But it became apparent in the development process that this function did not work properly. The suppliers had developed the system and used it widely in their own operations, where they had also tested it. But their own manufacturing environment was make-to-stock, and consequently they had not used the contract traceability function before in the quite different make-to-contract environment. Again, therefore, it appeared that Company B were the unwitting and unlucky guinea pigs.

(c) During the commissioning of initial modules in Company B, all sorts of major system bugs came to light. According to the MRP Project Manager, there were over 100 major faults. The time taken to fix even the most trivial faults was 9 months to a year, and sometimes the suppliers took no action. As a result, Company B were forced to take action themselves, in some cases (at the one site where they had the appropriate expertise) going into the source-code themselves and reprogramming it; in others fixing in 'patches' received from the supplier (which was US-based, as noted). Yet this package was reputedly a standard product some 10 years old!

This, of course, led Company B into a catch-22 situation. When they produced locally-customized solutions they subsequently had to undertake the ongoing commitment to maintaining the whole system themselves:

> the other problem we had was when we did customize the software, when [Supplier C] revised the software, they did not revise our revisions unless they had done the revisions.
> (Interview, MRP Project Manager, Company B, 1989)

The supplier company, whilst continually upgrading and enhancing their system, would not undertake to maintain local solutions. They did regularly provide enhancements, but these reflected the requirements and experiences of other users (and, of course, of themselves as users of their own system), essentially only where the users were sufficiently typical or important to force such attention. It seems that Company B were not. And so Company B, who did not want to become involved in extensive development of these systems (they were, after all, in the business of manufacturing motors and transformers, not software), had perforce to become involved, simply in order to get the system working.

(d) At a more technically superficial but nevertheless organizationally serious level, many of the screen forms and much of the documentation produced by the system did not suit Company B's requirements. They could not readily break into the source-code to alter these artefacts, especially at the site where they did not have appropriate programming expertise, and in any case it took six months to get a customized document back from their own computer services department. As a result, in some instances they moved over locally to using other PC-based programs, as indicated on Fig. 3.

(e) Finally, considerable effort was expended in getting their own local management procedures and operating practices into a form suitable for interfacing with the system:

> We have in-house procedures that we have developed to control what people do and they are not allowed to step outside the bounds of these procedures; and the procedures tell them how they do every single aspect of their part of the system. If it is booking materials into the system, there are certain rules and transactions that you use. If it is issuing materials to the shop floor: the shop floor people have to follow a particular procedure to get their material.
> (Interview, Computer Operations Manager, Company B, 1989)

These local procedures constituted other essential components of the CAPM configuration constructed, albeit non-technical ones. Existing management procedures had to be adapted and new ones devised. In addition, a considerable amount of knowledge and judgement about local operations (such as working out production times for particular contracts) was required to be captured into formal rules for the system. This highly contingent knowledge was distributed among various employees, in particular the production planning manager:

> I think I found myself in this position because, without being big-headed, I know the product, I know the job, I know how to go about building transformers and I think to a certain extent it has worked against me. I know too damn much about it and I mean I am put into a position here where they are using the information that is in my head and transferring it from my head into profiles such as this and putting it up.
>
> (Interview, Production Planning Manager, Company B, 1989)

This was not a straightforward matter, as the rules derived required checking and reformulating into terms which could be used by the system. Moreover, it was a matter of further judgement whether such rules were in fact fully adequate or the best available for running the operations. Later in the same interview, this manager, who was very much a traditional hands-on long-experienced production manager, expressed concern about the new formal system:

> Well, I feel we should be in a position that we should be able to get down into the shops and see what is going on. If the plan is not being followed or worked out, we should be in a position to say 'hey, look what have you got that guy on that job for? Get him off that and get him onto that and get that done!' And that worries me because I have not been proved wrong often over the years with the times that we've generated and I get a wee bit worried about the fact that for one we tend to be taking longer to do jobs now than we did two or three years ago and I think that is down to the fact that I have been off the shop floor for two or three years and there is not the same degree of control now as there was then.

Local tacit and contingent knowledge components were therefore being built into the new configuration, but were the results as efficient as previously, or were the observed shortfalls due to lack of experience with the new procedures?

After such a catalogue of disaster, and some five years of effort involving many people to varying extents, it was not surprising to find that Company B, as we looked at them, were planning to pull out all the software components of the grand CAPM system and to look afresh at the options available. These included a smaller network-based system from the same suppliers. To this end they had convened four committees to look further into four different areas: MRP, CAD, project management, and accounting systems.

► **DISCUSSION**

What then can be observed about this case in the light of the earlier discussion of implementation and innofusion? [In configurational development in particular, the processes of innovation and diffusion are collapsed into one another, in what might be called a process of 'innofusion'.]

First, it should be emphasized that this experience with implementation is not atypical. Many other studies of new technology implementation report high failure rates (e.g. [1,3]).[5] Not surprisingly, in general the greater the configurational nature of the technology (i.e. the more it is composed of selections of components to meet local requirements) the greater the chances of failure. Problems over communication between the supplier(s) and user, and areas of ignorance about the eventual configuration and how it should run, especially as regards integrating the various components with each other and the host organization, are intrinsic to the implementation process: implementation *is* highly uncertain. The economic separation and asymmetric distribution of knowledge between the users and suppliers are also typical of the situation in which implementation takes place, as is an asymmetry of power between users and developers [7]. As noted above, successful implementation is precisely a matter of finding a synthesis of local and generic knowledge adequate for meeting the particular application requirements. . . .

► **CONCLUSION**

In this paper, I have argued that innovation during the process of implementation is to be expected in the development of working configurations of CAPM, as an outcome of 'learning by trying'.[6] This analysis can help to make sense of events even in a case where attempts to employ a particular 'standard' system met with total failure. In conclusion, the following summary observations, oriented to policy considerations, can be made.

1 CAPM can be regarded as a configurational technology where each working configuration is made up of a particular selection of components, both technical and non-technical, geared to the needs of the individual organization.
2 Although particular elements, such as a materials requirements planning module, can be individually employed, it is their incorporation as components into wider operating configurations which ultimately determines their degree of success.
3 All components are potentially of importance, not just the purely technological ones, and certainly not just the high-tech developments. Bodies of management procedures and local knowledge, encapsulating

organization and product specific contingencies and arrangements, are essential for the effective operation of CAPM configurations. Indeed, they comprise distinct non-technical components of the overall operating configuration.

4 Existing 'old' technologies and conventional practices may continue to be important via their incorporation into emerging configurations, as for example with the incorporation of engineering drawings as 'parts' in the Bill of Materials.

5 A high variability among configurations of CAPM is therefore to be expected, and reflects the relative importance of organizational and market contingencies, rather than universally defined technological trajectories. This was nicely illustrated by the completely different configurations adopted by the two divisions of the one case study company.

6 The uncertain and evolutionary nature of the configurational development process implies a great amount of implementation effort within the user organization. It also gives rise to the opportunity, and indeed, even the necessity, for innovation during the implementation process itself. In the absence of such innovatory effort, even with purportedly standard technology such as MRP, failure is likely to result, as in fact happened in the case considered.

7 Substantial involvement in development on the part of users is consequently necessary to make the most of these opportunities. The value of the local knowledge held by users should be recognized as crucial for successful implementation, not least by the users themselves.

8 Successful suppliers of configurational technologies are likely to be those who recognize implicitly, if not explicitly, the need for a synthesis of global technological knowledge with local process knowledge. They are also likely to adopt appropriate methodologies for effectively coordinating supplier and user efforts in implementation, and for maximizing what can be learnt through feedback from the users about implementation and operation.

9 Nevertheless, particularly with smaller user companies who lack the confidence to value their own local knowledge and who lack the expertise to handle certain aspects of the generic knowledge (such as rectifying software faults), there will be a high degree of information asymmetry. This is likely to be exploited by suppliers; a tendency exacerbated by the one-off nature of many company-wide implementation programmes, although mitigated by the need for long-term interaction which limits the scope for such opportunism.

10 Finally, policy measures need to address these realities. Initiatives should aim at encouraging the overall industrial sector-based learning process, by facilitating the mobility of personnel, by providing avenues for the negotiation and formation of standards, and by encouraging user participation at all levels, from top management to the shop floor. In particular, some means of enabling smaller firms to fully appreciate the importance of their own local expertise, and of improving the valuation of such expertise more generally, could help in redressing any inhibiting information asymmetry. Other measures (such as demonstration

networks) should aim at reducing the costs of experimentation and at communicating the results (both positive and negative) of such experimentation as widely as possible, especially with respect to small and medium sized companies. In short, policy formulation should recognize that, in the case of the development of configurational technology, industry is the laboratory.

► NOTES

1 The collection of articles edited by Rhodes and Wield [9] also tackles implementation in the broad sense.
2 Winch and Twigg [15] critically examine various stage models for implementation, noting the evidence against a simple linearity.
3 This two way process of knowledge transfer is, of course, at the root of negotiations over the make-or-buy decision with respect to the supply and development of new technology.
4 Pace the Duhem-Ouine thesis, which suggests that it is always possible to save any hypothesis by making suitable adjustments to the background assumptions defining the context; in the present situation this could be done by redefining the notion of installation, for example, or by asserting that a 'standard' system includes all the customizations and adjustments required to cater for the particular firm's contingencies, which is of course rather departing from the gist of the quote and stretches the accepted meaning of standard.
5 There are, of course, a number of ways of defining failure (in order of increasing stringency): a lack of technical success; non-use in normal production; failure to achieve business success; failure to realize full expectations within a designated time horizon. Voss [13,14] discusses some of the issues. In the present case, despite partial achievements with particular subcomponents, failure at the level of the overall configuration was so total that definitional niceties did not arise.
6 This suggests the following taxonomy of learning processes: learning by doing (the industrial learning curve phenomena); learning by undoing (reverse engineering); learning by using (incremental improvements identified in the course of using a technology); learning by trying (improvements necessary before a new configuration can be implemented); learning by buying (the acquisition of knowledge by hiring in expertise); and learning by learning (the formation of expertise through in-house training programmes).

► REFERENCES

1 J. Bessant, The Integration Barrier: Problems in the Implementation of Advanced Manufacturing Technology. *Robotica* 3 (1985) 97–103.
2 F. P. Brooks, Jr., *The Mythical Man-Month* (Addison-Wesley Publishing Company. Reading, MA, 1982).
3 J. Fleck, The Introduction of the Industrial Robot in Britain. *Robotica* 2 (1984) 169–175.

4 L. Georghiou, J. S. Metcalfe, M. Gibbons, T. Ray and J. Evans, *Post Innovation Performance* (MacMillan, Basingstoke, 1986).

5 J. Harrington, Jr., *Understanding the Manufacturing Process: Key to Successful CAD/CAM Implementation* (Marcel Dekker, New York, 1984).

6 J. O. Hicks, Jr., *Information Systems in Business: An Introduction*, 2nd edn (West Publishing Company, Saint Paul, MN, 1990).

7 D. Leonard-Barton and W. A. Kraus, Implementing New Technology, *Harvard Business Review* 63 (Nov./Dec. 1985) 102–110.

8 D. C. McDermid, *Software Engineering for Information Systems* (Blackwell Scientific Publications, Oxford, 1990).

9 E. Rhodes and D. Wield (Editors), *Implementing New Technologies* (Basil Blackwell, Oxford, 1985).

10 R. G. Schroeder, *Operations Management: Decision Making in the Operations Function*, 3rd edn) (McGraw-Hill Book Company, New York, 1989).

11 R. G. Schroeder, J. C. Anderson, S. E. Tupy and E. M. White, A Study of MRP Benefits and Costs, *Journal of Operations Management* 2 (1) (1981) 1–9.

12 E. B. Swanson, *Information System Implementation: Bridging the Gap between Design and Utilization* (R. D. Irwin, Homewood, IL, 1988).

13 C. A. Voss, Implementation: A Key Issue in Manufacturing Technology: the Need for a Field Study, *Research Policy* 17 (2) (1988) 55–63.

14 C. A. Voss, Success and Failure in AMT, *Int. J. Technology Management* 3 (3) (1988) 285–297.

15 G. Winch and D. Twigg, The Implementation of Integrating Innovations: The Case of CAD/CAM, in: W. J. Vrakking and A. J. Cozijnsen (Editors), *Handbook of Innovation Management* (Basil Blackwell, Oxford, 1993).

16 J. P. Womack, D. T. Jones and D. Roos, *The Machine that Changed the World* (Rawson Associates, New York, 1990).

19 ▶ Working relations of technology production and use

Lucy Suchman

My starting place in this essay is recent feminist moves to reframe objectivity from an established body of knowledge to knowledges in dynamic production, reproduction and transformation. These reframings are relevant for our thinking about technologies insofar as technologies objectify knowledges and practices in new material forms.

For technology designers and developers, the basic change implied is from a view of design as the creation of discrete devices, or even networks of devices, to a view of systems development as entry into the networks of working relations – including both contests and alliances – that make technical systems possible. This represents a change insofar as the prevailing order of technology production is based not in acknowledgement and cultivation of these networks but in their denial in favor of the myth of the lone (male) creator of new technology on the one hand, and the passive recipients of new technology on the other. The fact that this myth belies the lived reality of systems development and use has so far gone largely unchallenged, as has the simple designer/user opposition that underwrites the myth. This paper develops the proposal that feminism offers a way to begin to replace the designer/user opposition – an opposition that closes off our possibilities for recognizing the subtle and profound boundaries that actually do divide us – with a rich, densely structured landscape of identities and working relations within which we might begin to move with some awareness and clarity of our own positions. . . .

To find an alternative stance for design I turn to Haraway's argument for a feminist objectivity, as a starting point for an alternative conception of what the responsible production and dissemination of new technical artifacts might be. Haraway writes:

> Feminist objectivity is about limited location and situated knowledge, not about transcendence and splitting of subject and object. In this way we might become answerable for what we learn how to see[1] . . . this chapter is an argument for situated and embodied knowledges and against various forms of unlocatable, and so irresponsible, knowledge claims. Irresponsible means unable to be called into account (1991, pp. 190–91).

Located accountability, then, is built on what Haraway terms 'partial, locatable, critical knowledges' (1991, p. 191). As she makes clear, the fact that our knowledge is relative to and limited by our locations does not in any sense relieve us of responsibility for it. On the contrary, it is precisely the fact that our vision of the world is a vision from somewhere – that it's inextricably based in an embodied, and therefore partial, perspective – which makes us personally responsible for it. The only possible route to objectivity on this view is through collective knowledge of the specific locations of our respective visions. By extension, the only possibility for the creation of effective objects is through collective knowledge of the particular and multiple locations of their production and use.

ASPECTS OF LOCATED ACCOUNTABILITY IN TECHNOLOGY PRODUCTION

To consider more concretely what technology production based on located accountability could mean, I offer some reflections on my own work experience in these terms. The vision of an alternative inspires my working relations, but is only partially realized in them. What follows is therefore a reporting in part on current practice, in part on desired transformations to it.

Relations of production

As members of a very large enterprise engaged in the production of new technologies we find ourselves enmeshed in an overwhelmingly complex network of social relations, for the most part made up of others we have never met and of whose work we are only dimly aware. The simple dichotomy of technology production and use masks (or indexes as we begin to respecify it) what is in actuality an increasingly dense and differentiated layering of people and activities, each operating within a limited sphere of knowing and acting that includes variously crude or sophisticated conceptualizations of the others. Within industrial research the distinctions are primarily disciplinary: computer science, electrical engineering, mathematics, cognitive psychology, linguistics, anthropology all orient not only to different problems but more significantly to different, sometimes incommensurate conceptions of the social/technical world. And as researchers we are all defined in contradistinction to enterprises of product design, development, manufacturing, finance, strategic planning, human resource management, marketing, sales and service, each of which in turn is itself a complex social world comprising distinctive concerns, accountabilities and working relations.

A central dilemma of our participation in these increasingly complex divisions of labor and professional specialization are the layers of mediation between each of us and the consequences of our work. In some real sense, no one of us *is* responsible for the outcomes of our collective labor. The possibility of invoking this reality as a rationale for abdicating responsibility for the products of technological labor are well known.[2] But the question concerns our responsibilities toward the *process* of technology production as well. Traditionally, the relations among disparate activities of technology production have been viewed as a series of hand-offs along a kind of multi-disciplinary assembly line. On this premise, for example, the role of research is to construct the technological foundations on top of which future devices will be built, including visions of how the future will be. A long-standing mutual dissatisfaction between research and product development arises from the failure of technologies and ideas to 'transfer' from one to the other, understood by one side as a failure of development to take advantage of the results of research, by the other as a failure of research to address the needs of development.

My own experience of this gap began in the early 1980s in grappling with the question of how an anthropology of technology might be made relevant to the design of machine interfaces. The first proposal was that, as ethnographers, we might mediate relations between designers and users. Increasingly, however, our reluctance to translate our practice directly into design terms was met with frustrations from the design community.[3] Our hesitation to produce such translations led to our characterization as recalcitrant social scientists, unwilling to roll up our sleeves and engage in the real work of design. For a time I at least was confused by this, feeling that to deliver design implications was indeed my responsibility but that I was unable to do so. I dwelled uncomfortably for several years within this gap between my practice and that of my design co-workers, seeing it not as a systemic discontinuity but as a personal shortcoming.

Gradually, however, we came to see that the problem lay neither in ourselves nor in our colleagues, but in the division of professional labor and the assumptions about knowledge production that lay behind it. The discontinuities among our traditions and our discursive practices meant that we could not simply produce 'results' that could be handed off to our colleagues. What we were learning was inextricably tied to the ongoing development of our own theorizing and practice, such that it could not be cut loose and exported elsewhere. Rather than feeling inadequate in the face of demands that our work produce design implications, we began to resist those demands. We resisted them not on grounds of scientific 'purity' or by denying our responsibility for design, but by rejecting assumptions on the basis of which the demands for our knowledge were being made. In place of the model of knowledge as a product that can be assembled through hand-offs in some neutral or universal language, we began to argue the need for mutual learning and partial translations. This in turn required new working relations not then in place.

At the same time, we began to find allies within the design community itself. Within the corporation, our colleagues who had spent much of their professional lives designing control panels for discrete, 'stand alone boxes' now were being told by their management to envision a future of devices that would be tied together through networks, with the functionality of the overall system distributed dynamically among them. Increasingly, our colleagues were finding that their traditional methods for generating design ideas (for example, with reference to prior products and the results of marketing focus groups), or establishing the adequacy of their designs (for example, through operability testing) were ineffective. Motivated first on our part by economic necessity (in particular, the necessity of obtaining funding from the design organization to support our anthropological re-search activities) and then increasingly by genuine affinity, a small network of working relations grew up across the divide.[4] Together, we realized, we might actually be able to bring our respective knowledges to bear on the shared problem of how to develop new grounds for technology design.

Our most recent attempt to develop these grounds has been to design a project intended deliberately to cut across the organizational boundaries that separate us, both from each other as design professionals within the corporation, and from the potential users of the technologies we are design-ing.[5] The goal of the project is to develop a work-oriented design practice that engages members of a specific site of potential technology use as collaborators in technology production. The site in this case is a large law firm, with its own highly elaborated and power differentiated network of working relations.

Relations of use

Just as the term 'designer' opens out, on closer inspection, onto an extended field of alliances and contests, so does the term 'user'. Organizations comprise multiple constituencies, each with their own professional identi-ties and views of the others. Our investigations of work at the law firm revealed the contested nature of members' representations of their own work and the work of others. Attorneys with whom we spoke described a status hierarchy within the firm comprising partners, associates, junior attorneys, paralegals, case assistants, and litigation support. The work of litigation support was quite literally invisible to the attorneys (being located on a lower floor of the firm, to which the attorneys never went). In addition, attorneys described this work to us as a mindless, routine form of labor, representing a prime target for automation or outsourcing as part of a general cost-cutting initiative within the firm.

Our direct investigations of the work of litigation support contradicted this view. In place of mindless workers we found a lively group of tempor-ary workers, most with bachelors degrees, supervised by a former paralegal with extensive experience in the maintenance and use of computerized databases. These 'document analysts,' as the supervisor called them, were engaged in carefully examining and encoding the thousands of documents used to assemble each case with the goal, vigorously instilled by their supervisor, of creating a valid and useful database for the attorneys. The litigation support supervisor expressed to us her belief that, given their familiarity with the document corpus, the document analysts could be responsible for certain other aspects of the document production process as well, now handled by junior attorneys. She also expressed her view that the attorneys underutilized the database, due to their ignorance of its capabilities and how to exploit them.

So we found ourselves cast into the middle of a contest over professional identities and practices within the firm, framed by the attorneys as a distinction between 'knowledge work' on the one hand and 'mindless labor' on the other, framed very differently by the workers within litigation support themselves. Our own observations of the work of the attorneys revealed no small measure of routine or tedious activities, all of which were, when brought into their awareness, acknowledged by them as inevitable if regrettable accompaniments of their professional practice. At the same time, the more we looked into the work of litigation support the more we saw the interpretive and judgmental work that the document coders were required to bring to it. We could not escape confronting directly these contrasting views as we realized that the work of document coding, which involved translations and transformations between paper and electronically-based media, was well-suited to our design agenda.[6] As a result, we decided to work with the supervisor of litigation support and her staff to prototype a redesigned document coding practice, incorporating some of our technologies. What interested us was the possibility of embedding bits of automation into the practice in a way that would relieve the tedium, while maintaining the level of interactive control over the process necessary for interpretation and the exercise of judgement.

After working for some time on the design of a document coding application, we coincidentally received a call from the firm's Director of Technology inquiring into the progress of our project. On hearing that, among other things, we were developing a proposal with respect to document coding, he responded that we should know that, in the interest of cost-cutting, the senior management of the firm were considering very seriously closing down the in-house coding operation altogether, and shipping the documents instead to the Philippines. He added that the supervisor of litigation support did not yet know the extent of this plan, and that he would appreciate it if we would not mention it to her.

This conversation placed us in an obvious dilemma, which we attempted to resolve at least partially in the following way. We arranged with the Director of Technology that we would provide him with an update on our work, including our observations and proposals regarding document coding. We then called the supervisor of litigation support and explained to her, without mentioning the offshore plan, that we were preparing a progress report for the Director of Technology and others, and that we would like to review with her what we planned to say to be sure that we were not misrepresenting her operation in any way, and to see whether she might have anything to add. In that way we hoped to speak at least in part on her behalf. We then attempted to construct our presentation in such a way that it called out the interpretive and judgemental work involved in document coding and its importance to the production of useful databases, as well as the impossibility of automating or outsourcing it without losing the value of that work.[7]

Our design efforts with respect to litigation support at the law firm ended with the research prototype. Nor do we have any illusions that our presentation alone could dissuade the management of the firm from pursuing outsourcing as a means of cost-cutting. Meanwhile, however, the litigation support staff took their own initiatives to increase the cost-effectiveness of their practice. At the time that our project ended, they had changed their practice to coding documents directly into the database rather than in two separate passes for document coding (on paper forms) and data entry (from forms into the database). At the same time, they had managed successfully to counter claims by outside sources to be able to do accurate database creation at a significantly cheaper rate. For the moment, then, their place within the firm seemed secure. We hope at least to have contributed to their efforts by seeing their work and acknowledging what we saw, both in our representations of it and our designing for it. . . .

▶ **CONCLUSION: WEBS OF CONNECTIONS**

In this paper I have tried to lay some groundwork for an approach to technology design informed by feminist theorizing and an awareness of the working relations of technology production and use. To return once again to Haraway:

> So, with many other feminists, I want to argue for a doctrine and practice of objectivity that privileges contestation, deconstruction, passionate construction, webbed connections, and hope for the transformation of systems and knowledge and ways of seeing (1991, pp. 191–92).

Haraway's argument is primarily an epistemological one. But insofar as the design of technical systems is a process of objectification, of the inscription of knowledges and activities into new material forms, her arguments point the way to transformation of technology design as well. Such transformation might entail at least the following:

1 Recognizing the various forms of visible and invisible work that make up the production/use of technical systems, locating ourselves within that extended web of connections, and taking responsibility for our participation;

2 Understanding technology use as the recontextualization of technologies designed at greater or lesser distances in some local site of practice;

3 Acknowledging and accepting the limited power of any actors or artifacts to control technology production/use;
4 Establishing new bases for technology integration, not in universal languages, but in partial translations;
5 Valuing heterogeneity and partial integration, achieved through practices of technology production/use, over homogeneity and domination.

Feminist scholars such as Cynthia Cockburn (1993) have argued compellingly that much existing technology systematically, and in manifold ways, incorporates masculinist assumptions and values, and that the relative absence of women from technical practice must be understood not only as the result of exclusion but as reflecting forms of resistance as well. In her book *Feminism Confronts Technology* (1991), Judy Wajcman suggests that really to understand these processes of exclusion and resistance, feminist scholars need to get inside the 'black box' of technology production: that there is room for an effective politics around gaining access to technological work and institutions, and that there are, as she puts it, 'opportunities for disruption in the engine rooms of technological production' (p. 164). Similarly Jane Flax, in a paper titled 'Postmodernism and Gender Relations,' outlines what she calls a four-fold task for feminist theorists:

> We need to (1) articulate feminist viewpoints of/within the social worlds in which we live; (2) think about how we are affected by these worlds; (3) consider the ways in which how we think about them may be implicated in existing power/knowledge relationships; and (4) imagine ways in which these worlds ought to and can be transformed (1990, p. 55).

I take Flax's four-fold tasks as both description and directive for constructing alternative forms of technology production and use. Technologies can be understood as materials whose stability relies upon the continuous reproduction of their meaning and usefulness in practice. There are two basic forms that technology stabilization can take. The first, prevailing form is stabilization through the handing-off of technologies across multiple, discontinuous worlds each of which stands as a black box for the others. Actors within these discontinuous worlds work to achieve enough coherence in the artifact that it becomes possible to hand it off to others. So product developers hand off a technology to marketers, whose work makes it possible to effect hand-offs to third party developers and system integrators, whose work makes it possible to effect hand-offs to purchasers, whose work makes it possible to effect hand-offs to local implementers, whose work in turn makes it possible to effect hand-offs to end-users. Two aspects of this process as currently constituted are crucial. It relies upon articulation work at each boundary crossing and that work, whether mythologized or denigrated, is largely invisible. The alternative form of technology production that I hope to have outlined here is built around a deepening awareness of and orientation to the work required to achieve technology stabilization, and of our own locations within the extended networks of working relations that makes technical systems possible.

▶ **NOTES**

1 And, I would add, for what we learn how to build.
2 Among the most dramatic are the arguments of those involved in the production of military technologies that it's not their work in particular that provides a weapon's deadliness.
3 For an account of these frustrations and proposals for their resolution see Blomberg *et al.* 1993.
4 The core of this network is a small affinity group comprising ten researchers and six product designers, distributed between research centers based on the West coast of the U.S. and in England, and development organizations on the East coast. Among us we draw on backgrounds in anthropology, sociology, computer science, industrial psychology, graphic design and product design. In addition, this core group has ties to a slightly larger network of approximately 50 researchers, designers and engineers within the company who have begun to meet periodically to exchange stories and provide mutual support.
5 For more on this project see Blomberg, Suchman and Trigg (1996).
6 Specifically, we were interested in exploring the potential usefulness of a class of image processing technologies, emerging from research and making their way into product development These technologies are aimed at supporting relations between paper and electronic documents, by turning marks made on paper into instructions to the machine at the point that a paper document is scanned. So, for example, the machine can 'recognize' a circled text region on a paper document and store just the circled text in a designated electronic file for subsequent reuse.
7 It is notable that when one is seen as designing hi-tech support for knowledge workers, the injunction is to capitalize the more expensive forms of labor. This in contrast to the goal of automating away the less expensive forms of labor.

▶ **REFERENCES**

Blomberg, Jeanette, Giacomi, Jean, Mosher, Andrea and Swenton-Wall, Pat (1993) Ethnographic Field Methods and their Relation to Design. In *Participatory Design: Principles and Practices*, eds D. Schuler and A. Namioka. Hillsdale, N.J.: Erlbaum, pp. 123–156.

Blomberg, Jeanette, Suchman, Lucy and Trigg, Randall (1996) Reflections on a Work-Oriented Design Project. *Human-Computer Interaction*, Volume 11, pp. 237–265.

Cockburn, Cynthia (1993) The Gender/Technology Relation: Taking Shape. Paper presented at the Conference on *Sex/Gender in Techno-Science Worlds*, University of Melbourne, Australia.

Flax, Jane (1990) Postmodernism and Gender Relations in Feminist Theory, in *Feminism/Postmodernism*, ed. Linda Nicholson, 39–62. New York: Routledge.

Haraway, Donna (1991) Situated Knowledges: the science question in feminism and the privilege of partial perspective. Chapter 9 in *Simians, Cyborgs, and Women*. New York, Routledge, pp. 183–201.

Wajcman, Judy (1991) *Feminism Confronts Technology*. University Park, PA: Pennsylvania State University Press.

PART THREE

▶ Reproductive technology

▶ Introduction

▶ TECHNOLOGICAL DETERMINISM AND REPRODUCTION

We are living in a time when it seems we have unprecedented technological options available to enhance our everyday lives. The impact of the revolution in information and biotechnologies is as great in the personal sphere of sexuality, consumption and leisure as it is in production. The home is increasingly portrayed as the centre of a web of technologies providing individuals and families with entertainment, shopping, global communication, and even paid work. Our bodies too are being reconstituted by advances in medical technologies and genetic engineering. The futuristic image of the cyborg, an amalgam of human and machine parts, is becoming less alien and more of a reality.

Reproductive technologies have held out particular promise for women, offering liberation from a previously ordained gender order. Reproduction here is taken to mean not only biological reproduction, but also the production and consumption that takes place in the home. It is in this area more than elsewhere that technological advances are seen as having directly transformed women's lives for the better.

In much early feminist writing, reproductive technology was seen as particularly progressive because it opened up the potential for finally severing the link between sexuality and reproduction. Writers such as Shulamith Firestone (1970) located women's oppression in their biology, and in the control of women's bodies, especially their sexuality and fertility, by men. The literature on reproductive technology has continued to be rife with technological determinist arguments which assume that changes in technology are the key to the massive social changes that have occurred for women's equality. The technologies of pregnancy and childbirth are said to have put an end to the dangerous and painful aspects of childbearing. The pill is seen as having caused the sexual revolution, smaller families and women's greater participation in the paid workforce. Advances in the technologies of fertility control are credited with making sexual equality possible.

Considerable optimism has also been attached to the possibility that technology may provide the solution to the tyranny of housework. Women's unpaid work in the home, servicing men, children and other dependants has for a long time been seen by feminists as the key to women's subordination. It is a widely held belief that domestic technologies have mechanized and rationalized housework to the point that drudgery has been largely eliminated. Technological wizardry and a multiplicity of appliances are seen as having brought an end to the sex war in the home.

However, there is no simple correlation between the development of these new technologies and the dawning of a new egalitarian era of intimate relationships. We hope to show in the following section that a social shaping perspective helps us understand why these technologies cannot be read simply as emancipatory. Let us first consider domestic technologies, after which we will move on to the technologies of human biological reproduction.

 ## DOMESTIC TECHNOLOGY

We do not usually think of the home as a hive of technological activity, yet domestic technology is of considerable economic significance. This fact has been eclipsed by the identification of work with paid work, an elision which is reflected in the gender-bias of both the sociology of work and the history of technology.

Over the past two decades, with the recognition of housework as work, research on domestic technology has grown rapidly, especially in the United States. A major contributor to this development is Ruth Schwartz Cowan, who set out to re-examine the conventional historical and sociological analysis of the industrial revolution's impact on the family. According to this view, the forces of technological change and the growth of the market economy have progressively absorbed much of the household's role in production. A classic formulation of the position is to be found in Talcott Parsons (1956) who argues that industrialization removed many functions from the family system, until all that remains is consumption, emotional support and bringing up children. Modern technology is seen as having either eliminated or eased the remaining household work, thereby freeing women to enter the labour force. This type of analysis, as Cowan puts it in Chapter 20, is 'a cultural artifact of vast importance' lending weight to popular rhetoric about the 'crisis' of the modern family. It assumes that housewives, with all their labour-saving devices, have nothing left to do.

Despite the massive technological changes in the home in America and other industrialized countries, such as running water, gas and electric cookers, electric irons, washing machines and refrigerators, household labour still accounts for approximately half of total working time (Robinson and Godley 1997). Furthermore, the time spent on housework remained remarkably constant between 1930 and the 1950s (Vanek 1974). During the period of the most intensive industrialization of the home (the 1920s and 1930s, with the widespread deployment of utilities and large appliances), middle-class women receiving these new domestic amenities did not, by and large, seek employment outside the home. The great promise of labour-saving technology had not been realized in the home.

So what was the relationship between these technological developments and household work? Cowan argues that the major changes in the pattern of household work during this period were not those that the traditional model predicts. Some technological systems do fit the production/consumption model, such as the move of aspects of the production of food, clothing and healthcare out of the home, others do not. Rather than household tasks simply being eased or eliminated, mechanization gave rise to a whole range of new tasks. Although not as burdensome physically, these jobs were as time-consuming as the jobs they had replaced.

The transportation system and its relation to changing consumption patterns exemplifies this argument. During the nineteenth century, household goods were often delivered, mail-order catalogues were widespread, and most people did not spend much time buying goods. With the spread of the motor car after the First World War, all this began to change. By 1930, the automobile had become the prime mode of transportation in the United States. Delivery services of all kinds began to disappear and the burden of providing transportation shifted from the seller to the buyer (Strasser 1982). As more and more business converted to the 'self-service' concept, households became increasingly dependent upon housewives to provide the service.[1] The time spent per week on shopping tasks, including the associated travel, expanded until today the average time spent by American housewives is almost the equivalent to an entire working day (Robinson and Godley 1997).

In the first essay reprinted here, Cowan stresses that there is no simple cause and effect relation between the mechanization of homes and changes in the nature of household work. Instead, the most significant change in the structure of household labour was the disappearance of domestic servants and the allocation of the entire job to the housewife herself. This, she argues, stimulated the mechanization of homes, which in turn hastened the disappearance of servants. As the job of the housewife changed, so did the accompanying ideology of the housewife-mother.

The development, in the early years of this century, of the domestic science movement, the germ theory of disease and the idea of 'scientific motherhood' led to new exacting standards of housework and childcare (Ehrenreich and English 1975). These trends were exploited and further promoted by advertisers in their drive to expand the market for domestic appliances. With home and housework acquiring heightened emotional significance, it became impossible to rationalize household production along the lines of industrial production.[2] Domestic technology has thus been designed for use in single-family households by a lone and loving housewife.

The social prevalence of the single-family household – with its assumption of the essentially unaided female homemaker, and its association with widespread goals of privacy and autonomy – has profoundly structured the form of technology that has become available. Domestic technologies that cross the boundaries of the single-family household have been invented, but have persistently failed, even though ownership by individual households is in many cases patently uneconomic in cost terms. The bias towards the individual household and individual housewife has had important design consequences. Papanek and Hennessey (1977: 27) write: 'Few tools in our society are designed for communal (or shared) ownership. If they were designed for sharing, rather than for individual use, we believe they would change structurally, mechanically and in material composition'.

GENDER SPECIALIZATION IN HOUSEHOLD TECHNOLOGY

We have so far examined how the social organization of the family household has structured the general form of domestic technology. However, it is also the case that particular domestic technologies are shaped by gender relations. Recently, more attention has been given to the innovation, development, and diffusion processes of specific technologies. The best study of this type is by Cynthia Cockburn and Susan Ormrod (1993) who trace the trajectory of the microwave oven from its conception right through to its consumption (see also Chabaud-Rychter 1994). The authors unravel the way that the sexual division of labour is mapped onto each stage in the journey of a domestic technology.

Like other domestic technologies, the microwave is designed by men in their capacity as engineers and managers, people remote from the domestic tasks involved, for use by women in their capacity as houseworkers. Where women do enter the picture, apart from on the production line, is primarily as home economists. Cockburn and Ormrod observe that the cooking expertise of the home economists is crucial to the successful design of the artifact. The women see themselves as doing 'a kind of engineering or science' but it is not acknowledged as such by the predominantly male culture of engineers. Their technical skills are undervalued because of the strong association of cooking with femininity. As a result, even at the one point when women enter the innovation process, they wield little influence over the development of new technologies.

What is so original about the microwave study is that it follows the gendering processes through the various stages of the artifact's life. Cockburn and Ormrod recognize that gendering does not begin and end with design and manufacturing. Domestic technologies are also encoded with gendered meanings during their marketing, retailing and appropriation by users (see Mackay and Gillespie 1992). While the technology is made into a physical object during production, the symbolic meanings attaching to it are continually being negotiated. The argument here is that domestic technologies are inscribed with specific ideologies that reflect the sexual division of domestic activities and men's and women's different relationship to machines.

The standard division of electronic technologies into 'white' goods (such as refrigerators, washing machines and microwaves) and 'brown' goods (such as televisions, video recorders, and music systems) is a case in point. Whereas white goods are portrayed as serviceable and simple to use, brown goods are often portrayed as complex, clever technologies that require skills in handling. The contrasting colours signify a gendered conception of household functions and thus a gendered construction of potential purchasers – those to do with domestic work as opposed to those to do with leisure and entertainment. Marketing and retailing play a key role in framing demand: 'there is an unclear dividing line between accurately *representing* the customer, *constructing* the customer and *controlling* the customer' (Cockburn and Ormrod 1993: 109). The market-place for music systems, for example, is peculiarly masculine, with young salesmen displaying their technical prowess in the pitch for certain products. For purchasers, the consumption of a domestic commodity is an activity of self-expression, and a marker of gender identity. Thus marketing and retailing and consumption are all part of the social shaping of technology.

These gendered inscriptions are reflected in the futuristic literature about 'home informatics' (Miles 1988; Negroponte 1995). Technology is at the heart of the house of the future. But it seems that the designers and producers of the technological home have little interest in housework. Anne-Jorunn Berg has studied prototypes for 'smart' houses to get a glimpse of what the high-technology dwellings of the future might be like. What does a utopian vision of a home humming with information technology hold for the family of the future? We might expect at least an attempt at the wholesale elimination of household labour.

Instead, innovations are concerned with the control of heating and lighting, security, information and messaging, entertainment and the environment. The core aim of automating the smart house turns out to be 'integration', that is, centralized control and regulation of all functions in a local network or 'house-brain'. Berg found that a labour-saving 'self-cleaning' house did not feature as an objective. The men who produce prototypes of the intelligent house of the future and design its key technologies ignore the fact that the home is a place of work – especially women's housework. They overlook women, whose domain they are in effect transforming. It is the technology as such, the way artifacts function in technical terms, that fascinates the designers. The target consumer is implicitly the technically-interested man, someone in their own image. The smart house is a deeply masculine vision of a house, rather than a home, somewhat like Corbusier's 'machine for living'. The famous house that Microsoft boss Bill Gates built is the latest example of this trend. Much has been made of the sophisticated audio-visual environment that he created, even featuring virtual artworks programmed to guest tastes. We are none the wiser about how the Microsoft magnate's home is serviced and sustained. The neglect of women's knowledge, experience and skills as a resource for technical innovation in the home is evidence of the gendering of the innovation process.

Berg's article illustrates well how masculine the images are of the future house. She correctly argues that many innovations which would simplify some tasks, like cleaning and ironing, have not been developed. More attention, however, needs to be paid to the physical capacity of machines and the inherent difficulties of automating housework. Berg wryly mentions the robot-butler that can serve drinks, but only after a person has poured them. Given the variety and material realities that household labour involves, there are limits to its mechanization. Even in the differently ordered world of paid work, robots perform only routine tasks in manufacturing, and personal service work has proved impossible to automate. It may be that even designers attuned to housework would have limited success in reducing it, given its nature.

 ## ALTERNATIVES TO INDIVIDUALIZED HOUSEWORK

Even the most forward looking of the futurists have us living in households which, in social rather than technological terms, resemble the households of today. A more radical approach would be to transform the social context in which domestic technology applies. Our final extract on household technology broadens the discussion to include the design of housing itself, and urban design generally.

How might housing be redesigned with alternative priorities in mind? As in the previous section, we need to ask this question in order to remind ourselves that things could be different. During the first few decades of this century there were a range of alternative approaches to housework being considered and experimented with. These included the development of commercial services, the establishment of alternative communities and cooperatives and the invention of different types of machinery. Drawing on Dolores Hayden's *The Grand Domestic Revolution*, Moyra Doorly describes a group of Victorian feminists who sought to redesign homes, neighbourhoods and cities. Recognizing that the exploitation of women's labour by men was embodied in the actual design of houses, these women believed that changing the entire physical framework of houses and neighbourhoods was the only way to free women from domestic drudgery. The object was to socialize household labour and childcare.

Two of the more influential women were Melusina Fay Peirce and Charlotte Perkins Gilman. In 1868 Melusina Fay Peirce outlined plans for cooperative residential neighbourhoods made up of kitchenless houses and a cooperative housekeeping centre. She suggested that women organize to perform their household tasks cooperatively, building communal kitchens, laundries and childcare centres as necessary. Freed from the domestic routine, they would then be able to develop other interests outside the home.

Charlotte Perkins Gilman saw the socialization of domestic work, rather than cooperation in its execution, as the means to economic independence for women. She envisaged a completely professionalized system of house-keeping which would free women from the ties of cooking, cleaning and childcare. Hayden describes these ideas as part of a 'lost feminist tradition' obscured by the dominance today of single-family residences and the private ownership of correspondingly small-scale amenities.

As Cowan so graphically puts it: 'Several million American women cook supper each night in several million separate homes over several million separate stoves' (Cowan 1979: 59; see also Cowan 1983). This is not an inevitable, immutable situation, but one whose transformation depends on the transformation of gender relations. Should such transformations occur, domestic technology, which presently reflects a man's world, would surely change also.

 ## THE TECHNOLOGIES OF BIOLOGICAL REPRODUCTION

Much of the recent literature on technoscience has concentrated on human biological reproduction – technologies for the body. Techniques such as in vitro fertilization, egg donation, artificial insemination, surrogacy and most recently the possibility of human cloning have generated considerable excitement, as well as anxiety, about their potential to disrupt the whole notion of genetic parenthood, thus placing conventional nuclear family relationships in jeopardy. At the same time, the new genetics is spawning explanations of phenomena as diverse as school failure, alcoholism, cancer, and even homosexuality in terms of our genetic make-up (Rose 1994: 172). These new technologies appear to offer fantastic opportunities for self-realization – we can literally redesign our bodies and commission designer babies.

What is obscured by this discourse of individual choice is the way our choices are highly constrained. We are in fact selecting from a very re-stricted range of technological options, which are themselves shaped by particular political and economic interests. This is well illustrated in our extracts on contraceptive technology. Here, a social shaping perspective reveals the profound influence of the professional interests of the medical establishment, and the ideology of marriage and motherhood, on the development of reproductive technologies.

TECHNOLOGY AND PROFESSIONALIZATION

Medical technology is moulded in no straightforward sense by a simple goal of efficacious healing. Prevailing medical theory, the social nature of the doctor-patient relationship, institutional frameworks such as hospitals, professionalism as a practice and ideology, the financial interests of the suppliers of drugs and other technologies, the place of medicine as an important area of economic activity in advanced societies, divides of gender, class and race, the role of the state – all these appear to have a place in shaping medical technology (see, for example, Reiser 1978; McKinlay 1981, 1982; Baker 1996). Within Western medicine, high technology activities are not only the key to power at the level of doctor-patient relations, but also to power within the profession.[3] Status, money and professional acclaim are distributed according to the technological sophistication of the medical speciality. Developing and expanding high technology procedures signals success in the competition for scarce resources – as between specialists, between hospitals, and between individuals. Medical specialization and technological innovation go hand in hand.

The evolution of in vitro fertilization and embryo transfer illustrate this process. They have both proceeded without sufficient work on establishing causes of infertility or improving other treatments. Given the low success rate of these two techniques, the level of physical discomfort, the health risk and the psychological distress that accompany them, why the current concentration on in vitro fertilization among infertility specialists? How does it happen that resources are allocated to this 'unsuccessful' technology?

While new technologies generally have a high failure rate until perfected, it is also the case that 'many roads' are not taken in science. Without doubt, professional interests explain a great deal about the development of these techniques (Pfeffer 1993). Before the introduction of in vitro fertilization and embryo transfer, the investigation and treatment of infertility had long been afforded low status in the medical hierarchy. Many of the procedures were carried out by general practitioners, as they required little special knowledge. In vitro fertilization provides gynaecologists with an exciting, high-status area of research as well as a technically complex practice. Status and substantial financial reward are to be had, as well as job satisfaction. Indeed, procreation has now become a highly profitable industry for the private medical sector and for the multinational pharmaceutical companies who supply fertility drugs. Professionalization strategies and commercialism are central to the development of biomedical technologies.

Equally important, but hidden from view, is the undertow created by the institution of the family and the ideology of motherhood. The political effect of the significance attached to maintaining the stability of this 'building block of society' should not be underestimated. Nowhere are sexual relations more profoundly formative of a set of technologies than in the sphere of contraception.

THE SEXUAL RELATIONS OF CONTRACEPTIVE TECHNOLOGIES

The perspectives from which most histories of fertility control are written are redolent of technological determinism. The conventional view is that in pre-industrial societies women were the victims of their own fecundity. It is assumed that earlier generations were prevented from practising birth control because they lacked the necessary technology. The pill, a technical invention, is credited both with enabling women for the first time to control their fertility, and with the massive social changes for women that accompanied its introduction.

On closer analysis it is apparent that birth control practices, and the form they take, are as much dependent on a society's attitude to sex, children and the status of women as on effective technology: 'For the use of birth control requires a morality that permits the separation of sexual intercourse from procreation, and is related to the extent to which women are valued for roles other than wife and mother' (Greenwood and King 1981: 169). Birth control has always been a matter of social and political acceptability as well as of medicine and technology.

The opening extract in this section describes how the first intrauterine contraceptive device (IUD) to be widely used, the Grafenberg Ring, was welcomed at the Sex Reform Congress of 1929. Anni Dugdale argues that we can understand the development and form of this contraceptive technology only in the context of the new discourse of equality in sexual desire. During the first quarter of this century, our ideas of a natural female body were dramatically transformed. The Victorian idea of women as non-sexual and passive gradually gave way to one which understood women's bodies as sexually active. In the new discourses of the sex reformers, men and women were positioned as having different but equal sexual desires. Women's sexual freedom was seen as the extension of the suffrage struggle from the public sphere into the sphere of personal relationships.

The characteristics of the Grafenberg Ring fitted in with the aspirations of the sex reform movement and the new model of femininity. As a completely internal device, it represented a refined approach to the flesh – it was not physically present during sex, so women felt comfortable with it. Further, it was represented as a scientific device that had to be inserted by a skilled physician and the association of IUDs with the knowledge and practices of midwives was thus broken. Finally, it was constituted as a contraceptive *choice* and so resonated with the discourse of the new sexually confident woman. The physical design of the IUD, and its institutionalization as a scientific method of birth control expressed the goals of women activists in the sex reform movement. In turn, the IUD participated in the material realization of a new body for women.

There has been much excellent feminist research on how scientific knowledge has represented the female body and female sexuality (Keller 1984; Jordanova 1989; Duden 1991). Dichotomies such as that between biology and culture situated women as natural and different, and have served to sustain women's oppression. The language of the biomedical sciences today is no less suffused with implicit assumptions about, and imagery of, sexual difference (Martin 1987). Most of this feminist literature is concerned with the analysis of scientific texts and theories, with the cultural production of images and meanings of sex and the body. Over the last decade or so, feminists have extended this linguistic work to argue that there is *now* no such thing as the natural, physiological body. One consequence of this work is that the conventional distinction between sex (natural) and gender (social) has been thoroughly contested and deconstructed. This has led to a major shift in the analysis of reproductive technologies. Earlier work on the impact of reproductive technologies assumed that the body is biologically given and fixed. Technologies, like science, are now seen as contributing to the stabilization of meanings of the body. With the rise of modern science, bodies have become objects that can be transformed with an increasing number of tools and techniques. Modern bodies are made and remade through science and technology; they too are technological artifacts.

The extract here by Nelly Oudshoorn illustrates the connection between the gendered discourses of the biomedical sciences and the institutionalization of medical techniques applied to women's bodies. Oudshoorn reminds us that the conceptualization of male and female bodies as essentially different, rather than similar, is a modern one, dating only from the eighteenth century. The identification of the female body as the Other resulted in positioning it as the quintessential medical object. Sex and reproduction were seen as the defining characteristics of women, and this was reflected in the establishment of gynaecology as a separate branch of medicine. With the rise of sex endocrinology in the 1920s and 1930s the notion of the female body as the reproductive body was integrated into the hormonal model. Women's bodies thus became set apart as the prime site for biomedical practices of the body. The idea that women are at the mercy of their hormones in a way that men are not is still influential today. Observe the vast new market for hormone replacement therapy to treat female menopause, which has no equivalent for men, as male menopause is not defined as a 'medical condition'.[4]

It was logical then for research on the first physiological contraceptive to be focused exclusively on women.[5] Oudshoorn shows how discourses about the natural body shaped the pill, and how the pill, in turn, constructed women's bodies as universal with respect to their reproductive functions. The scientists who were developing the pill attempted to design a universal 'one-size-fits-all' contraceptive technology, because they saw all women as being basically the same.

What is particularly interesting about Oudshoorn's account is that she shows how these scientists succeeded in literally 'making' women the same. It turns out that the design of the pill as a regime of medication to be taken for 20 days a month was shaped by moral considerations and notions of the natural body. Gregory Pincus, the American biologist who headed the research team, could have chosen any desired length for the menstrual cycle. He chose to make a pill that mimicked the 'normal' menstrual cycle. As a result, all pill users now have a regular cycle of four weeks and the variety in menstrual cycles among women has been diminished. Oud-shoorn argues that the pill thus literally homogenized women's repro-ductive functions on a mass scale.

As it happens, the pill has not been seen as universally applicable and appropriate for all women at all times in all places. Discourses of con-traception have therefore shifted from an assumption of sameness to a recognition of the diversity of women's bodies and needs. Witness the postmodern age of the designer contraceptive. However, postmodern dis-courses of difference sometimes overlap with modern racist discourses, especially those which centre on cultural difference. On the one hand, the 1980s saw the introduction of a variety of contraceptive methods and, for the first time, a male pill finally got onto the agenda.[6] Choices increased for white middle-class women. On the other hand, contraceptive hormonal implants (such as Norplant), developed by the Population Council in the 1970s, are mainly used by women in the Third World. The influence of population-control ideology is still central to modern birth control pro-grammes. Technological design and use may be more driven by difference but these differences are not as new as many postmodern authors would have us believe.

Recent analyses of reproductive technologies do not simply portray tech-nology as either progressive and life-enhancing, or as destroying life and the environment. We now recognize that harking back to the natural or bio-logical body, free of technological intervention, is misguided. There is no pure nature to return to, whether in the form of so-called natural childbirth or natural farming. In a world where sex, gender, bodies and sexuality can be altered, it makes even less sense to speak of the natural as opposed to the artifactual. Rather than fearing and rejecting technologies, we need to engage with the technology and argue for different kinds of technology based on social justice. In highlighting the way technologies do not have to be the way they are, the social shaping approach provides scope for human agency and political intervention in creating future technologies.

▶ **NOTES**

1 The idea that we are moving towards a self-service economy, in which paid services are increasingly substituted by the purchase of household commodities and self-servicing, has been developed by Gershuny (1983) who shares the popular belief that domestic technology will reduce the time spent on housework. For a discussion of these ideas, see Wajcman (1991a, Ch. 4).

2 For a broader account that locates the electrical modernization of the American home in the context of Franklin Roosevelt's New Deal, see Tobey (1996).

3 The decisive role of technology in the contest between female midwives and the emerging male-dominated medical profession as to who would have control over intervention in the birth process has been well documented. See, for example, Leavitt (1986).

4 Viagra, the first treatment for impotence that can be taken in tablet form, appears likely to change this situation drastically. Interestingly, its launch coincides with a redefinition of impotence as a biological, rather than a psychological, condition.

5 Research on the causes of infertility has similarly focused almost exclusively on the female body; see Pfeffer (1993).

6 Most of the research into medical contraceptive methods is still done on techniques for use by women, see Snow (1994).

20 ▶ The industrial revolution in the home

Ruth Schwartz Cowan

When we think about the interaction between technology and society, we tend to think in fairly grandiose terms: massive computers invading the workplace, railroad tracks cutting through vast wildernesses, armies of women and children toiling in the mills. These grand visions have blinded us to an important and rather peculiar technological revolution which has been going on right under our noses: the technological revolution in the home. This revolution has transformed the conduct of our daily lives, but in somewhat unexpected ways. The industrialization of the home was a process very different from the industrialization of other means of production, and the impact of that process was neither what we have been led to believe it was nor what students of the other industrial revolutions would have been led to predict.

Some years ago sociologists of the functionalist school formulated an explanation of the impact of industrial technology on the modern family. Although that explanation was not empirically verified, it has become almost universally accepted.[1] Despite some differences in emphasis, the basic tenets of the traditional interpretation can be roughly summarized as follows:

Before industrialization the family was the basic social unit. Most families were rural, large, and self-sustaining; they produced and processed almost everything that was needed for their own support and for trading in the marketplace, while at the same time performing a host of other functions ranging from mutual protection to entertainment. In these preindustrial families women (adult women, that is) had a lot to do, and their time was almost entirely absorbed by household tasks. Under industrialization the family is much less important. The household is no longer the focus of production; production for the marketplace and production for sustenance have been removed to other locations. Families are smaller and they are urban rather than rural. The number of social functions they perform is much reduced, until almost all that remains is consumption, socialization of small children, and tension management. As their functions diminished, families became atomized; the social bonds that had held them together were loosened. In these postindustrial families women have very little to do, and the tasks with which they fill their time have lost the social utility that they once possessed. Modern women are in trouble, the analysis goes, because modern families are in trouble; and modern families are in trouble because industrial technology has either eliminated or eased almost all their former functions, but modern ideologies have not kept pace with the change. The results of this time lag are several: some women suffer from role anxiety, others land in the divorce courts, some enter the labor market, and others take to burning their brassieres and demanding liberation.

This sociological analysis is a cultural artifact of vast importance. Many Americans believe that it is true and act upon that belief in various ways: some hope to reestablish family solidarity by relearning lost productive crafts – baking bread, tending a vegetable garden – others dismiss the women's liberation movement as 'simply a bunch of affluent housewives who have nothing better to do with their time.' As disparate as they may seem, these reactions have a common ideological source – the standard sociological analysis of the impact of technological change on family life.

As a theory this functionalist approach has much to recommend it, but at present we have very little evidence to back it up. Family history is an infant discipline, and what evidence it has produced in recent years does not lend credence to the standard view.[2] Phillippe Ariès has shown, for example, that in France the ideal of the small nuclear family predates industrialization by more than a century.[3] Historical demographers working on data from English and French families have been surprised to find that most families were quite small and that several generations did not ordinarily reside together; the extended family, which is supposed to have been the rule in preindustrial societies, did not occur in colonial New England either.[4] Rural English families routinely employed domestic servants, and even very small English villages had their butchers and bakers and candlestick makers; all these persons must have eased some of the chores that would otherwise have been the housewife's burden.[5] Preindustrial housewives no doubt had much with which to occupy their time, but we may have reason to wonder whether there was quite as much pressure on them as sociological ortho- doxy has led us to suppose. The large rural family that was sufficient unto itself back there on the prairies may have been limited to the prairies – or it may never have existed at all (except, that is, in the reveries of the sociologists).

Even if all the empirical evidence were to mesh with the functionalist theory, the theory would still have problems, because its logical structure is rather weak. Comparing the average farm family in 1750 (assuming that you knew what that family was like) with the average urban family in 1950 in order to discover the significant social changes that had occurred is an exercise rather like comparing apples with oranges; the differences between the fruits may have nothing to do with the differences in their evolution. Transferring the analogy to the case at hand, what we really need to know is the difference, say, between an urban laboring family of 1750 and an urban laboring family 100 and then 200 years later, or the difference between the rural nonfarm middle classes in all three centuries, or the difference between the urban rich yesterday and today. Surely in each of these cases the analyses will look very different from what we have been led to expect. As a guess we might find that for the urban laboring families the changes have been precisely the opposite of what the model predicted; that is, that their family structure is much firmer today than it was in centuries past. Similarly, for the rural nonfarm middle class the results might be equally surprising; we might find that married women of that class rarely did any housework at all in 1890 because they had farm girls as servants, whereas in 1950 they bore the full brunt of the work themselves. I could go on, but the point is, I hope, clear: in order to verify or falsify the functionalist theory, it will be necessary to know more than we presently do about the impact of industrialization on families of similar classes and geographical locations.

With this problem in mind I have, for the purposes of this initial study, deliberately limited myself to one kind of technological change affecting one aspect of family life in only one of the many social classes of families that might have been considered. What happened, I asked, to middle-class American women when the implements with which they did their everyday household work changed? Did the technological change in household appliances have any effect upon the structure of American households, or upon the ideologies that governed the behavior of American women, or upon the functions that families needed to perform? Middle-class American women were defined as actual or potential readers of the better-quality women's magazines, such as the *Ladies' Home Journal, American Home, Parents' Magazine, Good Housekeeping,* and *McCall's.*[6] Nonfictional material (articles and advertisements) in those magazines was used as a partial indicator of some of the technological and social changes that were occurring.

The *Ladies' Home Journal* has been in continuous publication since 1886. A casual survey of the nonfiction in the *Journal* yields the immediate impression that that decade between the end of World War I and the beginning of the depression witnessed the most drastic changes in patterns of household work. Statistical data bear out this impression. Before 1918, for example, illustrations of homes lit by gaslight could still be found in the *Journal*; by 1928 gaslight had disappeared. In 1917 only one-quarter (24.3 percent) of the dwellings in the United States had been electrified, but by 1920 this figure had doubled (47.4 percent – for rural nonfarm and urban dwellings), and by 1930 it had risen to four-fifths (80 percent).[7] If electrification had meant simply the change from gas or oil lamps to electric lights, the changes in the housewife's routines might not have been very great (except for eliminating the chore of cleaning and filling oil lamps); but changes in lighting were the least of the changes that electrification implied. Small electric appliances followed quickly on the heels of the electric light, and some of those augured much more profound changes in the housewife's routine.

Ironing, for example, had traditionally been one of the most dreadful household chores, especially in warm weather when the kitchen stove had to be kept hot for the better part of the day; irons were heavy and they had to be returned to the stove frequently to be reheated. Electric irons eased a good part of this burden.[8] They were relatively inexpensive and very quickly replaced their predecessors; advertisements for electric irons first began to appear in the ladies' magazines after the war, and by the end of the decade the old flatiron had disappeared; by 1929 a survey of 100 Ford employees revealed that ninety-eight of them had the new electric irons in their homes.[9]

Data on the diffusion of electric washing machines are somewhat harder to come by; but it is clear from the advertisements in the magazines, particularly advertisements for laundry soap, that by the middle of the 1920s those machines could be found in a significant number of homes. The washing machine is depicted just about as frequently as the laundry tub by the middle of the 1920s; in 1929, forty-nine out of those 100 Ford workers had the machines in their homes. The washing machines did not drastically reduce the time that had to be spent on household laundry, as they did not go through their cycles automatically and did not spin dry; the housewife had to stand guard, stopping and starting the machine at appropriate times, adding soap, sometimes attaching the drain pipes, and putting the clothes through the wringer manually. The machines did, however, reduce a good part of the drudgery that once had been associated with washday, and this was a matter of no small consequence.[10] Soap powders appeared on the market in the early 1920s, thus eliminating the need to scrape and boil bars of laundry soap.[11] By the end of the 1920s Blue Monday must have been considerably less blue for some housewives – and probably considerably less 'Monday,' for with an electric iron, a washing machine, and a hot water heater, there was no reason to limit the washing to just one day of the week.

Like the routines of washing the laundry, the routines of personal hygiene must have been transformed for many households during the 1920s – the years of the bathroom mania.[12] More and more bathrooms were built in older homes, and new homes began to include them as a matter of course. Before the war most bathroom fixtures (tubs, sinks, and toilets) were made out of porcelain by hand; each bathroom was custom-made for the house in which it was installed. After the war industrialization descended upon the bathroom industry; cast iron enamelware went into mass production and fittings were standardized. In 1921 the dollar value of the production of enameled sanitary fixtures was $2.4 million, the same as it had been in 1915. By 1923, just two years later, that figure had doubled to $4.8 million; it rose again, to $5.1 million, in 1925.[13] The first recessed, double-shell cast iron enameled bathtub was put on the market in the early 1920s. A decade later the standard American bathroom had achieved its standard American form: the recessed tub, plus tiled floors and walls, brass plumbing, a single-unit toilet, an enameled sink, and a medicine chest, all set into a small room which was very often 5 feet square.[14] The bathroom evolved more quickly than any other room of the house; its standardized form was accomplished in just over a decade.

Along with bathrooms came modernized systems for heating hot water: 61 percent of the homes in Zanesville, Ohio, had indoor plumbing with centrally heated water by 1926, and 33 percent of the homes valued over $2,000 in Muncie, Indiana, had hot and cold running water by 1935.[15] These figures may not be typical of small American cities (or even large American cities) at those times, but they do jibe with the impression that one gets from the magazines: after 1918 references to hot water heated on the kitchen range, either for laundering or for bathing, become increasingly difficult to find.

Similarly, during the 1920s many homes were outfitted with central heating; in Muncie most of the homes of the business class had basement heating in 1924; by 1935 Federal Emergency Relief Administration data for the city indicated that only 22.4 percent of the dwellings valued over $2,000 were still heated by a kitchen stove.[16] What all these changes meant in terms of new habits for the average housewife is somewhat hard to calculate; changes there must have been, but it is difficult to know whether those changes produced an overall saving of labor and/or time. Some chores were eliminated – hauling water, heating water on the stove, maintaining the kitchen fire – but other chores were added – most notably the chore of keeping yet another room scrupulously clean.

It is not, however, difficult to be certain about the changing habits that were associated with the new American kitchen – a kitchen from which the coal stove had disappeared. In Muncie in 1924, cooking with gas was done in two out of three homes; in 1935 only 5 percent of the homes valued over $2,000 still had coal or wood stoves for cooking.[17] After 1918 advertisements for coal and wood stoves disappeared from the *Ladies' Home Journal*; stove manufacturers purveyed only their gas, oil, or electric models. Articles giving advice to homemakers on how to deal with the trials and tribulations of starting, stoking, and maintaining a coal or a wood fire also disappeared. Thus it seems a safe assumption that most middle-class homes had switched to the new method of cooking by the time the depression began. The change in routine that was predicated on the change from coal or wood to gas or oil was profound; aside from the elimination of such chores as loading the fuel and removing the ashes, the new stoves were much easier to light, maintain, and regulate (even when they did not have thermostats, as the earliest models did not).[18] Kitchens were, in addition, much easier to clean when they did not have coal dust regularly tracked through them; one writer in the *Ladies' Home Journal* estimated that kitchen cleaning was reduced by one-half when coal stoves were eliminated.[19]

Along with new stoves came new foodstuffs and new dietary habits. Canned foods had been on the market since the middle of the 19th century, but they did not become an appreciable part of the standard middle-class diet until the 1920s – if the recipes given in cookbooks and in women's magazines are a reliable guide. By 1918 the variety of foods available in cans had been considerably expanded from the peas, corn, and succotash of the

19th century; an American housewife with sufficient means could have purchased almost any fruit or vegetable and quite a surprising array of ready-made meals in a can – from Heinz's spaghetti in meat sauce to Purity Cross's lobster à la Newburg. By the middle of the 1920s home canning was becoming a lost art. Canning recipes were relegated to the back pages of the women's magazines; the business-class wives of Muncie reported that, while their mothers had once spent the better part of the summer and fall canning, they themselves rarely put up anything, except an occasional jelly or batch of tomatoes.[20] In part this was also due to changes in the technology of marketing food; increased use of refrigerated railroad cars during this period meant that fresh fruits and vegetables were in the markets all year round at reasonable prices.[21] By the early 1920s convenience foods were also appearing on American tables: cold breakfast cereals, pancake mixes, bouillon cubes, and packaged desserts could be found. Wartime shortages accustomed Americans to eating much lighter meals than they had previously been wont to do, and as fewer family members were taking all their meals at home (businessmen started to eat lunch in restaurants downtown, and factories and schools began installing cafeterias), there was simply less cooking to be done, and what there was of it was easier to do.[22]

Many of the changes just described – from hand power to electric power, from coal and wood to gas and oil as fuels for cooking, from one-room heating to central heating, from pumping water to running water – are enormous technological changes. Changes of a similar dimension, either in the fundamental technology of an industry, in the diffusion of that technology, or in the routines of workers, would have long since been labeled an 'industrial revolution.' The change from the laundry tub to the washing machine is no less profound than the change from the hand loom to the power loom; the change from pumping water to turning on a water faucet is no less destructive of traditional habits than the change from manual to electric calculating. It seems odd to speak of an 'industrial revolution' connected with housework, odd because we are talking about the technology of such homely things, and odd because we are not accustomed to thinking of housewives as a labor force or of housework as an economic commodity – but despite this oddity, I think the term is altogether appropriate.

In this case other questions come immediately to mind, questions that we do not hesitate to ask, say, about textile workers in Britain in the early 19th century, but we have never thought to ask about housewives in America in the 20th century. What happened to this particular work force when the technology of its work was revolutionized? Did structural changes occur? Were new jobs created for which new skills were required? Can we discern new ideologies that influenced the behavior of the workers?

The answer to all of these questions, surprisingly enough, seems to be yes. There were marked structural changes in the work force, changes that increased the work load and the job description of the workers that remained. New jobs were created for which new skills were required; these jobs were not physically burdensome, but they may have taken up as much time as the jobs they had replaced. New ideologies were also created, ideologies which reinforced new behavioral patterns, patterns that we might not have been led to expect if we had followed the sociologists' model to the letter. Middle-class housewives, the women who must have first felt the impact of the new household technology, were not flocking into the divorce courts or the labor market or the forums of political protest in the years immediately after the revolution in their work. What they were doing was sterilizing baby bottles, shepherding their children to dancing classes and music lessons, planning nutritious meals, shopping for new clothes, studying child psychology, and hand stitching color-coordinated curtains – all of which chores (and others like them) the standard sociological model has apparently not provided for.

The significant change in the structure of the household labor force was the disappearance of paid and unpaid servants (unmarried daughters, maiden aunts, and grandparents fall in the latter category) as household workers – and the imposition of the entire job on the housewife herself. Leaving aside for a moment the question of which was cause and which effect (did the disappearance of the servant create a demand for the new technology, or did the new technology make the servant obsolete?), the phenomenon itself is relatively easy to document. Before World War I, when illustrators in the women's magazines depicted women doing housework, the women were very often servants. When the lady of the house was drawn, she was often the person being served, or she was supervising the serving, or she was adding an elegant finishing touch to the work. Nursemaids diapered babies, seamstresses pinned up hems, waitresses served meals, laundresses did the wash, and cooks did the cooking. By the end of the 1920s the servants had disappeared from those illustrations; all those jobs were being done by housewives – elegantly manicured and coiffed, to be sure, but housewives nonetheless (compare Figures 20.1 and 20.2).

If we are tempted to suppose that illustrations in advertisements are not a reliable indicator of structural changes of this sort, we can corroborate the changes in other ways. Apparently, the illustrators really did know whereof they drew. Statistically the number of persons throughout the country employed in household service dropped from 1,851,000 in 1910 to 1,411,000 in 1920, while the number of households enumerated in the census rose from 20.3 million to 24.4 million.[23] In Indiana the ratio of households to servants increased from 13.5/1 in 1890 to 30.5/1 in 1920, and in the country as a whole the number of paid domestic servants per 1,000 population dropped from 98.9 in 1900 to 58.0 in 1920.[24] The business-class housewives of Muncie reported that they employed approximately one-half as many woman-hours of domestic service as their mothers had done.[25]

In case we are tempted to doubt these statistics (and indeed statistics about household labor are particularly unreliable, as the labor is often transient, part-time, or simply unreported), we can turn to articles on the servant problem, the disappearance of unpaid family workers, the design of kitchens, or to architectural drawings for houses. All of this evidence reiterates the same point: qualified servants were difficult to find; their wages had risen and their numbers fallen; houses were being designed

Figure 20.1 The housewife as manager (*Ladies' Home Journal*, April 1918, by permission of Lever Brothers Co.)

without maid's rooms; daughters and unmarried aunts were finding jobs downtown; kitchens were being designed for housewives, not for servants.[26] The first home with a kitchen that was not an entirely separate room was designed by Frank Lloyd Wright in 1934.[27] In 1937 Emily Post invented a new character for her etiquette books: Mrs. Three-in-One, the woman who is her own cook, waitress, and hostess.[28] There must have been many new Mrs. Three-in-Ones abroad in the land during the 1920s.

Figure 20.2 The housewife as laundress (*Ladies' Home Journal*, August 1928, by permission of Colgate–Palmolive–Peet)

As the number of household assistants declined, the number of household tasks increased. The middle-class housewife was expected to demonstrate competence at several tasks that previously had not been in her purview or had not existed at all. Child care is the most obvious example. The average housewife had fewer children than her mother had had, but she was expected to do things for her children that her mother would never have dreamed of doing: to prepare their special infant formulas, sterilize their bottles, weigh them every day, see to it that they ate nutritionally balanced meals, keep them isolated and confined when they had even the slightest illness, consult with their teachers frequently, and chauffeur them to dancing lessons, music lessons, and evening parties.[29] There was very little Freudianism in this new attitude toward child care: mothers were not spending more time and effort on their children because they feared the psychological trauma of separation, but because competent nursemaids could not be found, and the new theories of child care required constant attention from well-informed persons – persons who were willing and able to read about the latest discoveries in nutrition, in the control of contagious diseases, or in the techniques of behavioral psychology. These persons simply had to be their mothers.

Consumption of economic goods provides another example of the housewife's expanded job description; like child care, the new tasks associated with consumption were not necessarily physically burdensome, but they were time consuming, and they required the acquisition of new skills.[30] Home economists and the editors of women's magazines tried to teach housewives to spend their money wisely. The present generation of housewives, it was argued, had been reared by mothers who did not ordinarily shop for things like clothing, bed linens, or towels; consequently modern housewives did not know how to shop and would have to be taught. Furthermore, their mothers had not been accustomed to the wide variety of goods that were now available in the modern marketplace; the new housewives had to be taught not just to be consumers, but to be informed consumers.[31] Several contemporary observers believed that shopping and shopping wisely were occupying increasing amounts of housewives' time.[32]

Several of these contemporary observers also believed that standards of household care changed during the decade of the 1920s.[33] The discovery of the 'household germ' led to almost fetishistic concern about the cleanliness of the home. The amount and frequency of laundering probably increased, as bed linen and underwear were changed more often, children's clothes were made increasingly out of washable fabrics, and men's shirts no longer had replaceable collars and cuffs.[34] Unfortunately all these changes in standards are difficult to document, being changes in the things that people regard as so insignificant as to be unworthy of comment; the improvement in standards seems a likely possibility, but not something that can be proved.

In any event we do have various time studies which demonstrate somewhat surprisingly that housewives with conveniences were spending just as much time on household duties as were housewives without them – or, to put it another way, housework, like so many other types of work, expands to fill the time available.[35] A study comparing the time spent per week in housework by 288 farm families and 154 town families in Oregon in 1928 revealed 61 hours spent by farm wives and 63.4 hours by town wives; in 1929 a U.S. Department of Agriculture study of families in various states produced almost identical results.[36] Surely if the standard sociological model were valid, housewives in towns, where presumably the benefits of specialization and electrification were most likely to be available, should have been spending far less time at their work than their rural sisters. However, just after World War II economists at Bryn Mawr College reported the same phenomenon: 60.55 hours spent by farm housewives, 78.35 hours by women in small cities, 80.57 hours by women in large ones – precisely the reverse of the results that were expected.[37] A recent survey of time studies conducted between 1920 and 1970 concludes that the time spent on housework by nonemployed housewives has remained remarkably constant throughout the period.[38] All these results point in the same direction: mechanization of the household meant that time expended on some jobs decreased, but also that new jobs were substituted, and in some cases – notably laundering – time expenditures for old jobs increased because of higher standards. The advantages of mechanization may be somewhat more dubious than they seem at first glance.

As the job of the housewife changed, the connected ideologies also changed; there was a clearly perceptible difference in the attitudes that women brought to housework before and after World War I.[39] Before the war the trials of doing housework in a servantless home were discussed and they were regarded as just that – trials, necessary chores that had to be got through until a qualified servant could be found. After the war, housework changed: it was no longer a trial and a chore, but something quite different – an emotional 'trip.' Laundering was not just laundering, but an expression of love; the housewife who truly loved her family would protect them from the embarrassment of tattletale gray. Feeding the family was not just feeding the family, but a way to express the housewife's artistic inclinations and a way to encourage feelings of family loyalty and affection. Diapering the baby was not just diapering, but a time to build the baby's sense of security and love for the mother. Cleaning the bathroom sink was not just cleaning, but an exercise of protective maternal instincts, providing a way for the housewife to keep her family safe from disease. Tasks of this emotional magnitude could not possibly be delegated to servants, even assuming that qualified servants could be found.

Women who failed at these new household tasks were bound to feel guilt about their failure. If I had to choose one word to characterize the temper of the women's magazines during the 1920s, it would be 'guilt.' Readers of the better-quality women's magazines are portrayed as feeling guilty a good lot of the time, and when they are not guilty they are embarrassed: guilty if their infants have not gained enough weight, embarrassed if their drains are clogged, guilty if their children go to school in soiled clothes, guilty if all the germs behind the bathroom sink are not eradicated, guilty if they fail to notice the first signs of an oncoming cold, embarrassed if accused of having body odor, guilty if their sons go to school without good breakfasts, guilty if their daughters are unpopular because of old-fashioned, or unironed, or – heaven forbid – dirty dresses (see Figures 20.3 and 20.4). In earlier times women were made to feel guilty if they abandoned their children or were too free with their affections. In the years after World War I, American

Figure 20.3 Sources of housewifely guilt: the good mother smells sweet
(*Ladies' Home Journal*, August 1928, by permission of Warner-Lambert Inc.)

women were made to feel guilty about sending their children to school in
scuffed shoes. Between the two kinds of guilt there is a world of difference.

Let us return for a moment to the sociological model with which this essay
began. The model predicts that changing patterns of household work will
be correlated with at least two striking indicators of social change: the

Figure 20.4 Sources of housewifely guilt: the good mother must be beautiful (*Ladies' Home Journal*, July 1928, by permission of Colgate–Palmolive–Peet)

divorce rate and the rate of married women's labor force participation. That correlation may indeed exist, but it certainly is not reflected in the women's magazines of the 1920s and 1930s: divorce and full-time paid employment were not part of the life-style or the life pattern of the middle-class housewife as she was idealized in her magazines.

There were social changes attendant upon the introduction of modern technology into the home, but they were not the changes that the traditional functionalist model predicts; on this point a close analysis of the statistical data corroborates the impression conveyed in the magazines. The divorce rate was indeed rising during the years between the wars, but it was not rising nearly so fast for the middle and upper classes (who had, presumably, easier access to the new technology) as it was for the lower classes. By almost every gauge of socioeconomic status – income, prestige of husband's work, education – the divorce rate is higher for persons lower on the socioeconomic scale – and this is a phenomenon that has been constant over time.[40]

The supposed connection between improved household technology and married women's labor force participation seems just as dubious, and on the same grounds. The single socioeconomic factor which correlates most strongly (in cross-sectional studies) with married women's employment is husband's income, and the correlation is strongly negative; the higher his income, the less likely it will be that she is working.[41] Women's labor force participation increased during the 1920s but this increase was due to the influx of single women into the force. Married women's participation increased slightly during those years, but that increase was largely in factory labor – precisely the kind of work that middle-class women (who were, again, much more likely to have labor-saving devices at home) were least likely to do.[42] If there were a necessary connection between the improvement of household technology and either of these two social indicators, we would expect the data to be precisely the reverse of what in fact has occurred: women in the higher social classes should have fewer functions at home and should therefore be more (rather than less) likely to seek paid employment or divorce.

Thus for middle-class American housewives between the wars, the social changes that we can document are not the social changes that the functionalist model predicts; rather than changes in divorce or patterns of paid employment, we find changes in the structure of the work force, in its skills, and in its ideology. These social changes were concomitant with a series of technological changes in the equipment that was used to do the work. What is the relationship between these two series of phenomena? Is it possible to demonstrate causality or the direction of that causality? Was the decline in the number of households employing servants a cause or an effect of the mechanization of those households? Both are, after all, equally possible. The declining supply of household servants, as well as their rising wages, may have stimulated a demand for new appliances at the same time that the acquisition of new appliances may have made householders less inclined to employ the laborers who were on the market. Are there any techniques available to the historian to help us answer these questions?

In order to establish causality, we need to find a connecting link between the two sets of phenomena, a mechanism that, in real life, could have made the causality work. In this case a connecting link, an intervening agent between the social and the technological changes, comes immediately to mind: the advertiser – by which term I mean a combination of the manufacturer of the new goods, the advertising agent who promoted the goods, and the periodical that published the promotion. All the new devices and new foodstuffs that were being offered to American households were being manufactured and marketed by large companies which had considerable amounts of capital invested in their production: General Electric, Procter & Gamble, General Foods, Lever Brothers, Frigidaire, Campbell's, Del Monte, American Can, Atlantic & Pacific Tea – these were all well-established firms by the time the household revolution began, and they were all in a position to pay for national advertising campaigns to promote their new products and services. And pay they did; one reason for the expanding size and number of women's magazines in the 1920s was, no doubt, the expansion in revenues from available advertisers.[43]

Those national advertising campaigns were likely to have been powerful stimulators of the social changes that occurred in the household labor force; the advertisers probably did not initiate the changes, but they certainly encouraged them. Most of the advertising campaigns manifestly worked, so they must have touched upon areas of real concern for American housewives. Appliance ads specifically suggested that the acquisition of one gadget or another would make it possible to fire the maid, spend more time with the children, or have the afternoon free for shopping.[44] Similarly, many advertisements played upon the embarrassment and guilt which were now associated with household work. Ralston, Cream of Wheat, and Ovaltine were not themselves responsible for the compulsive practice of weighing infants and children repeatedly (after every meal for newborns, every day in infancy, every week later on), but the manufacturers certainly did not stint on capitalizing upon the guilt that women apparently felt if their offspring did not gain the required amounts of weight.[45] And yet again, many of the earliest attempts to spread 'wise' consumer practices were undertaken by large corporations and the magazines that desired their advertising: mail-order shopping guides, 'product-testing' services, pseudo-informative pamphlets, and other such promotional devices were all techniques for urging the housewife to buy new things under the guise of training her in her role as skilled consumer.[46]

Thus the advertisers could well be called the 'ideologues' of the 1920s, encouraging certain very specific social changes – as ideologues are wont to do. Not surprisingly, the changes that occurred were precisely the ones that would gladden the hearts and fatten the purses of the advertisers; fewer household servants meant a greater demand for labor and timesaving devices; more household tasks for women meant more and more specialized products that they would need to buy; more guilt and embarrassment about their failure to succeed at their work meant a greater likelihood that they would buy the products that were intended to minimize that failure. Happy, full-time housewives in intact families spend a lot of money to maintain their households; divorced women and working women do not. The advertisers may not have created the image of the ideal American housewife that dominated the 1920s – the woman who cheerfully and skillfully set about making everyone in her family perfectly happy and perfectly healthy – but they certainly helped to perpetuate it.

The role of the advertiser as connecting link between social change and technological change is at this juncture simply a hypothesis, with nothing much more to recommend it than an argument from plausibility. Further research may serve to test the hypothesis, but testing it may not settle the question of which was cause and which effect – if that question can ever be settled definitely in historical work. What seems most likely in this case, as in so many others, is that cause and effect are not separable, that there is a dynamic interaction between the social changes that married women were experiencing and the technological changes that were occurring in their homes. Viewed this way, the disappearance of competent servants becomes one of the factors that stimulated the mechanization of homes, and this mechanization of homes becomes a factor (though by no means the only one) in the disappearance of servants. Similarly, the emotionalization of housework becomes both cause and effect of the mechanization of that work; and the expansion of time spent on new tasks becomes both cause and effect of the introduction of time-saving devices. For example, the social pressure to spend more time in child care may have led to a decision to purchase the devices; once purchased, the devices could indeed have been used to save time – although often they were not.

If one holds the question of causality in abeyance, the example of household work still has some useful lessons to teach about the general problem of technology and social change. The standard sociological model for the impact of modern technology on family life clearly needs some revision: at least for middle-class nonrural American families in the 20th century, the social changes were not the ones that the standard model predicts. In these families the functions of at least one member, the housewife, have increased rather than decreased; and the dissolution of family life has not in fact occurred.

Our standard notions about what happens to a work force under the pressure of technological change may also need revision. When industries become mechanized and rationalized, we expect certain general changes in the work force to occur: its structure becomes more highly differentiated, individual workers become more specialized, managerial functions increase, and the emotional context of the work disappears. On all four counts our expectations are reversed with regard to household work. The work force became less rather than more differentiated as domestic servants, unmarried daughters, maiden aunts and grandparents left the household and as chores which had once been performed by commercial agencies (laundries, delivery services, milkmen) were delegated to the housewife. The individual workers also became less specialized; the new housewife was now responsible for every aspect of life in her household, from scrubbing the bathroom floor to keeping abreast of the latest literature in child psychology.

The housewife is just about the only unspecialized worker left in America – a veritable jane-of-all-trades at a time when the jacks-of-all-trades have disappeared. As her work became generalized the housewife was also proletarianized: formerly she was ideally the manager of several other subordinate workers; now she was idealized as the manager and the worker combined. Her managerial functions have not entirely disappeared, but they have certainly diminished and have been replaced by simple manual labor; the middle-class, fairly well educated housewife ceased to be a personnel manager and became, instead, a chauffeur, charwoman and short-order cook. The implications of this phenomenon, the proletarianiz-ation of a work force that had previously seen itself as predominantly managerial, deserve to be explored at greater length than is possible here, because I suspect that they will explain certain aspects of the women's liberation movement of the 1960s and 1970s which have previously eluded explanation: why, for example, the movement's greatest strength lies in social and economic groups who seem, on the surface at least, to need it least – women who are white, well-educated, and middle-class.

Finally, instead of desensitizing the emotions that were connected with household work, the industrial revolution in the home seems to have heightened the emotional context of the work, until a woman's sense of self-worth became a function of her success at arranging bits of fruit to form a clown's face in a gelatin salad. That pervasive social illness, which Betty Friedan characterized as 'the problem that has no name,' arose not among workers who found that their labor brought no emotional satisfaction, but among workers who found that their work was invested with emotional weight far out of proportion to its own inherent value: 'How long,' a friend of mine is fond of asking, 'can we continue to believe that we will have orgasms while waxing the kitchen floor?'

 NOTES

1 For some classic statements of the standard view, see W. F. Ogburn and M. F. Nimkoff, *Technology and the Changing Family* (Cambridge, Mass., 1955); Robert F. Winch, *The Modern Family* (New York, 1952); and William J. Goode, *The Family* (Englewood Cliffs, N.J., 1964).

2 This point is made by Peter Laslett in 'The Comparative History of Household and Family,' in *The American Family in Social Historical Perspective*, ed. Michael Gordon (New York, 1973), pp. 28–29.

3 Phillippe Ariès, *Centuries of Childhood: A Social History of Family Life* (New York, 1960).

4 See Laslett, pp. 20–24; and Philip J. Greven, 'Family Structure in Seventeenth Century Andover, Massachusetts,' *William and Mary Quarterly* 23 (1966): 234–56.

5 Peter Laslett, *The World We Have Lost* (New York, 1965), passim.

6 For purposes of historical inquiry, this definition of middle-class status corre-sponds to a sociological reality, although it is not, admittedly, very rigorous. Our contemporary experience confirms that there are class differences reflected in magazines, and this situation seems to have existed in the past as well. On this issue see Robert S. Lynd and Helen M. Lynd, *Middletown: A Study in Contemporary*

American Culture (New York, 1929), pp. 240–44, where the marked difference in magazines subscribed to by the business-class wives as opposed to the working-class wives is discussed; Salme Steinberg, 'Reformer in the Marketplace: E. W. Bok and *The Ladies' Home Journal'* (Ph.D. diss., Johns Hopkins University, 1973), where the conscious attempt of the publisher to attract a middle-class audience is discussed; and Lee Rainwater *et al., Workingman 's Wife* (New York, 1959), which was commissioned by the publisher of working-class women's magazines in an attempt to understand the attitudinal differences between working-class and middle-class women.

7 *Historical Statistics of the United States, Colonial Times to 1957* (Washington, D.C., 1960), p. 510.

8 The gas iron, which was available to women whose homes were supplied with natural gas, was an earlier improvement on the old-fashioned flatiron, but this kind of iron is so rarely mentioned in the sources that I used for this survey that I am unable to determine the extent of its diffusion.

9 Hazel Kyrk, *Economic Problems of the Family* (New York, 1933), p. 368, reporting a study in *Monthly Labor Review* 30 (1930): 1209–52.

10 Although this point seems intuitively obvious, there is some evidence that it may not be true. Studies of energy expenditure during housework have indicated that by far the greatest effort is expended in hauling and lifting the wet wash, tasks which were not eliminated by the introduction of washing machines. In addition, if the introduction of the machines served to increase the total amount of wash that was done by the housewife, this would tend to cancel the energy-saving effects of the machines themselves.

11 Rinso was the first granulated soap; it came on the market in 1918. Lux Flakes had been available since 1906; however it was not intended to be a general laundry product but rather one for laundering delicate fabrics. 'Lever Brothers,' *Fortune* 26 (November 1940): 95.

12 I take this account, and the term, from Lynd and Lynd, p. 97. Obviously, there were many American homes that had bathrooms before the 1920s, particularly urban row houses, and I have found no way of determining whether the increases of the 1920s were more marked than in previous decades. The rural situation was quite different from the urban; the President's Conference on Home Building and Home Ownership reported that in the late 1920s, 71 percent of the urban families surveyed had bathrooms, but only 33 percent of the rural families did (John M. Gries and James Ford, eds, *Homemaking, Home Furnishing and Information Services*, President's Conference on Home Building and Home Ownership, vol. 10 [Washington, D.C., 1932], p. 13).

13 The data above come from Siegfried Giedion, *Mechanization Takes Command* (New York, 1948), pp. 685–703.

14 For a description of the standard bathroom see Helen Sprackling, 'The Modern Bathroom,' *Parents' Magazine* 8 (February 1933): 25.

15 *Zanesville, Ohio and Thirty-six Other American Cities* (New York, 1927), p. 65. Also see Robert S. Lynd and Helen M. Lynd, *Middletown in Transition* (New York, 1936), p. 537. Middletown is Muncie, Indiana.

16 Lynd and Lynd, *Middletown*, p. 96, and *Middletown in Transition*, p. 539.

17 Lynd and Lynd, *Middletown*, p. 98, and *Middletown in Transition*, p. 562.

18 On the advantages of the new stoves, see *Boston Cooking School Cookbook* (Boston, 1916), pp. 15–20; and Russell Lynes, *The Domesticated Americans* (New York, 1957), pp. 119–20.

19 'How to Save Coal While Cooking,' *Ladies' Home Journal* 25 (January 1908): 44.

20 Lynd and Lynd, *Middletown*, p. 156.

21 Ibid.; see also 'Safeway Stores,' *Fortune* 26 (October 1940): 60.

22 Lynd and Lynd, *Middletown*, pp. 134–35 and 153–54.

23 *Historical Statistics*, pp. 16 and 77.

24 For Indiana data, see Lynd and Lynd, *Middletown*, p. 169. For national data, see D. L. Kaplan and M. Claire Casey, *Occupational Trends in the United States, 1900–1950*, U.S. Bureau of the Census Working Paper no. 5 (Washington, D.C., 1958), table 6. The extreme drop in numbers of servants between 1910 and 1920 also lends credence to the notion that this demographic factor stimulated the industrial revolution in housework.

25 Lynd and Lynd, *Middletown*, p. 169.

26 On the disappearance of maiden aunts, unmarried daughters and grandparents, see Lynd and Lynd, *Middletown*, pp. 25, 99, and 110; Edward Bok, 'Editorial,' *American Home* 1 (October 1928): 15; 'How to Buy Life Insurance,' *Ladies' Home Journal* 45 (March 1928): 35. The house plans appeared every month in *American Home*, which began publication in 1928. On kitchen design, see Giedion, pp. 603–21; 'Editorial,' *Ladies' Home Journal* 45 (April 1928): 36; advertisement for Hoosier kitchen cabinets, *Ladies' Home Journal* 45 (April 1928): 117. Articles on servant problems include 'The Vanishing Servant Girl,' *Ladies' Home Journal* 35 (May 1918): 48; 'Housework, Then and Now,' *American Home* 8 (June 1932): 128; 'The Servant Problem,' *Fortune* 24 (March 1938): 80–84; and *Report of the YWCA Commission on Domestic Service* (Los Angeles, 1915).

27 Giedion, p. 619. Wright's new kitchen was installed in the Malcolm Willey House, Minneapolis.

28 Emily Post, *Etiquette: The Blue Book of Social Usage*, 5th ed. rev. (New York, 1937), p. 823.

29 This analysis is based upon various child-care articles that appeared during the period in the *Ladies' Home Journal, American Home*, and *Parents' Magazine*. See also Lynd and Lynd, *Middletown*, chap. 11.

30 John Kenneth Galbraith has remarked upon the advent of women as consumer in *Economics and the Public Purpose* (Boston, 1973), pp. 29–37.

31 There was a sharp reduction in the number of patterns for home sewing offered by the women's magazines during the 1920s; the patterns were replaced by articles on 'what is available in the shops this season.' On consumer education see, for example, 'How to Buy Towels,' *Ladies' Home Journal* 45 (February 1928): 134; 'Buying Table Linen,' *Ladies' Home Journal* 45 (March 1928): 43; and 'When the Bride Goes Shopping,' *American Home* 1 (January 1928): 370.

32 See, for example, Lynd and Lynd, *Middletown*, pp. 176 and 196; and Margaret G. Reid, *Economics of Household Production* (New York, 1934), chap. 13.

33 See Reid, pp. 64–68; and Kyrk, p. 98.

34 See advertisement for Cleanliness Institute – 'Self-respect thrives on soap and water,' *Ladies' Home Journal* 45 (February 1928): 107. On changing bed linen, see 'When the Bride Goes Shopping,' *American Home* 1 (January 1928): 370. On laundering children's clothes, see 'Making a Layette,' *Ladies' Home Journal* 45 (January 1928): 20; and Josephine Baker, 'The Youngest Generation,' *Ladies' Home Journal* 45 (March 1928): 185.

35 This point is also discussed at length in my paper 'What Did Labor-saving Devices Really Save?' (unpublished).

36 As reported in Kyrk, p. 51.

37 Bryn Mawr College Department of Social Economy, *Women During the War and After* (Philadelphia, 1945); and Ethel Goldwater, 'Woman's Place,' *Commentary* 4 (December 1947): 578–85.

38 JoAnn Vanek, 'Keeping Busy: Time Spent in Housework, United States, 1920–

1970' (Ph.D. diss., University of Michigan, 1973). Vanek reports an average of 53 hours per week over the whole period. This figure is significantly lower than the figures reported above, because each time study of housework has been done on a different basis, including different activities under the aegis of housework, and using different methods of reporting time expenditures; the Bryn Mawr and Oregon studies are useful for the comparative figures that they report internally, but they cannot easily be compared with each other.

39 This analysis is based upon my reading of the middle-class women's magazines between 1918 and 1930. For detailed documentation see my paper 'Two Washes in the Morning and a Bridge Party at Night: The American Housewife between the Wars,' *Women's Studies* (in press). It is quite possible that the appearance of guilt as a strong element in advertising is more the result of new techniques developed by the advertising industry than the result of attitudinal changes in the audience – a possibility that I had not considered when doing the initial research for this paper. See A. Michael McMahon, 'An American Courtship: Psychologists and Advertising Theory in the Progressive Era,' *American Studies* 13 (1972): 5–18.

40 For a summary of the literature on differential divorce rates, see Winch, p. 706; and William J. Goode, *After Divorce* (New York, 1956) p. 44. The earliest papers demonstrating this differential rate appeared in 1927, 1935, and 1939.

41 For a summary of the literature on married women's labor force participation, see Juanita Kreps, *Sex in the Marketplace: American Women at Work* (Baltimore, 1971), pp. 19–24.

42 Valerie Kincaid Oppenheimer, *The Female Labor Force in the United States*, Population Monograph Series, no. 5 (Berkeley, 1970), pp. 1–15; and Lynd and Lynd, *Middletown*, pp. 124–27.

43 On the expanding size, number, and influence of women's magazines during the 1920s, see Lynd and Lynd, *Middletown*, pp. 150 and 240–44.

44 See, for example, the advertising campaigns of General Electric and Hotpoint from 1918 through the rest of the decade of the 1920s; both campaigns stressed the likelihood that electric appliances would become a thrifty replacement for domestic servants.

45 The practice of carefully observing children's weight was initiated by medical authorities, national and local governments, and social welfare agencies, as part of the campaign to improve child health which began about the time of World War I.

46 These practices were ubiquitous, *American Home*, for example, which was published by Doubleday, assisted its advertisers by publishing a list of informative pamphlets that readers could obtain; devoting half a page to an index of its advertisers; specifically naming manufacturers and list prices in articles about products and services; allotting almost one-quarter of the magazine to a mail-order shopping guide which was not (at least ostensibly) paid advertisement; and as part of its editorial policy, urging its readers to buy new goods.

21 ▶ A gendered socio-technical construction: the smart house

Anne-Jorunn Berg

This chapter is about the shaping of the innovative home of the future and the importance of gender in that process. For several reasons, my study concentrates on a specific version of the home of tomorrow – the 'smart house'. First, it has information technology, *the* new technology, at its heart. Second, as publicly projected it resembles the popular scenarios of the 1980s in presenting new technologies as gender neutral. Third, several of the big international electronic corporations are already creating prototypes. The smart house is interesting because it is beyond the stage of unbridled imagination. It is already at pre-production stage – a serious IT-home-in-the-making. . . .

In our research project on the smart house I gathered information through the 'snowball' method, one source of information leading to another. The lack of empirical and theoretical social research on gender and innovation, particularly on the smart house, made such an exploratory approach appropriate. I interviewed designers and producers, systematically analysed advertisements and other kinds of written material and visited the three North American test houses described here.

I formulated three main questions, answers to which I hoped might reveal the relationship between gender and technology in the design of the smart house. These questions have their origins in the existing sexual division of labour. First, what material appliances are actually in the making today? Scenarios are not always to be trusted as a guide to the future. Second, what kind of household activities are the new artefacts or appliances meant for? Concretely, is housework taken into consideration during the design process? Third, who are the consumers the designers and producers see as their target group? For whom, exactly, are they making this new home? These three topics are addressed in the research in various ways with the aim of exposing to analysis how innovation can be said to be

a gendered process, how the smart house can be seen to be a gendered socio-
technical construction.

SMART HOUSE PROTOTYPES: MODELLING OUR FUTURES

Two of the three smart house prototypes I analysed are laboratory houses
financed respectively by Honeywell and the National Association of Home
Builders (NAHB). The third is a commercial show-case named Xanadu,
owned by private investors.

The Honeywell house

Honeywell is a multinational corporation producing control systems and
services, including thermostats, air cleaners, burglar and fire alarms. The
home is not Honeywell's only market, but its various control appliances are
already installed in more than 60 million one-family houses in the United
States.

Honeywell has been interested in home automation since 1979. Its first
laboratory house was built in natural surroundings in a residential area, but
it proved too difficult to test and change the infrastructure of the house in
such a location. The current Honeywell test house is therefore built inside
a laboratory. It embodies the Honeywell products and services linked together through one central programmable communication network: the
integration system. Honeywell aims to develop a flexible package that can
be adjusted to individual homes to suit different life situations and life
styles.

The house, as it was when I visited it, is a life-sized model of a typical
North American detached house. The 'Home Automation Test Laboratory'
(the house together with two environmental test chambers) is a large
research and development project within Honeywell Corporate Systems
Development Division. The six-room house is the test site for prototypes of
home automation products and the integrated house control systems
Honeywell are developing. Apart from the fact that the walls are 'open' to
simplify access by the engineers, the house looks like any ordinary house.
No fanciful details suggest that this is the home of the future. The R&D team
says this is deliberate: Honeywell plan to present home automation as
nothing out of the ordinary. Only the control panel, affording central
control of all electronic systems, reveals this to be a somewhat special
house.

The interior of the house is decorated in a studiedly 'ordinary' style.
People from the neighbourhood are sometimes invited along to consumer-
test Honeywell technologies. To make the test situation as natural as possible, say the producers, the house must resemble a typical home. Of course,

they have not fully achieved this, because the many small details and decorations that go to make a house a home are missing.

NAHB smart house system

The National Association of Home Builders (NAHB) is an association of producers and suppliers of different products for the home. The Association has about 150,000 members which makes them an organization to reckon with in the struggle over standardization to which we shall refer below. It has its own National Research Foundation which fostered the idea of developing a smart house. From 1986 they intensified and restructured their research effort, turning smart house R&D into an independent business, The Smart House Development Venture Inc. (SHDVI).

The NAHB demonstration house is located in a large long-distance truck, so that it can be moved from place to place. From the outside, it resembles less a house than a large caravan. The associated R&D work is carried on in a nearby building, where the house is also modelled in miniature. The house consists of entrance hall, kitchen, living room and bedroom. The rooms, to natural size, are arranged in a linear plan, one adjacent to the next. Each is in fact only half a room, but together they embody all the functions found in a normal house.

The main focus in the NAHB smart house system is on the communication network. The whole infrastructure of the home is going to change, say its designers. Their cable system integrates all kinds of power, independent of the energy source. NAHB are particularly competing to influence standards for signal transmission in networks made for homes.

Xanadu

Xanadu is located in Orlando, Florida. It is owned by private investors and used as a show-case for different suppliers to display and demonstrate their various products.

Unlike the other two houses, Xanadu embodies architectural innovations: its external appearance is unique. The unconventional form is supposed to express symbolically the novel thinking in the infrastructure of the house. One of its founding fathers, the architect Roy Mason, invented the term 'architronics', to signify the designed integration of building structure and information technology (Mason 1983).

Xanadu too has a central control unit that integrates various appliances. It is described as an analogue of the human brain, emphasizing differences in function between left and right hemispheres. The interior of the house seems unfinished; it has no comprehensive style. Each application stands alone and fails to blend into the futuristic unity promised in the brochures. Whereas in the Honeywell and NAHB houses the control network is designed for application to an existing structure, in Xanadu the net is integral with its innovatory structure and is thought of mainly as applicable to new construction.

TECHNOLOGICAL DEVELOPMENTS: WHAT DO DESIGNERS HAVE IN MIND?

Leaving aside the appliances designers liked to tell us about in futuristic terms, I chose to concentrate on new technologies actually visible or simulated here and now in the smart house prototypes.

Technologies in the Honeywell house

The substance of the Honeywell smart house concept is *integration*, the central programming of diversified control and regulation systems. The main R&D effort is directed towards a control system integrating: remote registration of outdoor temperature and humidity, regulation of indoor air quality, temperature and environmental control in specified zones and light regulation. It includes control of an advanced security system, with motion detectors. For instance, a video camera scans the main entrance and shows visitors on a monitor. It also involves service/diagnosis equipment and makes use of voice recognition and voice information. A safety device, for example, detects smoke and alerts the inhabitants, indicating on a video screen both the source of the fire and the appropriate action.

Honeywell pay particular attention to means of reducing energy consumption. With this in mind, they have invested considerable effort in a motion-activated system of light control in which lights switch on and off in response to information from motion detectors as to where people are currently located. (Besides, all lights are automatically switched on if the security alarm sounds.) Honeywell are not very happy with the results so far, however. To try out the system in a natural situation they had invited several people to the test house for dinner. As the guests entered the dining room, the lights obligingly went on. But when everyone had settled around the table and all was still the room was suddenly plunged into darkness. The Honeywell engineers had to ask their guests to flap their arms to activate the lights again. On consideration, Honeywell now feel voice activation may have more potential, in combination with infrared remote control.

To sum up, the technologies in the Honeywell version of the home of tomorrow are applied to light and heat regulation, to security control and alarm systems. Technologies that have anything to do with *housework* are notably absent.

Technologies in NAHB's smart house

In today's home we have different cables for different types of power or energy. In 1988 NAHB directed their R&D efforts to the development of a system to make all sources of power available with the same cable and through identical power-points. In such a multi-cable system the microwave oven, the washing machine, the home computer and the telephone may all be plugged into the same socket. This calls for purpose-designed appliances, of course, for only those equipped with the correct microchip will access the new power source. Signals from the microchip in each appliance will be sanctioned by a chip in the control unit.

SHDVI does not see its role as developing new appliances as such. They are, rather, projecting a future in which domestic technology is adjusted to their own project: the new network. The washing machine will signal on the television screen when the washing is ready to be moved to the tumble-dryer. The vacuum cleaner will be programmed to stop when someone is at the door or when the phone rings. When the temperature in the microwave indicates that dinner is almost ready it will signal to the hot-plate to warm the soup and to the stereo to provide the right background music for the meal. In addition to integrating in-house appliances in this way, the house net will also be connected to outside communication networks. It will support telework, telebanking and teleshopping as such services become available to private households.

The appliances displayed in this second smart house, then, are familiar technologies. The only thing that is new is their integration. Again we have to emphasize: the technological changes envisaged here are a long way from being consciously concerned with routine housework.

Technologies in Xanadu

As we saw, Xanadu has rooms with traditional functions: kitchen, bedroom, bathroom, etc. In every room appliances are presented that, claim the designers, are innovatory. In the book about the house, the new appliances are presented in a 'hi-tech' style, and to a certain extent this is carried through in the house itself. Xanadu is represented as a house you can talk to, a house that answers in different voices, where every room can be adjusted to your changing moods, a house that is servant, adviser and friend to each individual member of the household. Controlling all the functions in the home is Xanadu's house-brain.

Xanadu's technologies, with certain exceptions, are not dissimilar from those in the two prototypes already described, though their applications encompass a wider range of activities. These applications, however, have not all yet reached the stage of being prototypes. Some are simulations anticipating future possibilities.

Dreaming of integration

Our first question concerned the appliances in the home of the future. The list of what we found as testable prototypes – disregarding mere future possibilities – is neither extensive nor impressive. The innovations discussed above amount to control of:

- energy (heating and lighting);
- safety (security and fire alarms);
- communication (information and messaging, within the home and between the home and the outside world);

- entertainment (television, compact disc player, video recorder, computer games);
- environment (temperature, air pollution).

None of the technologies on this list differ radically from technologies already in existence (Miles 1988). All are available – albeit unintegrated – on the market today. All that is new about these 'smart houses' is the *integration* itself, linking different appliances in a central local network variously called a 'small area network' (SAN), 'homebus', *'domotique'* or 'house-brain'.

This is the designers' dream, therefore – integration, centralised control and regulation of all functions in the home. This is the core of the smart home as a socio-technical construction. Many different companies and organizations today are engaged in R&D projects for such home networks, and the battle over standards is preoccupying all the big electronic firms and other contenders such as the NAHB. Whoever wins the standardization battle will have a clear advantage in the future market for networks and new appliances.

 ## HOUSEWORK: OUT OF SIGHT, OUT OF MIND

Women and men traditionally have distinct and different work in the home. Housework, still mainly women's unpaid work, comprises the most repetitious and time-consuming tasks in the household – cooking, washing, cleaning, tidying, mending. Turning to the second topic in our research, then, the precise activities preoccupying the designers of smart houses, we must ask: Does housework feature in their thinking about the smart house? What do they seem to know about housework?

Honeywell: the house will 'do the job for you'

For Honeywell the main idea about the smart house is, to quote an R&D manager, that 'it does more things for you, the way you would like to have them done, than today's houses'. This means the overall purpose of the smart house is to help the owner so that he (I use the pronoun advisedly) will 'no longer have to think about how things are done'. All he has to be concerned with is 'whether the technologies are simple to use, increase comfort, are pleasant and affordable'. Housework is not mentioned in any way here – though one could imagine that technological solutions that increase comfort could relate to housework somehow. The designers responded to our question as to the *advantages* of the smart house with the following words – listed in order of the priority they felt they would have in the market: comfort, security, convenience, energy saving and entertainment. Labour-saving was not mentioned. Neither was it mentioned when they were asked about any *disadvantages* in the smart house. The main problem pointed out was a new kind of hazard: vulnerability to the 'house-break'. With all appliances integrated in a single electronic network, a failure due to technical breakdown or human interference would bring down the entire system.

It seems strange that in talk about 'a house that will do the job for you', 'job' does not refer to the actual work that is carried on in a house. Housework is no part of what this house will 'do' for you. This anomaly became yet more obvious when one of our informants went on to tell us how their prototype differed from Japanese home automation. According to him, Japanese systems are designed to leave the finishing of jobs to manual intervention. Japanese culture, he explained, is extremely service-minded. In Japanese households women render service to men. Even when technologically possible to eliminate it, the Japanese do not want to change this service relationship. North Americans, he said, by contrast, see the optimal use of technology as full automation of as many activities as possible. Yet he did not see the paradox of Honeywell's own lack of concern with housework.

NAHB: a house that will 'take care of me'

We found the same tendency at the NAHB. They present their smart house as, primarily, 'a house which will take care of me'. This implies the house will do 'anything you want to be done in your home today – and in the future'. Such a general concept could, of course, well include housework. On asking for more detail however we were told that it is first necessary to develop a communication network to modernize the basic infrastructure of the house. This network would then act as an invitation to suppliers of domestic technology to intensify their product development with the new network in mind. Housework as such was not NAHB's concern.

When asked more precisely what it means that a house will do 'anything you want done', the NAHB designers signalled as important that the house will be more comfortable, safer and easier to live in. Other advantages cited were better communication with the surrounding world, saving on energy, and enhanced entertainment. Things are vaguely glimpsed as being 'done' in a home, but housework as such is invisible and it has not occurred to the NAHB designers that users might like the house to carry out the time-consuming activities of housework.

Xanadu: a house to serve you

Xanadu is somewhat different. Integration in a house network is important here too, but in addition this more social, more imaginative version of the smart house does accommodate (albeit rather elliptical) references to housework. The title page in the book about Xanadu, for example, has a picture of a robot serving mother breakfast. The accompanying text reads: 'We are not replacing Mommy with a robot. We are presenting ideas on how to design, build and use a home in new ways that can reduce drudgery while

Figure 21.1 Xanadu: a house to serve you

increasing comfort, convenience and security' (Mason 1983: 1). Housework is not mentioned explicitly, but it is easy to interpolate: drudgery will be reduced. The robot serving mother is a technical solution that serves the traditional server.

Family life is often mentioned, and the house is designed to be a place in which people can live happily together. 'What is really futuristic about an architronic house like Xanadu . . . is not the way it looks but the way it works. In this sense, the house of the future will be more like the houses of the past than like the houses of today' (Mason 1983: 43). The house is intended to work in such a way as to give the family more opportunity to spend time together, but the designers say nothing about how work time can be saved to make this a possibility.

Xanadu's message is therefore ambiguous. Women as mothers and housewives are shown as playing an important role in the home. But we are left unclear as to whether it is women's work they mean when claiming the house will reduce drudgery.

Ignorance of housework

During the interviews we pressed the housework issue. When making appointments with informants we told them explicitly that housework would be our main focus. Nevertheless, when it came to it, they seemed surprised to be asked about it, and their answers were imprecise. NAHB said housework was not their concern – they left that to the white goods producers. At Honeywell they said they had paid some attention to housework, but when prompted they gave unconvincing examples such as the automatic light switch. They justified this as facilitating housework by arguing that a housewife entering a room with her hands full of wet clothes would not have to put them down to turn on the light.

In Xanadu one example of a housework appliance was the 'robutler'. This device was said to serve drinks. In fact, someone had to fill the glasses first and place them on exactly the right spot for the robutler to collect them. The machine then required guidance by remote control in serving them. There seemed to be little saving in manual intervention and, besides, serving drinks can hardly be said to be a burdensome task in the average home.

Another instance of Xanadu's insubstantial inventions is the 'gourmet autochef'. It sounds interesting and looks fancy but on closer inspection is found to do nothing except suggest a menu for the dinner party. The housewife (a woman is shown in the picture) is still responsible for planning, shopping, cooking and all the rest of the work that goes into preparing a meal. The gourmet autochef, when the chips are down, is no more than a computer program.

When reflecting on what goes on in a home the Xanadu designers say 'today home is often little more than a place to sleep, eat a meal or two and store possessions' (Mason 1983: 16). This is a highly misleading description of what takes place in a home (Cowan 1983). A housewife would certainly ask 'Does nobody change the sheets? Is that meal not cooked and are the dishes not washed up afterwards? Are those possessions never dusted?'

The designers' knowledge of such material realities of housework is scanty indeed. First, the variety and necessity of today's housework is quite overlooked. 'Once household chores were regarded as inescapable duties – like tending animals or crops on a farm. But today they are more often resented as impositions that everyone in the family would like to avoid.' They continue: 'As a result a host of new household appliances are appearing that require less time and physical effort to do unpleasant jobs.' Yet they clearly have no idea of housework's centrally time-consuming tasks for they cite quite peripheral appliances: 'the toilet-bowl cleaner that fits inside the tank, the in-sink garbage disposal, trash compactor etc.' (Mason 1983: 19–20).

What is more, many of the appliances described in connection with Xanadu, technologies that might seem to promise to reduce household labour, are as yet no more than future possibilities. There are plans, for instance, for a closet that could make the washing machine superfluous. Vapour is distributed inside the closet during the night to clean the clothes where they hang. Despite the fact that a woman, Frances GABe, pioneered such a concept some years ago in her 'self-cleaning house' (GABe 1983), this cleaning-cupboard does not exist in Xanadu even as a prototype – its control panel is no more than a simulation, the innovation no more than a thought.

In printed publicity about these smart homes, the housewife is sometimes pictured with new technologies. She smiles happily by the computer in her kitchen. Our interviews with the men behind the publicity, however, convinced us that they did not know much about this woman's work and visions for the future.

▶ **WOMEN AS A SOCIAL GROUP: RELEVANT BUT ABSENT**

I move now to the third question to ask: just whom, then, do the designers have in mind as their target consumer? 'Relevant social group' is a term used by Trevor Pinch and Wiebe Bijker to denote 'institutions and organizations . . . as well as organized or unorganised groups of individuals' for whom an artefact has a shared 'set of meanings'. They emphasise specifically that the social group of 'consumers' or 'users' fulfils such a requirement and should be included in the analysis of a technological development (Pinch and Bijker 1987: 30).

Women are a relevant social group in the development of the smart house in at least two ways. First, women possess important skills for and know-ledge about the home that should be a resource in the design process. Second, since the home is women's traditional domain, women could be seen as an important target in the marketing of the smart house.

Often, when the user-producer relationship is discussed in connection with technical design it refers to a relation where the user's competence, based on task-related experience, knowledge and skills, could guide the development of a new tool or machine for factory or office. In a similar way women's competence in housework could constitute an important innovative resource for the development of home-oriented information technology.

When asked about the relevance of users in the design process, producers said they found it 'an interesting idea'. Such an interest would seem self-evident: after all, how can one expect a product to sell except by ensuring it corresponds to consumer needs and demands? Yet we found evidence of only one actual contact by these designers with potential users: the instance already described when Honeywell invited guests to dinner and left them in the dark. Women's housework skills are being entirely neglected as a design resource.

Who exactly do the producers see as the target purchaser of their smart house? It proved difficult to pin them down. 'Anyone and everyone' seemed to be the answer. NAHB was the most specific: they had at least decided to concentrate on the one-family house. The others had only vague pictures of potential consumers. Honeywell see the user as 'the owner' synonymous with the man of the house. It is 'the owner' who will no longer have to think about 'how things are done'. The 'things' they rank as potentially most important in the house market are male activities and the most important consumer group (they say when pressed) is the technically-interested male.

In Xanadu too the user/consumer is difficult to identify. When the *user* is discussed at all it is in connection with specific appliances. Alleviation of household labour is mentioned as a potential demand in the market but the robutler and the gourmet autochef typify the lightweight response.

When we asked about the *consumer* the designers had in mind, we were answered several times with rather similar stories of individuals that had built and equipped their own houses with new technologies. One example was a Norwegian engineer living in Texas who had built his own private smart house. A detail that fascinated the storytellers was the lighting system he had devised: the lights would dim along the corridor when, for example, the children went to their rooms to sleep. This was the kind of consumer the producers liked and speculated about, one who would share their fascination with electronic or technological gadgets. That this man had been unsuccessful in his attempt to sell his house seemed not to dampen their enthusiasm. It is the technology-as-such, the way artefacts function in technical terms, that fascinates the designers. Again, the target consumer is implicitly the technically-interested man, not unlike the stereotype of the computer hacker.

In summary, then, the men (and it is men) producing prototypes of the intelligent house of the future and designing its key technologies have failed to visualize in any detail the user/consumer of their innovation. In so far as they have one in mind, it is someone in their own image. They have ignored the fact that the home is a place of work (women's housework) and overlook women, whose domain they are in effect transforming, as a target consumer group.

 ## GENDERED INNOVATION PROCESS, GENDERED TECHNOLOGY

Smart house prototypes resemble their literary forerunners, those scenarios of the 1980s that presented home-oriented information technologies as gender neutral. We have seen just how far from reality this is. Nothing could be clearer evidence of the gendering of the technological innovation process than the absence of women we have uncovered in this design and development process – their invisibility, the waste of their housework knowledge.

Smart house prototypes are one of several kinds of attempt to create today the technological home of tomorrow. The nature of that future home has serious implications for women. Technology as an element in social action has the power to change, or to preserve, today's gender relations, including the sexual division of labour. This particular socio-technical construction is transparently not intended to change gender inequality.

To say the smart house is a masculine construct and leave it at that, however, is unnecessarily defeatist. Technology should not be understood as ready-made artefacts whose use is non-negotiable. A technology's impacts are never entirely determined by its designers' and producers' intentions or inscribed visions (Akrich 1992). Rather, technology should be seen as *process* – open to flexible interpretation by its various user groups. To look at the eventual application of a technology, to see what users make of it for themselves, is often to dissipate the pessimism. Unfortunately, in the case of the smart house still at prototype stage, we cannot yet see it in use.

Despite the non-fixed nature of a technology, however, to observe its gendering in those early stages before it reaches the user is of vital importance for understanding what happens subsequently. The smart house is a typical case of 'technology push', in contrast to 'consumer pull'. Its inspiration lies not in the practices of everyday life but in a fascination with what is technically possible.

The gender implications of this are clear. Technology is traditionally a masculine domain and an interest in technology is seen as constitutive of masculinity (Lie 1991). When technological possibilities lead, as they do in the socio-technical construction of the smart house, the house that results is somewhat like Corbusier's 'machine for living' – a highly masculine concept.

Conversely, decor and style are traditionally a feminine domain, and creative flair in home-making has been described as an important part of feminine identity (Prokopi 1978, Gullestad 1989). There is a crucial difference between a house and a home. It is women, in the main, whose work and skills make the former into the latter. Decor and style have no place in these prototypes. The smart house is no home (Miles 1991).

As we saw, historical studies of housework and domestic technology have disproved the idea that new technologies have reduced the time spent on housework (Boalt 1983, Cowan 1983, Hagelskjær 1986). None the less, the popular conception of domestic technology remains one of 'time-saving' (Vanek 1974), and it seems reasonable to expect to find such a conception among the designers of the smart house. Instead, they manifest neither interest in nor knowledge of housework. The home is acknowledged as an important area of everyday life, yet the work that sustains it is rendered invisible.

To summarize, then, the integrative technology that is the core of the smart home project appears unlikely to initiate any developments that would substitute or save time in housework. As a result we may anticipate that the producers will experience serious difficulties in selling their ideas. This socio-technical construct reflects a male idea of the home and responds to male activities in it. It is gendered in what it leaves out – its lack of support for changes in the domestic sexual division of labour.

There has been no actual user participation in the innovation process of the smart house to date, nor is the anticipated user described in more than hazy outline. Behind the shadowy notion however we can see the consumer these designers really have in mind: a technically-interested man in their own image. Women and men could, at different stages in the development process, 'negotiate'. Here such a negotiation has not taken place. And in women's exclusion from the imagination of its designers as a 'relevant social group' we see how deeply gendered the nature of an innovation process can be.

Is our finding on the smart house cause for hope or fear? For women who may have reckoned on saving themselves time and labour the smart house looks like being a cheat. For techno-freaks who hope for some really interesting and significant inventions the smart house will likewise be a disappointment. On the other hand, for those who may have feared 'technology is taking command' the evidence assembled here should be reassuring. Nothing much, it seems, is going to change because of the smart house – at least not in terms of gender.

▶ **REFERENCES**

Akrich, Madeleine (1992) 'The de-scription of technical objects', in Bijker, W. E. and Law, J. (eds) *Shaping Technology/Building Society: Studies in Sociotechnical Change*. Cambridge, MA: MIT Press.

Boalt, Carin (1983) 'Tid för hemarbete. Hur lång tid då?' (Time for housework, but how much time?), in Åkerman, Brita (ed.) *Den okända vardagen – om arbetet i hemmen* [*Unknown Everyday Life – On Housework*]. Stockholm: Akademilitteratur.

Cowan, Ruth Schwartz (1983) *More Work for Mother. The Ironies of Household Technology from the Open Hearth to the Microwave*. New York: Basic Books.

GABe, Frances (1983) 'The GABe self-cleaning house', in Zimmerman, J. (ed.) *The Technological Woman: Interfacing with Tomorrow*. New York: Praeger.

Gullestad, Marianne (1989) *Kultur og Hverdagsliv* [*Culture and Everyday Life*]. Oslo: Universitetsforlaget .

Hagelskjær, Elin (1986) *Teknologiens Tommeliden: Moderne Tider i Husholdningen* [*The Tom Thumb of Technology: Modern Times in Households*]. Aalborg: Aalborg Universitetsforlag.

Lie, Merete (1991) *Technology as Masculinity*. Trondheim: SINTEF-IFIM.

Mason, Roy (1983) *Xanadu: The Computerised Home of Tomorrow and How it Can Be Yours Today!* Washington, DC: Acropolis Books.

Miles, Ian (1988) *Home Informatics. Information Technology and the Transformation of Everyday Life*. London: Pinter Publishers.

Miles, Ian (1991) 'A smart house is not a home?', in Sørensen, K. and Berg, A.-J. (eds) *Technology and Everyday Life: Trajectories and Transformations*. Report No. 5: Proceedings from a Workshop in Trondheim, May 1990. Oslo: NAVF, NTNF, NORAS.

Pinch, Trevor and Bijker, Wiebe E. (1987) 'The social construction of facts and artefacts: or how the sociology of science and the sociology of technology might benefit each other', in Bijker, W. E., Hughes, T. P. and Pinch, T. (eds) *The Social Construction of Technological Systems*. Cambridge, MA and London: MIT Press.

Prokopi, Vlrike (1978) *Kvindelig livssammenherg* [*Female Life Situations*]. Copenhagen: GMT.

Vanek, Joann (1974) 'Time spent in housework', *Scientific American*, Vol. 231, No. 5.

22 ▶ A woman's place: Dolores Hayden on the 'grand domestic revolution'

Moyra Doorly

In the last half of the 19th century and the first quarter of this one, there existed in the United States a remarkable school of feminist thought which tied together architecture and economics in a cogent social theory. The most basic cause of women's inequality, they argued, was the economic exploitation of women's labour by men. Women suffered from two of the fundamental characteristics of industrial capitalism: the physical separation of household space from public space and the economic separation of the domestic economy from the political economy.

These women – 'material feminists,' as they are dubbed in Dolores Hayden's classic study of their ideas – demanded a grand domestic revolution.* They wanted wages for housework. They set up new kinds of neighbourhood organisations – such as housewives' cooperatives which would undertake housework for payment. Most significant of all, they chivvied architects into exploring radical new types of building. They pushed architects and town planners into looking more intently at the effects of design on family life.

The central object of their campaigning was the need to socialise domestic work. They wanted all household labour and child care to become social labour, in home-like, nurturing neighbourhoods. They wanted neighbourhoods planned to provide laundry facilities, dining and cooking services and extensive child care facilities. In her book, *The Grand Domestic Revolution*, Dolores Hayden records their belief 'that women must create feminist homes with socialised housework and child care before they could become truly equal members of society.'

Two of the more influential women were Melusina Fay Peirce, and

* *The Grand Domestic Revolution*, subtitled *A History of Feminist Designs for American Homes, Neighbourhoods And Cities*, by Dolores Hayden is published by the MIT Press.

Figure 22.1 Melusina Fay Peirce, patent for apartment house, Chicago, 1903

Charlotte Perkins Gilman. Melusina Fay Peirce laid out her proposals for cooperative housekeeping in 1868. She loved Cambridge, Massachusetts and after six years of marriage to a Harvard lecturer she described the 'costly and unnatural sacrifice' of her wider talents to 'the dusty drudgery of house ordering.' Her idea was that 'groups of 12–50 women would organise cooperative associations to perform all their domestic work collectively and charge their husbands for these services. Through membership fees a group would purchase a building to serve as its headquarters, furnish it with appropriate mechanical equipment for cooking, baking, laundry and sewing, and supply a cooperative store with provisions.' She goes on to describe how people with particular skills such as cooking would be hired and how workers would be paid wages equivalent to those paid to skilled men. Women would also be able to develop outside interests and careers.

These proposals had wide implications for neighbourhood planning and house design. She outlined plans for cooperative residential neighbour-hoods made up of kitchenless houses and one cooperative housekeeping centre. She proposed apartment houses with communal kitchens, laundries, dining facilities, rest rooms etc., family hotels in other words. She advised women to 'gather in towns and cities, elect their own officers and set up women's committees to deal with public issues such as education, health and welfare.' . . .

Dolores Hayden describes how in 1916 a self-educated architect, Alice Austin, drew up plans for a feminist socialist city that 'developed the urban infrastructure necessary for cooked food delivery and laundry services and carried Howard's proposals for cooperative housekeeping to their ultimate conclusion in terms of urban design.'

A group of farmers and urban workers planned to build this alternative to the suburban sprawl of Los Angeles at Llano del Rio, California. Austin first developed plans for kitchenless houses. She criticised the waste of time, strength and money which traditional houses with kitchens required and the 'hatefully monotonous drudgery' of preparing 1,095 meals in the year and cleaning up after each one. In her plans a network of underground tunnels linked each home to the centre and cooked food, laundry and other necessities would be delivered in railway cars. Dirty dishes could be returned to the central kitchen for washing by machine. All gas, water, electricity and telephone lines could be placed underground in the same tunnel as the residential system. The expense would be met by the savings in fuel and machinery, made by cutting down on the need for these activities to be carried out in each home. Public delivery services would also bring shopping to each home.

The city was never built. Sufficient funds couldn't be raised. But many other attempts to organise housing around communal facilities were tried out in New York and Chicago. Indeed, in the half century up to 1917, about 5,000 women and men took part in feminist experiments to socialise domestic work.

As Dolores Hayden points out, these experiments reflected an approach to the organisation of urban life similar to that advocated by Robert Owen and his French contemporary, Charles Fourier. As owner of a textile mill at New Lanark, in Scotland, between 1800 and 1824, Robert Owen included an attempt at developmental care for the children of working mothers. Owen also purchased land in New Harmony, Indiana and with his architect produced a model of an ideal settlement which included community kitchens, a childcare centre and a women's association.

This project was never completed but Charles Fourier, who identified the private dwelling as 'the greatest obstacle to improving the position of women in civilisation,' inspired about 30 Associations or Phalanxes in America beginning in the 1840s. At the North American Phalanx, a community of about 125 members established in New Jersey in 1843, a communal kitchen, laundry and bakery were contained in the same building as private, kitchenless apartments. Members were also allowed to own homes with kitchens on the domain.

Like Owen and Fourier, the material feminists held a place in middle ground between those two great 19th century social movements, socialism and feminism. They criticised industrial capitalism for its effects on human work, and saw the need to build communities which would give equal work to domestic and industrial labour. But they went beyond Owen in their proposals that the entire physical design of towns and cities must reflect equality for women.

Why did the grand domestic revolution 'fizzle out'? Dolores Hayden describes its ideas as part of a 'lost feminist tradition'. It flourished particularly during a period when urban populations were becoming increasingly tightly packed into the new industrial cities. Houses with multiple occupation became more common – tenements for the poor, apartments for the better off.

At the same time, industrialisation and mass production changed the quality of women's lives. More paid work for women meant that more middle class housewives had to learn to cope with fewer servants. More goods which had to be bought – instead of home-produced, as in the pre-industrial economy – made women more conscious of their lack of cash. 'The growth of manufacturing,' records Dolores Hayden, 'meant that while the rest of society appeared to be moving forward to socialised labour, the housewife, encased in women's sphere, slowly became more isolated from her husband, who now worked away from home; her children, who attended school all day; and the rural social networks of kin and neighbours which were disrupted by migration to the growing urban centres.'

But by the 1920s and 1930s, the industrial cities of both Britain and America which had previously attracted vast populations from the countryside began spilling them out again to the new suburbs. Dolores Hayden marks 1931 as the end of the feminists' campaign. For it was in this year that the Hoover Commission report, Home Building And Home Ownership, was published in America. It advocated single family home ownership and 'eventually led to the development of 50 million low technology, single family homes, housing three quarters of American families.'

The retreat to the suburbs left the housewife more isolated than she had been in the new cities. At the same time, the pressures of Depression and high unemployment kept women firmly out of the job market. Meanwhile, appliance manufacturers were beginning to produce in huge quantities washing machines and other domestic gadgets which were small enough to justify individual usage in the single family kitchen.

The Victorian feminists might have been able to dream of communal kitchens and laundries as cheaper than the work of the individual housewife, as well as more sociable: with the growth of mass produced domestic appliances, the economic logic of the idea was shot away. The average home was soon to be filled with enough manufactured equipment to service a primitive village. Today, architecture is still very much a man's world. An advertisement for psychotropic drugs in the US recently showed a frowning housewife, with the caption, 'You can't change her environment but you can change her mood.' Perhaps architects could do more to improve the quality of women's lives if they rediscovered the revolutionary ideas of Charlotte Gilman and Melusina Fay Peirce.

Inserting Gräfenberg's IUD into the sex reform movement

Anni Dugdale

At the 1929 Sex Reform Congress, Dr Ernst Gräfenberg from Germany won medical support for the Grafenberg Ring, the first intrauterine device (IUD) to be widely accepted and used. The Sex Reform Congress was a discursive space in which women and men struggled for a new sexual ethics and morality founded on the recognition of their mutual equality.[1] Women such as Dora Russell, Vera Brittain, Helene Stocker, Marie Stopes, Stella Browne, Naomi Mitchison, Elsie Ottesen-Jensen, and Dr Hannah Stone were respected leading members of the Sex Reform Movement. For these reformers, women had been excluded from the position of sexual subject in the sex economy of male domination under the 'old order', and now was the time to reconstitute women as agents in a new discourse of sexual freedom.

How could this be achieved? It required nothing less than the construction of something new to count as the natural woman's body. It was not simply a matter of insisting on a new representation of a natural woman's body as possessed of sexual desires, a body freed from the distorted Victorian sexual mores that had insisted on the passive and non-sexual but compliant nature of women's sexual bodies. The women active in the sex reform movement naturalized a new corporeality for all women through the contraceptive technologies that they took up and distributed. This chapter is about the Grafenberg Ring and its insertion into the sex reform movement of the late 1920s. How did this device, one of many contraceptive strategies being touted to conference participants, win their support and become one of the contraceptive methods promoted in sex reform movement clinics from New York to Melbourne? I will argue that in order for this IUD to change hands and pass from Gräfenberg's own clinic into the hands of the sex reform movement it acquired certain characteristics. These characteristics enabled the Grafenberg Ring to carry the desires of the sex

reform movement. This was the case because the Grafenberg Ring configured the bodies of its users in a way that manufactured the new women's
body envisaged by the sex reformers.

The Grafenberg Ring was derived from the stem pessary. Stem pessaries
were widely used to treat all kinds of uterine complaints in the nineteenth
century. The womb had special significance particularly in the first half of
the nineteenth century. The concentration of death in the hospital and the
practice of pathological anatomy opened up the mysteries of women's
bodies to the medical gaze. By the mid-nineteenth century the uterus had
became localized as the essence of femininity and the source of women's
ailments. Stem pessaries were a class of device much used for non-surgical
intervention into the uterus. They consisted of a part that was pushed
through the cervix into the uterus and a stem that protruded through the
cervix. Different models were made from all kinds of materials, some soft
and some hard, and were all kinds of shapes, sometimes with a hollow stem
and sometimes with a solid stem, and were used to treat all kinds of
conditions from infertility to uterine prolapse (a common complaint in
the nineteenth century), and to prevent pregnancy. These devices were
readily available to women through catalogues, from peddlers, and in
'rubber shops' and were inserted not only by doctors but by midwives, and
sometimes by abortionists seeking to prevent the continuation of a pregnancy.

The design of the contraceptive IUD appears to have been an accidental
affair. Its 'discovery' and development owe much to fortuitous and ad hoc
conjunctions. A Dr Richter, as early as 1908, published a paper in support of
the insertion of a simple ring of silk gut into the uterus. It is thought that
this device was inspired by his noting that the part of a broken stem pessary,
the part that sat inside the womb, was still working as a contraceptive even
though the parts designed to impede the flow of sperm into the uterus had
been expelled. Some stem pessaries, such as the Pust pessary, consisted of a
ring of silk gut that was inserted into the uterus. This held in place a thick
string suspended from it into the cervix that hindered the progress of any
sperm that escaped the occlusive action of the button-like cap attached to
the string and covering the entrance to the cervix. Even when the suspended string and button were missing, such devices seemed to remain
effective as contraceptives.

To say that Richter discovered the IUD would be to ignore the contributions of those, including Gräfenberg, without whose work such devices
may well have disappeared from use along with the class of stem pessaries.
Gräfenberg did not cite Richter and these two researchers may have
developed their contraceptive rings independently and at about the same
time. Many doctors of both liberal and leftist politics were associated with
genuine mass movements seeking to alleviate the misery of working people
in Weimar, Germany, through the liberalization and improvement of
contraception and abortion services.[2] Gräfenberg, being Jewish, had to
overcome prejudice and exclusionary measures in order to gain access
to gynaecological and obstetric training. At the municipal Berlin hospital
where he was head of the gynaecological department, he extended access

to contraceptives and abortion to the wives and daughters of the factory workers and tradesmen who consulted him, access already enjoyed by wealthy patients at his fashionable and thriving private practice.[3]

Gräfenberg tinkered with contraceptive devices for inserting into the uterus, experimenting with silk gut, silver thread and gold wire, star shapes and coils. He eventually settled on a circular coil design made from inert silver wire which could be compressed for ease of insertion but readily sprang back into shape inside the uterus. He had first experimented with circles when his star shapes tended to be expelled. He wound the rings made from silkworm gut with silver wire, perhaps as a means of making them opaque to X-rays. It is possible that the efficacy of his devices was improved by an unknown and idiosyncratic characteristic. The composition of the silver wire he used was rather impure and had a high copper content but the efficacy of copper as a contraceptive would not be noted by science for another half century.[4]

Many studies of the social shaping of design focus on the making of technologies as a series of contested decisions. The outcome, that has come down to us in the design of the artifact, could have been otherwise and thus could have resulted in the technology being something different. Social constructivists study how such decisions are shaped by the interests and motives of designers, thus locating the process of design in the social contexts in which they occurred. It might be possible to impute motivations and interests to Gräfenberg and to study how these influenced his decision-making processes, and the origins of his choice of the IUD as a method of contraception, rather than one of the many other methods that were beginning to be developed. However, the kind of documents that would enable such a study are not readily available. What we do have are the devices themselves and the scientific papers making knowledge claims on their behalf. Whilst such documents cannot tell us whether Gräfenberg purpose-designed his devices for the new woman, or with her characteristics and desires in mind, they can be studied in terms of how they configured women as the users of the IUD. My aim is not to impute to Gräfenberg shared interests with the women of the sex reform movement, nor to demonstrate that such a movement was able to bring pressure to bear on the design of the devices. Rather, by exploring the Grafenberg Ring as a heterogeneous technology involving specific modes of distribution and particular ways of presenting knowledge claims, I want to analyse how the new woman's body was *materialized*. My focus is on how the arrangement of the various elements constituting the Grafenberg Ring as a working technology in a field of contraceptives materially inscribed the new women's bodies and desires, and thereby constituted the Grafenberg Ring as a contraceptive that could bear the goals of the sex reformers.

Have I then fallen into a version of technological determinism, arguing that the design of the Grafenberg Ring determined the new woman or caused more equitable intimate relations? Such arguments are often made about contraceptives, particularly about the oral contraceptive pill. The pill is seen as having caused the sexual revolution, increased promiscuity, smaller families and women's greater participation in paid work. But the much earlier availability of the IUD and its failure to cause such widespread social reorganization suggests that such forms of technological determinism give far too much weight to technological change in explanations of social change. Moreover, as some feminists have recently argued, contraceptive technologies can be seen not as emancipating women but as being consumed in ways that absorbed the technologies into existing sexual politics. Women were not only to be available to men to relieve their sexual urges, but became responsible for preventing any inconvenient outcomes such as pregnancy. Rather than see-sawing between the two poles of social shaping or technological determinism, I want to explore how technologies, bodies and new patterns of sociality are made together.

Following World War I, at a mass level, both women's and men's bodily desires were repositioned against the discourses of consumption. According to the capitalist ideology of consumption, the market allowed the pursuit of comforts and pleasures to become every citizen's right without threatening the social order, indeed as the social glue that would replace social ranking based on one's station of birth. So too, in the discourses of the sex reformers, the old patriarchal hierarchy could be dismantled in the home. Men and women were positioned alongside each other in a new pattern of difference and equality – each could freely express their different but equal sexual desires. This would replace the old order in which men's active sexual desire was both privileged over and construed as complementary to women's lack of desire. The social bond, the suffrage struggle and the family became associated in new ways. Social coherence would be maintained precisely because of the heterosexual bonds and happy families that would result from a corporeal organization that constituted women as having an equal right to sexual pleasure, and as desiring sexually fulfilling lives within loving partnerships. This was represented as the extension of the suffrage struggle from the victories in the public sphere into the private realm of the home.

The freedom of choice to satisfy individual sexual desires (at least within the bounds of heterosexual unions) came to stand for the emancipated woman's body. But sexual pleasures were to be enacted by women as a rational activity, guided by science and founded on sex education, including knowledge of contraception. Thus women's sexual urges, freed from the bounds of male domination, were to escape that long-forecast fate of destroying civilization. Social order, under the new regime of equality between the sexes, was to be guaranteed because men and women following their sexual desires would form the bonds of social cohesion. Here was a context which, in contrast to the mainstream medical organizations, was most welcoming of new contraceptive technologies. Just as the free market would supposedly ensure the equitable distribution of goods and services, thereby creating a modern world without class divisions, so too would freeing sexual desires from any constraints but those of the new scientific contraception create a world without the bitter sexual divisions that this generation saw as having characterized the suffrage struggles. Gräfenberg contextualized the Grafenberg Ring, his IUD, within this new discourse of equality in sexual desire. In order to couple together the new women with his devices, a particular kind of IUD crystallized at the 1929 Sex Reform Congress in London.

The local knowledges and practices constituting the IUD in Germany in the second and third decades of the twentieth century were circulated in the international context through the presentation of papers by Dr Ernst Gräfenberg and Dr Hans Lehfeldt at the 1929 Sex Reform Congress.[5] There are three sets of arrangements patterning Gräfenberg's IUD that seem to me to be crucial to the success of making the Grafenberg Ring appear as a solution to the problems posed by the sex reform discourse. These arrangements positioned the Grafenberg Ring in the place of the social standards whereby women's lack of interest in sexual pleasures controlled fecundity in civilized society. The device reinscribed women's happiness, sexual freedom and fertility control.

First, Gräfenberg severed his wholly intrauterine devices from the class of stem pessaries. This was an important rhetorical move. His intrauterine device, he insisted, was a *new* device, and would prove itself to be much safer than the earlier devices because it broke the link between the organisms of the vagina and the sterile uterus. This was a pattern of disassociation which was also carried by the arrangement of the parts of the IUD, by its physical design. The Grafenberg Ring had no tail and consisted only of a coil of wire that had to be inserted and removed by instruments in the hands of a skilled physician. This linkage between rhetoric, design and skill brokered the IUD as a device for use by women of refinement. Sexual pleasure could be enjoyed even by those women of the higher and middle classes with a civilized approach to matters of the flesh. As Gräfenberg wrote, here was a device for women with a subjective reaction of 'repugnance' to contraceptive methods that involved 'deliberate preparation on each occasion'.[6]

The second set of associations important to the translation of the Grafenberg Ring into the international arena of the sex reform movement concerned the positioning of the device as scientific. Gräfenberg supported his claim for the very high efficacy of his coils as contraceptives with a table of figures derived from his own patients' experience with IUDs. This translation of his patient records from a bundle of cards confined to the privacy of his clinic to a table of figures, alongside his insistence that this was a physician-administered contraceptive, located the Grafenberg Ring within a modern, scientific practice of contraception. Moreover, the tabulation of his clinical practice with the Grafenberg Ring constituted it as safe for use in the context of the doctor-patient relationship – when used in medical practice, the devices had 'done no harm'. Connections were broken between the IUD and women's association with midwives and abortionists. These were the bearers of what Gräfenberg saw as the murky feminine world of tradition, knowledge based not on science but on questionable practices that did not measure up to the high standards of morality and cleanliness of the medical profession. Gräfenberg insisted that medical practitioners should be the *only* inserters of his devices. He sought to put in place a distribution system for his IUDs that did not rely on door-to-door peddlers, but on the clinic.

The final pattern of importance is the presentation of the Grafenberg Ring as one possible contraceptive choice among many, its alignment alongside other scientific contraceptives such as the Messinga diaphragm. There was no talk of this being the ideal contraceptive for women incapable of exerting rational control over their sexual passions. It was a matter of personal preference, of aesthetic taste, whether a woman preferred to use the Grafenberg Ring, the diaphragm, or one of the many foams and other spermicides being tested by pharmaceutical companies. This positioning of the intrauterine device within a context of 'choice', whether conscious or accidental, fitted well with the discourse of the sexually confident new women. Taken together, these patterns of association and substitution enabled the intrauterine device user to be contextualized in the international network of clinics (often run by the women activists of the sex reform movement) as a choosing individual.

Many studies argue that at a time when most physicians were still hostile to birth control, doctors who were involved with early contraceptive innovations sought support among the radicals of the sex reform movement.[7] What I have pointed out is how, in order to link the IUD with the sex reform movement, a particular set of agencies had to be attributed to the device which at least enabled a set of emancipatory meanings to be associated with it. I have shown how these attributed characteristics were relationally produced in the rhetorical moves made by Gräfenberg and circulated at the 1929 Sex Reform Congress. But this should not be reduced to mere language, a matter of how things were symbolized. The arrangement of the physical parts of the IUD, the numerical record of its use, its distribution through doctors and the positioning of the IUD in a field of other scientific contraceptives, all contributed to constituting the Grafenberg Ring as a device that could carry forward the goals of the Sex Reform Movement.

Gräfenberg's IUD thus participated in the materialization of a new corporeal order. This is sometimes seen as ushering in a new age of liberated sexuality for women, as ending the repressive regulation of women's sexuality characteristic of the Victorian period. But the Proceedings of the Third Congress of the World League for Sexual Reform (1930) evidence the mandating of a highly regulated and organized coital heterosexuality. Women's sexual desires were not the free expression of an unfettered natural sexuality, although they were no longer organized according to a binary division of women into the two mutually exclusive categories of 'damned whores' and 'God's police'.[8] The new woman's body was just as much a fabrication, a matter of ordering. Contraceptive planning and sex education on contraceptive techniques became the means of ordering and inscribing women's bodies and desires in a new economy of exchange between mutual sexual enjoyment, marital happiness and social health.

▶ **NOTES**

1 N. Haire (1930) *Proceedings of the Third Congress of the World League for Sexual Reform*. London: Kegan Paul, Trench, Trubner & Co. Ltd.

2 A. Grossmann (1983) Satisfaction is domestic happiness: mass working-class sex reform organisations in the Weimar Republic, in M. N. Dobkowski and I. Wallman (eds) *Towards the Holocaust: The Social and Economic Collapse of the Weimar Republic*. Westport, CT: Greenwood Press.

3 H. Lehfeldt (1975) Ernst Gräfenberg and his Ring. *Mt. Sinai Journal of Medicine*, 42: 345–51.

4 L. Margulies (1975) History of intrauterine devices. *Bulletin of the New York Academy of Medicine*, 51(5): 662–7; C. Tietze (1965) History and statistical evaluation of intrauterine devices. *Journal of Chronic Diseases*, 18: 1147–59.

5 E. Gräfenberg (1930) The intra-uterine method of contraception; H. Lehfeldt (1930) Contraceptive methods requiring medical assistance. Both in *Proceedings of the Third Congress of the World League for Sexual Reform*. London: Kegan Paul, Trench, Trubner & Co. Ltd.

6 E. Gräfenberg (1930) The intra-uterine method of contraception. *Proceedings of the Third Congress of the World League for Sexual Reform*. London: Kegan Paul, Trench, Trubner & Co. Ltd, p. 611.

7 See L. Gordon (1990) *Woman's Body, Woman's Right; Birth Control in America* (revised and updated). New York: Penguin.

8 A. Summers (1975) *Damned Whores and God's Police: The Colonization of Women in Australia*. Ringwood, Victoria: Penguin.

24 ▶ The decline of the one-size-fits-all paradigm, or, how reproductive scientists try to cope with postmodernity

Nelly Oudshoorn

Have you ever heard of andrology? The very fact that andrology, the medical specialism concerned with the reproductive functions of men, is still a cinderella profession compared to its bigger sister, gynaecology, is one of the striking examples of the institutional and discursive processes of othering in the biomedical sciences. Feminist discourses of the last two decades have provided major challenges to these 'othering processes of scientific discourse'. In this chapter I want to show how major changes in the dominant paradigm of subject–object dichotomies emerged in one specific area of the biomedical sciences: the reproductive sciences. I begin by describing how the identification of 'woman' as 'the other' eventually resulted in setting the female body apart in a separate branch of medicine. The emergence of gynaecology and sex endocrinology in the late nineteenth and early twentieth centuries established a discursive practice in which sex and reproduction became considered as 'more fundamental to Woman's than Man's nature' (Moscucci 1990; Oudshoorn 1994). Next, I compare the discourses on contraceptive technologies of the 1950s and 1980s. I will describe a major transformation in the reproductive paradigm, in which the emphasis on the universality of women and their bodies – the ultimate consequence of the process of othering – became replaced by a discourse that acknowledged the diversity of human bodies. I will show how this shift meant a break in the process of othering that had dominated the reproductive sciences since the nineteenth century.

THE INSTITUTIONALIZATION OF WOMAN AS THE OTHER

The institutional process of othering in medicine has a surprisingly recent history. For our postmodern minds it is hard to imagine that, for two thousand years, male and female bodies were not conceptualized in terms of differences. Medical texts from the ancient Greeks until the late eighteenth century described male and female bodies as fundamentally similar. Women had even the same genitals as men, with one difference: 'Theirs are inside the body and not outside it.' In this approach, characterized by Thomas Laqueur as the 'one-sex model', the female body was understood as a 'male turned inside herself' – not a different sex, but a lesser version of the male body.

It was only in the eighteenth century that biomedical discourse began to conceptualize the female body as the Other: a body that was to be considered as essentially different from the male body. The long established tradition that emphasized bodily similarities over differences began to be heavily criticized. In the eighteenth century, anatomists increasingly focused on bodily differences between the sexes and argued that sex was not restricted to the reproductive organs; or, as one physician put it, '[t]he essence of sex is not confined to a single organ but extends, through more or less perceptible nuances, into every part' (Schiebinger 1989: 189). The first part of the body to become sexualized was the skeleton. If sex differences could be found in 'the hardest part of the body', it would be likely that sex penetrated 'every muscle, vein, and organ attached to and molded by the skeleton' (Schiebinger 1989: 191).

In nineteenth-century cellular physiology, the medical gaze shifted from the bones to the cells (Laqueur 1990: 6, 215). By the late nineteenth century, medical scientists had extended this sexualization to every imaginable part of the body: bones, blood vessels, cells, hair and brain (Schiebinger 1989: 189). Only the eye seems to have no sex (Honegger 1991: 176). Biomedical discourse thus shows a clear shift in focus from similarities to differences. This shift seems to have been caused by epistemological and socio-political changes rather than by scientific progress. In *Making Sex*, Laqueur described this shift in the context of changes in the political climate. The French Revolution and new liberal claims in the eighteenth century led to new ideals about the social relationships between men and women in which the complementarity between the sexes was emphasized. This theory of complementarity 'taught that man and woman are not physical and moral equals but complementary opposites. Women now became viewed as fundamentally different from, and thus incomparable to, men' (Laqueur 1990: 32, 216, 217). The theory of sexual complementarity was meant to keep women out of competition with men, designing separate spheres for men and women. In this theory, which came to be known as the 'doctrine of the two spheres', the sexes were expected to complement, rather than compete with, each other.

The female and the male body now became conceptualized in terms of opposite bodies with 'incommensurably different organs, functions, and feelings' (Laqueur 1990: viii). This change is visible in medical language as well. 'Organs that had shared a name, ovaries and testicles, were now linguistically distinguished. Organs that had not been distinguished by a name of their own, the vagina, for example, were given one' (Laqueur 1990: 149).

Following this shift, the female body became the medical object *par excellence* (Foucault 1981), emphasizing woman's unique sexual character. Medical scientists now started to identify the ultimate cause of woman's otherness. The medical literature of this period shows a radical naturalization of femininity in which scientists reduced woman to one specific organ. In the eighteenth and nineteenth centuries, scientists set out to localize the essence of femininity in different places in the body. Until the middle of the nineteenth century, scientists considered the uterus as the seat of femininity. This conceptualization is reflected in the statement of the German poet and naturalist Johann Wolfgang Goethe (1749–1832): 'Der Hauptpunkt der ganzen weiblichen Existenz ist die Gebaermutter' ['The uterus is the essence of the whole female existence'] (Medvei 1983: 213). In the middle of the nineteenth century, medical attention began to shift from the uterus to the ovaries, which came to be regarded as largely autonomous control centres of reproduction in the female animal, while in humans they were thought to be the essence of femininity itself (Gallagher and Laqueur 1987: 27). In 1848, Virchow (1817–1885), often portrayed as the founding father of physiology, characterized the function of the ovaries thus:

> It has been completely wrong to regard the uterus as the characteristic organ . . . The womb, as part of the sexual canal, of the whole apparatus of reproduction, is merely an organ of secondary importance. Remove the ovary, and we shall have before us a masculine woman, an ugly half-form with the coarse and harsh form, the heavy bone formation, the moustache, the rough voice, the flat chest, the sour and egoistic mentality and the distorted outlook . . . in short, all that we admire and respect in woman as womanly, is merely dependent on her ovaries.
>
> (Medvei 1983: 215)

In the late nineteenth century, the search for the cause of woman's otherness eventually led to setting women's bodies apart in a medical specialism: gynaecology. In her fascinating account of the rise of the 'Science of Women', Moscucci has described how 'the belief that the female body is finalised for reproduction defined the study of "natural woman" as a separate branch of medicine.' With the emergence of gynaecology, women became identified as 'a special group of patients' (Moscucci 1990: 2). The turn of the century witnessed the founding of societies, journals and hospitals specifically devoted to the diagnosis and treatment of the female body. 'Woman' thus became set apart in the discursive and institutional practices of the biomedical sciences. The growth of gynaecology was not paralleled by the establishment of a complementary 'science of masculinity'. 'As the male was the standard of the species, he could not be set apart on the basis of his sex' (Moscucci 1990: 32).

This institutional process of othering was continued and reinforced by the rise of sex endocrinology, a discipline devoted to the study of sex hormones that emerged in the 1920s and 1930s. In *Beyond the Natural Body* I have described how the very existence of gynaecology facilitated a situation in which the new science of sex endocrinology focused almost exclusively on the female body. The by then established gynaecological practices had transformed the female body into an easily accessible supplier of research materials, a convenient guinea pig for tests and an organized audience for the products of sex endocrinology. Both laboratory scientists and pharmaceutical firms depended on these institutional practices to provide them with the necessary tools and materials to transform the hormonal model of the body into a new set of disease categories, diagnostic tools and drugs. Sex endocrinologists integrated the notion of the female body as a reproductive body into the hormonal model, but not without thoroughly changing it. They provided the medical profession with tools to intervene in features that had been considered inaccessible prior to the hormonal era. The introduction of diagnostic tests and drugs enabled the medical profession to intervene in the menstrual cycle and the menopause, thus bringing the 'natural' features of reproduction and ageing into the domain of medical intervention.

The introduction of the concept of sex hormones not only changed the medical treatment of the female body but also redefined the existing social configurations structuring medical practice. The field of sex endocrinology generated a set of social relationships that did not exist prior to its emergence. What changed in this episode was the question of who was entitled to claim authoritative knowledge about the female body. The hormonal model enabled gynaecologists to draw the female body more and more deeply into the gynaecological clinic. Gynaecologists, however, had to share their increased medical authority with another professional group: the laboratory scientists. With the introduction of the concept of sex hormones, scientists explicitly linked women's diseases with laboratory practice. The study of woman as the other thus became extended from the clinic to the laboratory and thereby firmly rooted in the heart of the life sciences.

▶ **ONE SIZE FITS ALL**

Bearing in mind this short history of the process of othering in the biomedical sciences, it will be no surprise that the development of the first physiological means of contraception focused exclusively on women. The history of the contraceptive pill indicates how the process of othering required an emphasis on similarities among women.[1]

Remarkably, this time the choice to focus on women, rather than men, was not made by the medical profession or laboratory scientists but by an 'outsider', Margaret Sanger, a women's rights activist and pioneer for birth control in the United States of America. Sanger, arrested and jailed for opening the first birth-control clinic in New York in 1916, believed that the most important threat to women's independence came from unwanted and unanticipated pregnancies. She advocated birth control as a basic precondition to the liberation of women (Christian Johnson 1977: 1). In 1951, at the age of 72, Sanger approached Gregory Pincus, an American reproductive biologist specializing in the study of hormones, and persuaded him to start research on contraceptives. She was very explicit about what type of contraceptive had to be developed: it had to be a 'universal contraceptive' that could be used by all women, regardless of colour, class, age, or educational background (Christian Johnson 1977).

These early ideas on contraception set the stage for the reproductive paradigm of the 1960s and early 1970s. The adage 'One Size Fits All' became the major cornerstone of R&D in contraceptives. The quest for universal contraceptives can be considered as the ultimate consequence of the process of othering. Classifying woman as the other directs the attention to similarities among women. Consequently, the design of medical technology does not have to take into account the diversity of its users. The history of the pill therefore reads as an intriguing story of how scientists tried to construct similarities between women. This is very obvious in the texts that Pincus and his colleagues published reporting the clinical trials of the pill. A perusal of these publications reveals a very telling picture: the women participating in the clinical trials have disappeared from the stage. They were replaced, quite simply, by the number of treated menstrual cycles. In the 1958 publication of one of the first large-scale clinical trials, Pincus concluded that 'in the 1279 cycles during which the regimen of treatment was meticulously followed, there was not a single pregnancy' (Pincus 1958: 133). In the 1959 publication in *Science*, which described all four field trials of the pill, it was reported:

> We have recently collected and analyzed the data (to November 1958) from these four projects and present here the outstanding findings derived from these data; 830 subjects took the medication for a total of 8133 menstrual cycles, or 635 woman years.
>
> (Pincus 1959: 81)

A popular writer adopted this representation strategy as well. In *The Hormone Quest*, the author concludes:

> By 1960 1600 women at the Caribbean Field Trial Centers had used Enovid as a contraceptive for from a few months to nearly four years. Their experience, as a group, with the new steroid covered nearly 40,000 menstrual cycles or – as medical statisticians prefer to put it – about 3000 woman-years of exposure to the possibility of pregnancy.
>
> (Maisel 1965: 134)

This representational strategy clearly emphasizes the similarities between women. The use of such categories as 'cycle' replaces the individual subject by the group, suggesting a continuity that did not exist in the trials. That suggestion simultaneously affirms continuity while obscuring discontinuity by framing new scientific categories for data measurement. A representation in terms of cycles implies an abstraction from the bodies of individual women to the universal category of a physical process. Here we see how scientific texts are not simply a reflection of the proceedings of research. Texts are a far stronger tool than that: they are a representation which creates a new reality. The discourse of pill researchers constructed women's bodies as universal with respect to their reproductive functions.

The construction of similarities between women is not just a matter of discourse. During the testing of the pill, similarities were literally created by the introduction of a specific regimen of medication. In one of the first clinical trials, women were quite distressed when they noticed that their menstruations ceased during the treatment with oral progestins (Maisel 1965: 119). If these women were distressed, Pincus reflected, it would be very likely that women taking progestins for contraceptive purposes would experience similar reactions to cessation. A contraceptive that suppresses menstruation did not meet the requirements of a 'universal' contraceptive. Pincus therefore changed the medication. The pills should be taken for twenty days, starting on the fifth day after menstruation, as was the practice in the hormonal treatment of menstrual irregularities in the 1940s (McLaughlin 1982: 110). This suggestion set the standard for the administration of progestins in all later trials and eventually for the use of the contraceptive pill in the 1960s.

The choice of this regimen of medication was shaped by moral objections to any drugs that would interfere with menstruation. Pincus was directly confronted with this norm by Searle, the pharmaceutical firm which put the pill on the market. Searle's director of biological research let Pincus know that he did not want to take part in the development of any compound that might interfere with the menstrual cycle (McLaughlin 1982: 111). In later publications, Pincus presented the effect of progestin on menstruation as a way of mimicking nature: women would still have their menstrual periods. In 1958, Pincus legitimated the regimen of medication as follows:

> Actually, in view of the ability of this compound to prevent menstrual bleeding as long as it is taken, a cycle of any desired length could presumably be produced. We had chosen our standard day 5 through day 24 regime in the expectation that 'normal' cycle length would occur.
>
> (Pincus 1958: 1338)

This quotation illustrates that concepts such as 'normality' and 'similarity' are medical constructs, rather than rooted in nature. Pincus could have made a menstrual cycle of any desired length. He chose to make a 'normal' menstrual cycle that subsequently became materialized in the pill. This diminished the variety in menstrual patterns among women: all pill-users have a regular cycle of four weeks. The pill thus literally created similarities in women's reproductive functions.

The next step toward creating similarities in women's reproductive functions was to adjust women to the demands of the new technology. Women had to learn to follow the relatively complicated instructions of 'one tablet a day, beginning on day 5 of the menstrual cycle until one vial of 20 tablets was consumed, i.e. through day 24 of the cycle' (Pincus 1958: 1335). During the trials in Puerto Rico, the control to ensure a strict adherence to this regimen of medication was assigned to a team of trained social workers. In one of the reports of this trial, published in the *American Journal of Obstetrics and Gynaecology* in 1958, Pincus described this regimen as follows:

> A schedule of visits by a trained social worker was arranged so that in every medication cycle each subject was seen shortly after she should have taken the last tablet. Initially only one vial was distributed to each woman. This was replaced on the social worker's visit. Since, in a number of instances, the housewives were not at home when she called, it was decided, after a few months, to leave two vials with each subject so that the continuity of the regime of medication (day 5 through day 24) might not be broken.
>
> (Pincus 1958: 1335)

During the visits of the social worker, the women were asked to cooperate in interviews as well.

> At each consultation, information was elicited concerning the length of the menstrual cycle, the occurrence of side effects, the frequency of coitus, and the number of missed tablets. A rough check, in some instances, on the number of tablets omitted was made by counting those remaining in the vial.
>
> (Pincus 1958: 1335)

This strategy to ensure that women would adhere to the required test protocols did not work in all cases. In the publications of this trial, Pincus reported 17 pregnancies due to what he described as 'patient failure': these women had missed some days of tablet-taking (Pincus 1958: 1335, Pincus 1959: 81). Obviously only women can fail, not the technology.

The most obvious illustration of scientists' emphasis on similarities, however, is their choice of test subjects. The quotations cited earlier show that the testing of hormones as contraceptives did not take place in the continental USA (where the laboratory research took place) but in the Caribbean Islands. In the late 1950s and early 1960s, four large-scale field trials were organized (three in Puerto Rico and one in Haiti) in which more than 1,600 women participated. So it was women of colour, especially in former colonial settings, who entered this history as the guinea pigs of one of the most revolutionary drugs in the history of medicine. Most importantly, the choice to test hormones on women of colour could only be made because scientists did not recognize any fundamental differences between women.

The emphasis on similarities in the development of medical technologies such as the pill is not unproblematic. The concept of similarity functions as the cornerstone for the development of universal technologies, technologies that can be used by women all over the world. The theoretical assumption underlying the idea of universal technologies is that technologies can be made to work everywhere because scientific knowledge is universal by nature. The case study of the pill exemplifies the failure of this claim. Despite all the emphasis on similarities, the pill has not developed into a universal technology. The dream of making the ideal contraceptive for any woman, regardless of her specific background, was not fulfilled. The main acceptance of the pill had been among middle- and upper-class women in the Western, industrialized world, with one exception: China. Most women in countries of the South had adopted sterilization and intra-uterine devices as means of contraception (Seaman and Seaman 1978: 76). Scientists may explain this failure by saying that women are to blame. My argument is that, if anything is to blame, it is the technology. I suggest that we may be able to understand the failures of science and technology by adopting a social constructivist approach that emphasizes the contextual nature of science and technology. In this perspective, every technology contains a configured user.[2] Consequently, technologies cannot simply be transported elsewhere.

The case of the contraceptive pill illustrates the complications that emerge when Western technologies are introduced into developing countries. Although the pill was developed as a universal, context- independent contraceptive, it nevertheless contained a specific user: a woman, medicalized enough to take medication regularly, who is accustomed to gynaecological examinations and regular visits to the physician, and who does not have to hide contraception from her partner. It goes without saying that this portrait of the ideal pill user is highly culturally specific (with varieties even within one culture). This user is more likely to be found in Western, industrialized countries with well-developed health-care systems. From this perspective it can be understood that the pill has not found a universal acceptance. Actually, the user-specificity of the envisioned universal contraceptive pill was already manifest during the clinical testing in the Caribbean Islands. The early trials witnessed a high percentage of drop-outs. Disciplining women to the conditions of the tests was not always successful. Many women did not participate in the gynaecological examinations, 'forgot' to take the pills, or simply quit the programme because they preferred other contraceptive methods, particularly sterilization. The Caribbean trials provided the pill researchers with information indicating that the pill did not meet with universal acceptance. These test conditions were, of course, a much heavier burden than the conditions of using the pill after it had been approved by the Food and Drug Administration. Two conditions remained the same, however: frequent visits to a physician (the pill was only available by prescription) and regular gynaecological examinations (women using the pill had to have regular gynaecological examinations, including blood-pressure tests and vaginal smears to check for adverse health effects).

The making of the pill into a successful contraceptive technology thus required a specific context, in which:

- there exists an easily accessible, well-developed health-care infrastructure;
- people are accustomed to taking prescription drugs (many countries in the Two-Third's World mainly use 'over the counter' drugs, which people can buy in shops);
- women are accustomed to regular medical controls; and
- women and men are free to negotiate the use of contraceptives.

The pill could only be made into a universal contraceptive if its producers put great effort into mobilizing and disciplining people and institutions to meet the specific requirements of the new technology. Needless to say, many of the required transformations were beyond the power of the inventors of the pill.

 THE CAFETERIA DISCOURSE

In the 1970s, scientists concluded that the development of 'a magic bullet' – that is, a perfect contraceptive for everyone – had failed (Harper 1983: 212). The previous twenty years of experience with the pill had made it clear that this method had significant limitations, not only with respect to acceptability but also to safety and continuity of use (Greep *et al.* 1976: 3). In the 1970s, the safety issue became of central concern because feminist health advocates and physicians reported serious side-effects of both oral contraception and intra-uterine devices (Seaman and Seaman 1978; Boston Health Collective 1971). In this period we see the emergence of a totally different type of contraceptive discourse. In *Reproduction and Human Welfare* (1976), the first extensive review of the reproductive sciences and contraceptive development initiated and funded by the Ford Foundation in the United States of America, the authors evaluated the experiences with birth control methods as follows:

> Thus, current technology cannot be regarded as adequate to meet individual or societal needs in either industrial or developing nations . . . The heterogeneity of personal, cultural, religious, and economic circumstances of human life, as well as the varying needs of individuals at different stages in the life cycle, impose diverse demands upon the technology. It is thus likely that there will never be an 'ideal' contraceptive for all circumstances.
>
> (Greep *et al.* 1976: 4)

The bulky report on the state of the art of contraceptive technology concluded with a list of recommendations addressed to 'biomedical scientists and those who act for the general public on Capitol Hill and in the

White House' (Greep *et al.* 1976: xvii) which opened with the following statement:

> A variety of safe and effective methods of fertility regulation beyond those now available is urgently needed by the world's diverse population living under different conditions and circumstances. This requires increased efforts ranging from fundamental research on reproductive processes to targeted activities in contraceptive development.
>
> (Greep *et al.* 1976: 25)

This drastic shift in the reproductive paradigm coincided with broader cultural changes in the late 1970s: the collapse of the dreams of modernity. The declining belief in grand theories and ideologies to understand and control the world led to a situation in which locality and individuality became of central concern in Western culture. The notion of differences became an important theme. Feminists, among others, acknowledged the importance of the vast differences among women's experiences and characteristics in different cultural settings (hooks 1982). The crisis in modernity eroded the belief in one technological fix to improve the human condition (Smart 1992). Reproductive scientists readily adopted the postmodern acknowledgement of differences, not least because it enabled them to expand their research programme. They used the voiced need for a wider variety of contraceptive methods to negotiate an increase in financial support for fundamental research in the reproductive sciences. To industry, the recognition of diversity among users indicated a variety of new markets. To quote Adele Clarke's felicitous phrase, 'In postmodernity, capital has fallen in love with difference' (Clarke 1995: 10).

In the 1980s, contraceptive R&D resulted in the introduction of a wider variety of contraceptive methods. On the one hand, scientists put great effort into differentiating the existing methods. The pill was now produced in many different types consisting of varying sorts and doses of hormones, tailored to the various physiological and psychological reactions of the user. Similarly, industry started marketing new types of intra-uterine device (IUD). One of these new IUDs nicely exemplifies industry's adjustment to the changing paradigm: Organon advertised an IUD which is able to adjust to differences in size of the uterus as 'the flexible alternative'. Besides this differentiation of existing methods, scientists developed new contraceptive methods such as long-acting injectable contraceptives, subdermal implants (devices which have to be inserted under the skin) and vaginal rings. The most recent innovations include the possibility of developing vaccines against pregnancy and nonsurgical methods of sterilization. In *Birth Control Technologies: Prospects by the Year 2000*, the author aptly describes the new reproductive paradigm for the last decades of the twentieth century:

> Family planning programs need to have available for consumers a variety of safe and effective methods, so that in a 'cafeteria' style, self-selection by consumers will lead to greater individual motivation to use

any particular method and ensure continued widespread use of birth control methods.

(Harper 1983: 9)

Most remarkably, the acknowledgement of the need 'to modify technology to fit people, rather than modifying people to fit technology' (Marshall 1977: 65) broadened contraceptive R&D to include a new group: men. Since the introduction of the condom and sterilization, both methods that date from the nineteenth century, no new means of contraception for men had been developed. A review of the state of the art of contraceptive development in 1977 concluded that,

although there were as many as 14 methods available for limitation of fertility, only 3 of these were ones that could be used by men: coitus interruptus, condoms and vasectomy.

(Schearer 1977: 178)

At present, it is estimated that worldwide only 21 per cent of contraceptive-using couples rely on male methods – that is, condoms and vasectomies (Hatcher 1990). Reproductive scientists suggest that research on reproductive functions of the male has lagged at least fifteen years behind that on the female (Greep *et al.* 1976: 15), one of the most striking consequences of the process of othering in the reproductive sciences. Until the late 1970s, scientists and culture at large – including feminists! – considered the female as the sex responsible for contraception. Rather ironically, reproductive biologists have argued that, in terms of population control, it would have been more efficient to choose men as the major target for controlling fertility because men have a much longer fertile life than women (Spilman *et al.* 1976: 2, 3).

History repeats itself. As in the case of the pill for women, the request for developing new male contraceptives came from outside the scientific community. In this case, social pressures came from two different sides: feminists in the Western, industrialized world; and governments of the Southern world, most notably China and India. Feminists demanded that men share the responsibilities and health hazards of contraception, whereas governmental agencies urged the inclusion of 'the forgotten 50% of family planning' as a target for contraceptive development (Handelsman 1991: 230; Wu 1988: 443). In the 1970s and 1980s, the question 'What about a male pill?' appeared at regular intervals in newspaper headlines, particularly in periods during which the serious health risks of the female pill were reported (*The Lancet* 1984: 1108). Although research in male reproduction has increased due to these pressures, the pill's 'male twin' has not yet appeared on the market. Reproductive scientists estimate that it is very unlikely that new male contraceptives will be available before well into the twenty-first century (Mastroianni *et al.* 1990: 483). Nevertheless, the male has definitely been put on the scientific and political agenda. The United Nations' International Conference on Population and Development, held recently in Cairo, addressed male responsibilities in family planning as one of its goals.

The shift in the reproductive paradigm toward acknowledging differences among users thus coincided with an erosion of the gendered subject–object relations in scientific discourses in which men traditionally possessed the subject position and women were the target. For the first time in the history of the reproductive sciences, male scientists are testing contraceptive compounds on their own sex.[3] The reverse also happened: women have increasingly adopted the subject position. Since the 1980s, women are represented in substantial numbers in decision-making bodies dealing with contraception, and the number of women scientists has increased as well (Djerassi 1989: 358).

▶ ## CONTROL OR CHOICE?

After this short retrospective view of the history of contraceptives, it is time to reflect on the meaning of the shift in the reproductive paradigm. Is the cafeteria discourse really so different from the one-size-fits-all discourse? I have argued that there is an important difference: the cafeteria discourse acknowledges the diversity of users, whereas the one-size-fits-all discourse emphasizes the universality of women and their bodies, a discourse which largely erased diversities. We may therefore conclude that the shift in discourses does matter a lot. It meant a break in the process of othering that had dominated the reproductive sciences since the nineteenth century. The universal category of 'woman' became replaced by human bodies in all their diversities.

Nevertheless, if you make such neat distinctions, continuities will inevitably be discovered between these discourses as well. Adele Clarke recently described another distinctive feature of modern approaches to human reproduction: their quest to achieve control over reproductive processes (Clarke 1988, 1995). If we compare the cafeteria discourse with the one-size-fits-all discourse from the perspective of control, the two discourses seem to conflate. In the discourse on the pill, control was of central concern. In his own account of what made him decide to begin contraceptive research, Pincus cited 'two overtly ascertainable factors: a visit from Mrs Margaret Sanger in 1951, and the emergence of the appreciation of the importance of the population explosion' (Pincus 1965: 5). The pill was thus called into existence as a technological fix to solve what was perceived as 'the population problem'. From the viewpoint of control, the cafeteria discourse is more ambiguous. Here reproductive scientists frequently adopt double talk: their reports are couched in terms of individual choice while they simultaneously emphasize the need to control population growth. In this type of discourse, individual control is on uneasy terms with population control by the state. The aim to lower birth rates has to be attuned to the individual's reproductive health needs. The volume mentioned earlier, *Reproduction and Human Welfare*, exemplifies my point. Discussing the problem of the 'current unprecedented rate of world population growth', the authors concluded:

Reproductive research aimed at improved methods of fertility regulation thus links the search for solutions of the personal problems of individuals and the social problems of societies with some of the most critical overarching problems facing the world as a whole.

(Greep *et al.* 1976: 2)

This ambiguity of the cafeteria discourse is clearly visible in one of the recently developed contraceptive methods, contraceptive implants. Contraceptive implants (like Norplant) are long-acting methods (five years) which are inserted under the skin of a woman's upper arm. Although this contraceptive method is presented as a technology to increase the freedom of women to choose the contraceptive they prefer, it facilitates a situation in which the individual's control over her fertility is replaced with control by family-planning organizations or the state. Similarly, as in the case of the pill, contraceptive implants are not a 'neutral' device; they contain a specific type of user – a woman who is considered likely to forget to take her contraceptives. By introducing long-acting implants, the continuity of use is guaranteed because it is embedded in the technology itself. The contraceptive is now delivered in a form that ensures that women will continue contraception over a specific period. This type of contraceptive is a nice example of 'technical delegates': artefacts that are designed to compensate for the perceived deficiencies of their users.[4] Feminist health activists have noted the danger of abuse of this type of contraceptive because the user depends on the assistance of health workers to remove the device if she wants to get pregnant (Mintzes 1992; Mintzes *et al.* 1993). The new generation of provider-dependent long-acting contraceptives aims at 'minimizing user failure', but simultaneously 'minimizes women's control' (Hartmann 1992: 6; Hardon 1992).

The cafeteria discourse can thus be portrayed as simultaneously modern and postmodern. This is in line with Clarke's argument which questions the robustness of the boundary between modern and postmodern approaches and points to 'the simultaneity in time and space of modern and postmodern approaches' (Clarke 1995: 3). However, in reproductive discourses the modern and postmodern conflate only in time, not in space. If we look more closely at the cafeteria discourse, we can see an important differentiation between that part of the population deemed worthy of a greater individual choice, and that part in need of a stricter fertility control. The rhetoric of individual choice seems to be addressed to users all over the world, whereas the rhetoric of population control is more exclusively centred on the countries of the South. To quote Hartmann:

We in the industrialized countries have the right to voluntary choice as to whether and when to bear children, but these rights are subordinate to the overriding imperative of population control.

(Hartmann 1987: 14)

Modernistic tendencies are thus still very much alive in reproductive discourses concerned with countries of the Two-Third's World. Although the cafeteria discourse has disrupted the former scientific representations of the gendered subject–object relations, it has reinforced the othering of people of colour. The focus on 'woman' as the responsible sex has been replaced with a focus on a specific category of women and men who are considered more responsible than others. In the present political climate, in which population control in countries of the South is deemed a precondition for the solution of environmental problems, this politics of othering will remain a crucial issue in the biomedical sciences.

▶ NOTES

1 See Oudshoorn 1994 for a more extended analysis of the development of the pill in the 1950s and the 1960s.
2 The concept of the configured user has been introduced by Steve Woolgar. See Woolgar 1991.
3 With one exception: one of the first clinical trials that Pincus performed to investigate the contraceptive activity of hormones included eight men, all patients from a mental institution. Despite the fact that the hormone preparations had a definite contraceptive effect in these male patients, men were not included in later trials due to the occurrence of side-effects (Oudshoorn 1994).
4 Madeleine Akrich, 'Comment décrire les objets techniques', *Technique et Culture*, vol. 5, 1987, pp. 49–63; Bruno Latour, 'Mixing Humans and Non-humans Together: The Sociology of a Door-Closer', *Social Problems*, vol. 35, 1988, pp. 298–310.

▶ REFERENCES

Boston Health Collective (1971) *Our Bodies, Ourselves*, Simon & Schuster, New York.
Christian Johnson, R. (1977) 'Feminism, Philanthropy and Science in the Development of the Oral Contraceptive Pill', *Pharmacy in History*, vol. 19, no. 2, pp. 63–79.
Clarke, A. (1988) 'The Industrialization of Human Reproduction, 1889–1989', Keynote Address, Conference on Athena Meets Prometheus: Gender and Technoscience, University of California – Davis.
Clarke, A. (1995) 'Modernity, Postmodernity and Reproductive Processes, 1890–1990 or, "Mommy, Where Do Cyborgs Come From Anyway?"', in C. H. Gray, J. Figueroa-Sarriera and S. Mentor, *The Cyborg Handbook*, Routledge, New York.
Djerassi, C. (1989) 'The Bitter Pill', *Science*, vol. 245, pp. 356–61.
Foucault, M. (1981) *The History of Sexuality Volume 1: An Introduction*, trans. Robert Hurley, Penguin Books, Harmondsworth.
Gallagher, C. and T. Laqueur, eds (1987) *The Making of the Modern Body: Sexuality and Society in the Nineteenth Century*, University of Chicago Press, London, Berkeley and Los Angeles.
Greep, R. O., M. Koblisky and F. Jaffe (1976) *Reproduction and Human Welfare: A*

Challenge to Research. A Review of the Reproductive Sciences and Contraceptive Development, MIT Press, Cambridge, Mass. and London.

Handelsman, D. J. (1991) 'Bridging the Gender Gap in Contraception: Another Hurdle Cleared', *The Medical Journal of Australia*, vol. 154, no. 4, pp. 230–33.

Hardon, A. (1992) 'Norplant: Conflicting Views on its Safety and Acceptability', in H. B. Holmes, ed., *Issues in Reproductive Technology: An Anthology*, Garland Publishing Inc., New York and London, pp. 11–31.

Harper, M. J. K. (1983) *Birth Control Technologies: Prospects by the Year 2000*, University of Texas Press, Austin.

Hartmann, B. (1987) *Reproductive Rights and Wrongs: The Global Politics of Population Control and Contraceptive Choice*, Harper & Row, New York.

Hartmann, B. (1992) 'Contraceptive Choice: A Multitude of Meanings', in H. B. Holmes, ed., *Issues in Reproductive Technology: An Anthology*, Garland Publishing Inc., New York and London, pp. 3–9.

Hatcher, R. (1990) *Contraceptive Technology: 1990–1992*, Irvington, N.Y.

Honegger, C. (1991) *Die Ordnung der Geschlechter: Die Wissenschaften vom Menschen und das Weib*, Campus Verlag, Frankfurt and New York.

hooks, b. (1982) *Ain't I a Woman: Black Women and Feminism*, Pluto Press, London.

The Lancet (1984) 'Gossypol Prospects', vol. 8386, pp. 1108–9.

Laqueur, T. (1990) *Making Sex: Body and Gender from the Greeks to Freud*. Harvard University Press, Cambridge, Mass. and London.

Maisel, A. Q. (1965) *The Hormone Quest*, Random House, New York.

Marshall, J. F. (1977) 'Acceptability of Fertility Regulating Methods: Designing Technology to Fit People', *Preventive Medicine*, vol. 6, pp. 65–73.

Mastroianni, L., P. F. Donaldson and T. T. Kane (1990) 'Development of Contraceptives – Obstacles and Opportunities', *The New England Journal of Medicine*, vol. 322, 15 February, 482–5.

McLaughlin, L. (1982) *The Pill, John Rock and the Church: The Biography of a Revolution*, Little Brown, Boston, Mass. and Toronto.

Medvei, V. C. (1983) *A History of Endocrinology*, MTP Press, The Hague.

Mintzes, B., ed. (1992) *A Question of Control. Women's Perspectives on the Development and Use of Contraceptive Technologies*, Report of an International Conference held in Woudschoten, The Netherlands, April 1991, WEMOS Women and Pharmaceuticals Project and Health Action International, Amsterdam.

Mintzes, B., A. Hardon and J. Hanhart (1993) *Norplant: Under Her Skin*, Eburon, Delft.

Moscucci, O. (1990) *The Science of Woman. Gynaecology and Gender in England, 1800–1929*, Cambridge University Press, Cambridge.

Oudshoorn, N. (1994) *Beyond the Natural Body. An Archeology of Sex Hormones*, Routledge, London and New York.

Pincus, G. (1958) 'Fertility Control with Oral Medication', *American Journal of Obstetrics and Gynaecology*, vol. 75, pp. 1333–47.

Pincus, G. (1959) 'Progestational Agents and the Control of Fertility', *Vitamins and Hormones*, vol. 169, pp. 307–25.

Pincus, G. (1965) *The Control of Fertility*, Academic Press, New York and London.

Schearer, S. B. (1977) 'Pharmacological Approach to Contraception in Men', *Drug Therapy*, vol. 5, no. 2, pp. 72–7.

Schiebinger, L. (1989) *The Mind Has No Sex? Women in the Origins of Modern Science*, Harvard University Press, Cambridge, Mass. and London.

Seaman, B. and G. Seaman (1978) *Women and the Crisis in Sex Hormones. An Investigation of the Dangerous Uses of Hormones: From Birth Control to Menopause and the Safe Alternatives*, Harvester, Brighton, Sussex.

Smart, B. (1992) *Modern Conditions, Postmodern Controversies*, Routledge, London and New York.

Spilman, C. H., T. J. Lobl and K. T. Kirton (1976) *Regulatory Mechanisms of Male Reproductive Physiology*, Sixth Brook Lodge Workshop on Problems of Reproductive Biology (Amsterdam), Excerpta Medica, Oxford.

Woolgar, S. (1991) 'Configuring the User: The Case of Usability Trials', in J. Law, ed., *A Sociology of Monsters: Essays on Power, Technology and Domination*, Routledge, London.

Wu, F. C. W. (1988) 'Male Contraception: Current Status and Future Prospects', *Clinical Endocrinology*, vol. 29, pp. 443–65.

PART FOUR

▶ Military technology

▶ Introduction

▶ TECHNOLOGICAL DETERMINISM AND WEAPONRY

Military technology has contributed centrally to the shaping of our world. One of its central features, the dominance of the countries of the 'North' over those of the 'South', is in part the legacy of innovations in military technology in Europe. The gun-carrying, ocean-going sailing ship was a vital implement in creating European empires (Cipolla 1965), and the military-technological dominance of the North remained, and remains, a vital factor in nineteenth- and twentieth-century politics and international economics (see, for example, Headrick 1981; Kaldor 1982: 131–68). From 1939 onwards, achieving technological improvements to weaponry became a central concern in most industrialized countries: at the height of the arms race, as much as 40 per cent of research and development effort worldwide was devoted to military technology (Barnaby 1981). In the post-war United Kingdom, for example, over half of government-funded research and development was typically military (see, for example, Norman 1979).

The combination of the shaping role of military technology and the immense resources devoted to improving it meant that during the cold war informal thinking about war and weapons tended to have a strongly technologically determinist air. The eminent British science policy maker Lord Zuckerman, for example, wrote that 'the pure or fundamental science of today inexorably becomes the applied science and technology of tomorrow, with unforeseeable consequences, either immediate or remote, flowing from its exploitation' (1980: 246). Frank Barnaby, former director of the justly renowned Stockholm International Peace Research Institute, similarly wrote that 'the military technological tail wags the political dog' (1980). And though his thought was complex enough to defy simple classification, the historian E. P. Thompson's resonant warning (1982) that the social systems of East and West were in the grip of 'exterminism' had a technologically determinist flavour.

Since the end of the cold war, much less attention has been paid to military technology, but it remains a vitally important topic. The thousands of nuclear weapons that the world's arsenals still contain are amply sufficient to cause a human catastrophe without historical precedent. The collapse of the Soviet Union may in practice mean that its arsenal is more, not less, dangerous, as previously stringent political controls erode. The very 'victory' of the West, which has left the USA with unquestioned worldwide supremacy in conventional weaponry, increases the incentive for other nations to turn to weapons of mass destruction rather than embark on the hopeless task of matching American conventional might. More mundanely, military procurement remains a major component of state expenditure, and the resultant industries, though much shrunk in size, remain large employers, especially of scientists and technologists.

Military concerns influence civilian technologies as well as overtly military ones, as we noted in our overall introduction (p. 15). An interesting example, and our reason for including the extract from the work of Janet Abbate, is the Internet, which evolved from an earlier computer network, ARPANET, named after its military sponsor, the US Department of Defense Advanced Research Projects Agency (ARPA). An important and distinctive feature of both ARPANET and the Internet is their topology. Unlike most previous computer networks, which had a star configuration (a central node and single connections radiating out to every other node), ARPANET had no 'centre': it was a 'distributed' network, with a denser web of connections ensuring that there were at least two routes from any node to any other (see Figure 1). ARPA took this idea even further in designing the Internet, a worldwide network of networks originally built around the ARPANET. While the ARPANET was centrally managed, the Internet was specifically designed to allow decentralized operation and expansion of the member networks. This permitted the Internet to develop in an almost anarchic way.

Although the idea of a large-scale technology that is not based on centralized control has appealed to antiestablishment groups of all kinds, Abbate reveals that the distributed topology had quite different origins. One concern that it reflected was the desire by the US Air Force for a communications network that could survive nuclear war. A star configuration was plainly a poor choice, since nuclear attack on the central node could paralyse the entire network. Paul Baran of the defence think-tank, the Rand Corporation, proposed as a better solution a distributed network and what has come to be called 'packet switching', in which messages would be split into segments, and each node in the network would read the addresses of incoming messages and forward them appropriately towards their destination.

Cold war concerns, however, form only part of the story told by Abbate. Donald Davies, a computer scientist at the British National Physical Laboratory, independently developed the idea of packet switching. Davies was working in a quite different context, one in which military concerns were in

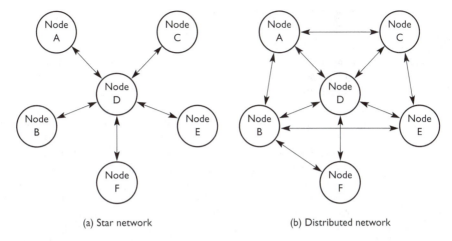

(a) Star network   (b) Distributed network

Figure P4.1 Nodal configurations (courtesy Janet Abbate)

the background and one influenced by the apparent commitment of Harold Wilson's 1960s Labour government to forging a new Britain in the 'white heat of technology'. Davies's ambitious plan for a user-friendly, packet-switched national computer network foundered on lack of resources and technical scepticism: it is interesting to speculate on the extent to which Britain's technological history might have been different if he had succeeded!

Abbate shows that Davies's practical design for a packet switching network was an important intermediate step between Baran's work and the ARPANET. ARPA was not closely bound to a particular military mission and had remarkable freedom to pursue an ambitious and technologically risky project (telephone engineers, for example, feared that packet switching was too complex to be practical) with little certainty of immediate practical pay-off. In the process of building the ARPANET, the research community took the technology in new directions, making it a medium for social interaction in the form of electronic mail. But military concerns and priorities still shaped the development of the ARPANET's successor, the Internet (see Abbate 1998), and ARPA has continued to play a strong role in computer network research and development in the USA. The episode is also of more general interest in showing that (despite the conventional wisdom that formed during the 1980s) government action can play a crucial creative role in technological development, if that action is properly resourced and not heavy-handed or too focused on short-term commercial pay-off.

▶ GENDER AND MILITARY TECHNOLOGY

Of all the spheres of technology, none is more stereotypically masculine than war and weaponry are seen as male affairs, and the imagery surrounding them is typically either an exclusively male imagery (the army as male brotherhood, the missile as phallus) or one in which women are relegated to the sidelines as the wife/girlfriend/daughter that the soldier is fighting to 'defend'. Some authors have tried to root the male fascination with weapons and war in biology and psychology, arguing that men need a substitute for the babies they cannot bear (Easlea 1983). To us, however, it seems more plausible to follow authors such as Cynthia Enloe (1983) and see the maleness of weaponry as the outcome of a process of exclusion in which both masculinity and technology are constructed.

Enloe emphasizes that the maleness of militarism is an appearance sustained only by the systematic denial of certain awkward facts and tendencies. For example, as weapons become more and more heavily based on electronics, the role of traditional male military 'virtues' is diminished – no enemy is ever seen in the flesh, much less confronted physically – while simultaneously the salience of women's labour becomes ever greater. The electronics industry is largely a women's industry, at least as far as production is concerned. Enloe (1983: 195) gives the example of a modern navy cruiser, an archetypally male artifact – but in actuality one built in large part by women. Forty per cent of the cost of such a vessel is accounted for by its electronics, and without those electronics it would be useless in modern naval warfare.

Discussion of the relations between gender and the military often has a strong symbolic focus, and one reason for including the extract from Rachel Weber's work is that it reminds us that exclusionary processes can take straightforward material forms. She focuses on the design of aircraft cockpits, showing how the physical dimensions of traditional cockpits excluded the majority of women, because they were set with men's typically larger bodies in mind. Paradoxically, though, Weber shows that feminists have been able to challenge this physical gendering of technology more successfully in the case of military aircraft (where explicit political intervention is possible); the more diffuse processes shaping civilian cockpit design are harder to influence.

▶ OTHER ASPECTS OF THE SOCIAL SHAPING OF MILITARY TECHNOLOGY

Part of what makes Weber's study distinctive is that feminist pressure has made the gendering of the design of cockpits explicit. More often, it is harder empirically to show specific influences of gender because of the extent of the exclusion of women from the arenas in which military technology is shaped (see the discussion on p. 22 of this methodological problem). The influences that are more immediately apparent are organizational ones, and our extracts from the work of Fallows and of Armacost exemplify these processes.

Fallows demonstrates the capacity of a set of military organizations, the

US Army's ordnance corps, to shape American rifle design. The ordnance corps to a large extent successfully resisted a technical innovation from an outside inventor – the lightweight Armalite AR-15 rifle – and imposed on the US Army a rifle design, the M-16, which proved disastrous in combat. Fallows's story can be presented as simply one of an aberrant abuse of power. But purged of some of its extraordinary features, such as the alleged deliberate and conscious rigging of test results, it is perhaps of some generality. In the military, as elsewhere, organizations develop technological styles and preferences which are embodied in the criteria they use to evaluate technical changes. In effect, organizations can come to hold socially specific definitions of what it is for a technology to 'work'. In the case of the ordnance corps, the definition of what it was for a rifle to work was influenced strongly by the 'gravel-bellies', the sharpshooters whose key criterion was the accuracy of a weapon in controlled, deliberate, long-range firing on the best range. Given that it is impossible to test any technology for all possible conditions, or to shape it to satisfy all possible demands on it, an organization must select criteria. The criteria it prioritizes will depend in part on its history and its relations to other competing organizations.

The feature of *competition between organisations* has, indeed, become the focus for the best-developed approach to the shaping of military technology, the *bureaucratic politics* model. This model, which has been applied to the whole range of military (and sometimes other) decisions, not simply those concerning technology, is a distinctive contribution of post-1950 American political science. In this approach, it is argued that it is misleading to see the actors on the military or diplomatic stage as being countries. 'America', say, decides nothing, plans nothing, intends nothing. Policy is an *outcome* of a process of bargaining and competition, one in which the key actors are often neither individuals nor political parties, but bureaucratic organizations. While all those involved may see their goal as being the enhancement of national security (no necessarily cynical view of individual motivation is implied), those with different institutional locations will put a very different slant on how to achieve that goal.

In the USA, to whose defence politics this approach has been most frequently applied, key institutional actors include not simply the presidency, Congress and the office of the secretary of defense, but also the armed services. Nothing could be more mistaken, suggest the proponents of the bureaucratic politics model, than to imagine that the armed services simply obey orders passed down from their political superiors. They lobby actively for their goals, and the relationship between the US Air Force, say, and the presidency is more appropriately described as 'bargaining' than as political command (see, for example, Ball 1980). An individual concerned with his or her career has to reckon with the fact that armed services remain while presidents come and go, and against the president's formal right of command the Air Force can counterbalance extensive influence over Congress, with whom the president will also be involved in a complex bargaining process.

From a range of studies of military technology reflecting in various degrees the influence of the bureaucratic politics model (for example,

Sapolsky 1972; Greenwood 1975; Beard 1976), we have chosen an edited extract from an early study that has become a classic, Armacost's *The Politics of Weapons Innovation: The Thor–Jupiter Controversy* (1969). Armacost's focus is on the conflict in the 1950s between the US Air Force and Army over the intermediate range ballistic missile. Both services coveted the right to design and deploy such a weapon, and produced rival missile plans, Thor in the case of the air force and Jupiter in that of the Army. One interesting feature of Armacost's study is that he shows that the proprietorial interests of the two services influenced technical design. The Army sought to design a mobile missile, in the belief that the analogy between such a missile and existing field artillery would strengthen their claim for operational control of the intermediate range ballistic missile. The Air Force, on the other hand, held that true mobility was neither technically feasible nor necessarily desirable, and argued for the strategic necessity of a weapon tightly integrated into the overall nuclear command and control network which they dominated.

A further useful feature of Armacost's study (one not always to be found in 'bureaucratic politics' studies) is that he points to the involvement in the Thor–Jupiter controversy of the industrial producers of weapons as well as their military users. The conflict between Thor and Jupiter was not simply between Air Force and Army, but also between the powerful, established Aircraft Industries Association and Chrysler, an automobile company, experienced in the supply of vehicles to the Army, that was trying to break into the missile field.

The existence of such links between weapons industries and the armed forces has led some authors to follow the lead of President Eisenhower and identify a 'military-industrial complex' fuelling the arms race, especially in America (Eisenhower's 1961 'Farewell to the Nation' speech on the topic is reprinted in Melman 1970: 235–9). While some claims for the coherence and influence of the military-industrial complex are exaggerated (see Sarkesian 1972; Rosen 1973), there can be little doubt that the nature of military technology is profoundly affected by the institutions in which it is produced as well as by those in which it is used.

Perhaps the most thought-provoking writer on this topic is Mary Kaldor. Her 1982 book, *The Baroque Arsenal*, together with the related work of authors such as Fallows, and Gansler (1982), presented a powerfully argued analysis of military technology that found an audience even within the 'defence establishment'. The armament sector of Western economies, argues Kaldor, is unusual in that it contains essentially one customer, the state, and a small number of competitive suppliers of weapons (international arms sales are economically important, but do not remove the dependence of most large weapons suppliers on their states). Competition between suppliers, and the need to justify new weapons politically, dictates that the performance characteristics of a new aircraft, say, be 'better' than those of its predecessor; but the essential conservatism of military institutions freezes judgements of what constitutes 'better' into clearly defined moulds (greater ballistic missile accuracy, for example). This institutional context thus produces a distinctive style of technological innovation:

weapons systems become more complex, more elaborate, more sophisti-cated and grotesquely more expensive.

What Kaldor points to is a process of the shaping of technology in which economic factors play a major role: defence contractors are surely no less motivated by profit considerations than other firms. However, because those economic factors operate within a different framework from that more common in civilian markets, they produce markedly different effects. Because of the peculiar nature of military contracting, where a firm's total outlay traditionally was met by the state with a percentage profit on top,[1] firms had until recently little incentive to focus innovation around reduc-ing the cost of the *process* by which weapons are produced. Competitive pressure has acted to focus innovation on 'improving' the *product*, even at the expense of greatly increased production costs.

In the extract from her work we have chosen, Kaldor develops these ideas into an ambitious model of the shaping of military technology. She identifies two characteristics of the organizations supplying this tech-nology: whether or not they are 'sovereign', that is, responsible for their own financial affairs; and whether or not they are heavily dependent on military contracts. The traditional cold war Western pattern was the sovereign, dependent supplier, and for the reasons discussed above Kaldor suggests that this gave rise to the pattern of over-elaborate improvement she calls 'baroque'. The former Soviet Union was characterized by non-sovereign, dependent suppliers, which gave rise to a more conservative pattern of technical change. 'Revolutionary' changes in military tech-nology – those that fundamentally alter weapons platforms and weapons themselves rather than simply improving them – are, Kaldor suggests, typically the product of a third kind of supplier: sovereign, independent organizations.

 ## UNINVENTING NUCLEAR WEAPONS

Our final reading is from the work of one of us, and is intended to tie the topic of military technology to two of the central overall threads of this book: theories of the relationship of technology and society; and the politics of technology. MacKenzie considers the possibility of the world-wide abolition of nuclear weapons, and asks what theories of technical change might contribute to an analysis of its feasibility: specifically, what they might imply for ways of making the abolition of nuclear weapons permanent and irreversible. He suggests, for example, that the conventional wisdom that nuclear weapons cannot be uninvented is misleading. If some of the knowledge needed to design and produce such weapons is tacit knowledge (analogous to our capacity to ride a bicycle without being able to say precisely how we do it), then it can plausibly be lost if there is a sufficiently long hiatus in design and production. Of course, nuclear weapons could then be reinvented, but 'reinvention' might be the right word: the process might be significantly harder than conventional wisdom allows.

MacKenzie also applies Tom Hughes's system perspective and the actor-network perspective (both discussed in Part One of this book) to the abolition of nuclear weapons. At its simplest, the system perspective suggests that we should look for ways of 'unplugging' those weapons from the large technical systems that produce them and that they are part of. The actor-network emphasis on contingency and reversibility might seem to make it a voice of caution and pessimism, in that an abolition agreement that is broken would be extremely dangerous, but MacKenzie explores ways in which more optimistic lessons might be drawn from actor-network theory.

We do not expect all readers to feel that abolishing nuclear weapons is a desirable goal (and MacKenzie's chapter deliberately focuses on the goal's feasibility, not its desirability), but we make no apologies for ending this book with this explicitly political chapter. The social studies of technology are impoverished if they remain an entirely academic pursuit, but there are strong pressures to focus their political application on promoting technological innovation. That is often a sensible and laudable focus, but the field will lose its critical edge if it is restricted to it. Not all technological change is desirable, and a fully-fledged politics of technology must tangle with questions of the elimination of technologies – a process that, to be sure, will involve technological innovation – as well as with their promotion.

 NOTE

1 *Explicit* 'cost-plus' contracts of this kind have been frowned upon for many years, but Gansler (1982) reveals a range of strategies whereby firms can make contracts *de facto* cost-plus. Only very recently has procurement reform really begun to alter the entrenched cost-plus culture of defence contracting.

▶ Cold war and white heat:
the origins and meanings of
packet switching

Janet Abbate

Of all the technical innovations featured in the ARPANET, forerunner of
the Internet, perhaps the most celebrated was packet switching. Packet
switching was an experimental, even controversial method for transmitting data across a network. Its proponents claimed that it would increase
the efficiency, reliability and speed of data communications, but it was
quite complex to implement, and some experts argued that the technique
would never work. Indeed, one reason the ARPANET became the focus of
so much attention within the computer science community was that it
was the first large-scale proof of the feasibility of packet switching. The
successful use of packet switching in the ARPANET and other early networks paved the way for the technique's widespread adoption, and at the
end of the twentieth century packet switching dominates networking
practice.

To many computer professionals, packet switching appears to have
obvious technical advantages over alternative methods for transmitting
data, and they have tended to treat its widespread adoption as a natural
result of these advantages. In fact, however, the success of packet switching
was not automatic: it had to be socially constructed. For many years there
was no consensus on what packet switching actually was – what its defining
characteristics were, what advantages it offered, how it should be implemented. Before the technique could achieve legitimacy in the eyes of data
communications practitioners, its proponents had to prove that it would
work by marshalling the resources to build demonstration packet switching
networks. The wide disparity in the outcomes of these early packet switching experiments demonstrates that the concept could be realized in very
different ways, and that, far from being a straightforward matter of superior

technology winning out, the 'success' of packet switching depended greatly on how it was interpreted.

Packet switching was invented independently by two computer researchers working in very different contexts: Paul Baran at the Rand Corporation in America and Donald Davies at the National Physical Laboratory in England. Baran was the first to explore the idea, around 1960; Davies came up with his own version of packet switching a few years later and subsequently learned of Baran's prior work. Davies was instrumental in passing on the knowledge of packet switching that he and Baran had developed to Lawrence Roberts, who oversaw the creation of the ARPANET. However, while Baran's and Davies's versions of packet switching had some basic technical similarities, their conceptions of what defined packet switching and what it was good for were very different. Much of this difference was due to the strong political pressures that were brought to bear on computing research in Britain and the United States. Large computer projects in both countries were developed in a context of government funding and control, and national leaders saw computers as a strategic technology that was vital to pursuing important political goals. But in the very different policy contexts of the United States and United Kingdom, packet switching took on different meanings for Baran, Davies and Roberts. Packet switching was never adopted on the basis of purely technical criteria, but always because it fitted into a broader socio-technical understanding of how data networks could and should be used.

 ## NETWORKING DR STRANGELOVE: THE COLD WAR ROOTS OF PACKET SWITCHING IN THE UNITED STATES

In 1964, movie theatres across the United States presented Stanley Kubrick's brilliant black comedy of cold war paranoia, *Dr Strangelove*. The film, though humorous, highlighted a serious problem for American defence strategists: the vulnerability of communications channels to disruption by a Soviet attack, which might make them unavailable just when they were needed most. In the movie, a psychotic air force commander named Jack D. Ripper sets in motion a nuclear holocaust by invoking a strategy of mutually assured destruction called 'Plan R.' Plan R – which allows Ripper to circumvent the president's authority to declare war – is specifically designed to compensate for a wartime failure in command, control and communications. As the movie dramatizes, this plan is hardly ideal, since it allows Ripper to launch a 'retaliatory' attack even though no Soviet first strike has actually occurred. In reality (as the film's disclaimer states), the air force never had any such strategy. In fact, the air force was at this time exploring a very different solution to the problem: building a communications system that would be able to survive an enemy attack, and thus maintain 'proper command and control.'

The need for 'survivable communications' was a generally recognized problem by the early 1960s, and among those intent on solving it was a researcher at the US air force's premier think-tank, the Rand Corporation. Founded by the air force in 1946 as an outgrowth of operations research efforts initiated in World War II, Rand (or RAND, an acronym for Research and Development) was a non-profit corporation dedicated to research on military strategy and technology. Rand was primarily funded by contracts from the air force, though it served other government agencies as well. The corporation attracted talented minds through a combination of high salaries, relative autonomy for researchers, and the chance to contribute to policy decisions of the highest importance.

In 1959 a young engineer named Paul Baran joined Rand's computer science department. Immersed in a corporate culture focused single-mindedly on the cold war, Baran soon developed an interest in survivable communications, which he felt would decrease the temptation of military leaders to launch a pre-emptive first strike:

> Both the US and USSR were building hair trigger nuclear ballistic missile systems . . . If the strategic weapons command and control systems could be more survivable, then the country's retaliatory capability could better allow it to withstand an attack and still function; a more stable position. But, this was not a wholly feasible concept because long distance communications networks at that time were extremely vulnerable and not able to survive attack. That was the issue. Here a most dangerous situation was created by the lack of a survivable communication system.[1]

Baran was able to explore this idea without an explicit contract from the air force, since Rand had a considerable amount of open-ended funding that researchers could use to pursue projects they deemed relevant to the American defence concerns.[2]

Baran began in 1959 with a plan for a minimal communications system that could transmit a simple 'Go/No go' message from the president to the commanders using AM radio broadcasts. When Baran presented this idea to military officers, however, they immediately insisted that they needed greater communications capacity. So he went back to the drawing board and spent 1960–2 formulating ideas for a new system that would combine survivability with high communications capacity.[3] Baran envisioned a system that would allow military personnel to carry on ordinary voice conversations or use teletype, facsimile, or low-speed computer terminals under wartime conditions.

The key to this new system was a technique Baran called 'distributed communications'.[4] In a conventional communications system such as the telephone network, switching is concentrated and hierarchical. Calls go first to a local office, then to a regional or national switching office if they need to connect beyond the local area. Each user is connected to only one local office, and each office serves a large number of users. This means that destroying a single local office would cut off many users from the network. By contrast, a distributed system would have many switching nodes and

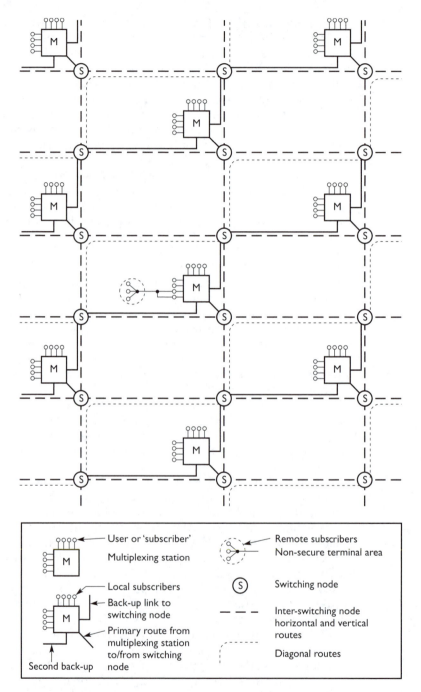

Figure 25.1 Baran's distributed network of switching nodes and multiplexing stations (Paul Baran 1964a *On Distributed Communications*, vol. VIII, Figure 1)

many links attached to each node. Such redundancy would make it harder to cut off service to network users.

Baran described this new type of network in a series of eleven reports called *On Distributed Communications*. His proposed system (a small segment of which is shown in Figure 25.1) consisted of a set of several hundred switching nodes, each connected to other nodes by up to eight lines. There was also a set of several hundred multiplexing stations that provided the interface between the users and the network. Each multiplexing station would be connected to two or three switching nodes and up to 1024 users with data terminals or digital telephones. The switching was distributed among all the nodes in the network, so that the enemy could not disable the whole network by targeting a few important centres. To make the system even more secure, Baran planned to locate the nodes in areas remote from the network users, since population centres were considered military targets, and he designed the multiplexing stations with a wide margin of excess capacity, on the assumption that enemy attacks would cause some equipment to fail. He also added military features such as cryptography and a priority/precedence system that would allow high-level users to pre-empt messages from lower-level users.[5]

To move the data through the network, Baran adapted a technique known as 'message switching' or 'store and forward' switching. A common example of message switching is the postal system. In a message switching system, each message (such as a letter) is labelled with its origin and destination addresses, and it is then passed from node to node through the network. The messages are temporarily stored at each node (such as a post office) until they can be forwarded to the next node or the final destination. Each successive node uses the address information to determine the next step of the route. In the 1930s, message switching began to be used in telegraph systems, with messages being stored on paper tape at each intermediate station before being transmitted to the next station. At first the messages were switched manually by the telegraph operators, but in the 1960s telegraph offices began to use computers to store and route the messages.[6]

The postal and telegraph systems adopted message switching because it was more efficient than transmitting messages or letters directly from source to destination. Letters are stored temporarily at a post office so that a large number can be gathered for each delivery route. In the case of the telegraph, message switching also addressed the problem of uneven traffic flow on the expensive long-distance telegraph lines. If traffic was light, the lines would be underused and the excess capacity wasted; on the other hand, if the lines were overloaded, there would be a risk that some of the messages would be lost. Storing the messages at intermediate stations made it possible to even-out the traffic flow: if a line was busy, messages could be stored at the switch until the line was free. In this way message switching increased the efficiency, and hence the economy, of long-distance telegraph transmission.

Baran appreciated the efficiency offered by message switching, but he also saw the technique as a way to make his system more survivable. Since the nodes act independently in processing the messages and there are no pre-set routes between nodes, the nodes can adapt to changing conditions by picking the best route at any given moment. As Baran described it, 'There is no central control; only a simple local routing policy is performed at each node, yet the over-all system adapts'.[7] This increases the ability of the system to survive an attack, since the nodes can re-route messages around non-functioning parts of the network. Baran realized that survivability depended on more than just having redundant links; the nodes must be able to make use of those extra links: 'Survivability is a function of switching flexibility'.[8] Therefore, his proposed network would be characterized by distributed routeing as well as distributed links.

Departures from other contemporary systems

Baran was not the first to propose either message switching or survivable communications systems to the military. Both types of system already existed or were in development, and there was one large-scale distributed communications network being constructed in the early 1960s. This was the AUTOVON network, designed and operated for the Defense Department by the US telephone giant AT&T. In 1961 AT&T had provided the army with a communications network called SCAN (Switched Circuit Automatic Network), and in 1963 they provided a similar network for the air force called NORAD/ADS (North American Air Defense Command/ Automatic Dial Switching). The Defense Communications Agency, which was charged with coordinating the provision of communications services throughout the armed services, decided to integrate these networks into a new system called CONUS AUTOVON (Continental United States Automatic Voice Network). AUTOVON was not a message switching system but rather a special military voice network that was built on top of the existing civilian telephone network. AUTOVON had ten switching nodes and came on line in April 1964.[9]

While AUTOVON was an example of distributed communications, Baran's approach differed from AT&T's in two crucial ways. First, although AUTOVON had nodes distributed throughout the system, control of those nodes was concentrated in a single operations centre, where operators monitored warning lights, analysed traffic levels, and controlled system operations. If traffic needed to be re-routed, it was done manually: operators at the control centre would make the decision and then contact the operators at the switching nodes with instructions to change routes.[10] In Baran's network, control was fully distributed, as noted above. Nodes would be individually responsible for determining routes, and would do so automatically, without human intervention: 'The intelligence required to switch signals to surviving links is at the link nodes and *not* at one or a few centralized switching centers'.[11] Clearly such a system would be more survivable than one dependent on a single operations centre, which, Baran noted, 'forms a single, very attractive target in the thermonuclear era'.[12]

An implication of Baran's design was that the nodes would have enough 'intelligence' to perform their own routeing; they would need to be computers, and not just telephone switches. This brings us to Baran's second departure: unlike AT&T (and earlier message switching systems), he envisioned an all-digital network, with computerized switches and digital transmission across the links. Computers would be needed at the nodes to handle the complexity of routeing messages, since the switches had to be able to determine, on their own, the best path to any destination, and would need to update that information as network conditions changed. Such computerized switches had never been designed before. As Baran acknowledged, 'These problems place difficult requirements on the switching. However, the development of digital computer technology has advanced so rapidly that it now appears possible to satisfy these requirements by a moderate amount of digital equipment'.[13] Digital transmission lines would be needed to preserve the clarity of the signal. One consequence of having a distributed network was that a connection between any two endpoints would typically be made up of many short links rather than a few long ones, with messages passing through many nodes on the way to their destination. Having many links in a route posed a problem when transmitting ordinary analogue signals, because the signal degenerated slightly whenever it was switched from one link to another, and distortion accumulated with each additional link. Digital signals, on the other hand, could be regenerated at each switch; thus digital transmission would allow the use of many links without cumulative distortion and errors. Digital transmission was still a novelty at the time; Bell Labs had only begun developing its T1 digital trunk lines in 1955, and they would not be ready for commercial service in the Bell System until 1962.[14]

The requirements of Baran's system would push switching and transmission technology to their limits, so it is understandable that contemporary experts reacted sceptically to his claims. The engineers in AT&T's Long Lines division, which ran the long distance telephone service and the AUTOVON system, tended to be familiar only with analogue technology, and they were sceptical of Baran's claims that an all-digital system could transcend the well-known limitations on the number of links per call.[15]

Baran's system departed from traditional telephone company practice in other ways that show the effect of cold war military considerations on his design assumptions. For instance, the phone company tried to increase the reliability of the system as a whole by making each component as reliable as possible, and for an additional fee would provide computer users with lines that were specially conditioned to have low error rates. Baran chose instead to make do with lower-quality communications links and to provide redundant components to compensate for failures. Conditioned lines would be too expensive for a system with so many links, and in any case, the reliability of individual components could not be counted on in wartime conditions. As Baran observed, 'Reliability and raw error rates are secondary. The network must be built with the expectation of heavy damage anyway'.[16]

Packet switching in Baran's system

Baran's proposed network began as a distributed message switching system. His final innovation was to alter message switching to create a new technique: packet switching. A message in his system could be anything from digitized speech to computer data, but the fact that these messages were all sent in digital form – as a series of binary numbers or 'bits' – meant that the information could be manipulated in new ways. Baran proposed that rather than sending messages of varying sizes across the network, messages should be divided into fixed-size units that he called 'message blocks'. The multiplexing stations that connected users to the network would be responsible for dividing outgoing messages into uniform blocks of 1024 bits. A short message could be sent as a single block, while longer messages would require multiple message blocks. The multiplexer would add to each block a header that specified the addresses of the sending and receiving parties as well as other control information. The switching nodes would use the header information to determine what route each block should take to its destination; since each block was routed independently, the different blocks that made up a single message might be sent on different routes. When the blocks reached their destination, the local multiplexer would strip the header information from each block and reassemble the blocks to form the complete message. Baran's message block idea would eventually be widely adopted for computer networks. In these later systems the message blocks would be called 'packets', and the technique would become known as packet switching.[17]

For all its eventual significance, the decision to transmit data as packets was not the original focus of Baran's work. As the title of his eleven volume system description, *On Distributed Communications*, indicates, he began with the idea of building a distributed network, an idea that had already been identified with survivability by people working in military communications. In describing the system, Baran tended to stress the idea of link redundancy, rather than other elements such as packet switching. But as he developed the details of the system, the use of message blocks emerged as a fundamental element. By the time he wrote the final volume of the series, Baran had changed the name he used to refer to the system to reflect the new emphasis: 'While preparing the draft of this concluding number, it became evident that a distinct and specific system was being described, which we have now chosen to call the "Distributed Adaptive Message Block Network", in order to distinguish it from the growing set of other distributed networks and systems'.[18] What, then, was so important about packet switching? What did it mean to Baran and his sponsors?

Note that transmitting packets, rather than complete messages, imposed certain costs on the system. The interface computers had to perform the work of dividing users' outgoing messages into packets and reassembling incoming packets into messages. There was also the overhead of having to include address and control information with each packet (rather than once per message), which increased the amount of data that had to be transmitted over the network. And since packets from a single message could take different routes to their destination, they might arrive out of sequence, which meant that there had to be provisions for reassembling them in the proper order. All of this made the system more complex and presented more opportunities for failure. For Baran, these costs were outweighed by his belief that packet switching would solve some pressing problems and support some fundamental goals of the system.

Packet switching offered a variety of benefits. Baran was determined to use small, inexpensive computers for his system, rather than the huge ones he had seen in other message switching systems, and he was aware that, given the state of the art at the time, the switching computers would have to be simple in order to be both fast and inexpensive. Using fixed-size packets simplified the design of the switching node. Another advantage for the military was that breaking messages into packets and sending them along different routes to their destination would make it harder for enemy spies to eavesdrop on a conversation. But the biggest potential reward was efficient and flexible transmission of data. 'Most importantly', wrote Baran, 'standardized data blocks permit many simultaneous users, each with widely different bandwidth requirements[,] to economically share a broad-band network made up of varied data rate links'.[19] In other words, packet switching allowed a more efficient form of multiplexing, the sharing of a single communications channel between many users.

In the conventional telecommunications systems of the time, the usual form of multiplexing was by frequency division: each caller would be assigned a particular frequency band for their exclusive use for the duration of their connection. If the caller did not talk or send data continuously, the idle time would be wasted. An alternative method was called time division multiplexing, in which time is divided into short intervals, and each user in turn is given a chance to transmit data for the duration of one interval. Only users with data to transmit are offered time slots, so no slots go idle as long as anyone has data to transmit; time division multiplexing is thus more efficient for usage situations where bursts of information alternate with idle periods. Since computer data tends to have this 'bursty' characteristic, Baran felt that time division was a more 'natural' form of multiplexing for data transmission.[20] And since the time slot accommodated a fixed amount of data, Baran believed that having fixed-size message blocks was a prerequisite for time division multiplexing. Thus he associated packet switching with time division multiplexing and its promise of efficient data transmission.

Packet switching would also make it easier to combine links with different data rates in the network. The data rate is the number of bits per second that can be transmitted on a given link. In the conventional telephone system, each caller is connected at a fixed data rate, and data that flows into a switch must flow out at the same rate. With packet switching, data flowing into a switch can be divided among the outgoing links in a variety of ways, rather than having to be sent out at fixed data rates. This would make it easier for devices transmitting at different data rates (computers and digital telephones, for instance) to share a common link to the network. The system could also take advantage of new media, such as low-cost microwave, that had different data rates from the standard phone company circuits. Thus, while packet switching increased system complexity in some ways, in other ways it made the system simpler and more economical to build.

In sum, packet switching was important to Baran because it furthered some key requirements of a survivable military system. Cheaper nodes and links made it economically feasible to build a highly redundant (and thus robust) network. Efficient transmission made it possible for commanders to have the higher communications capacity they wanted. Dividing messages into packets also increased security by making it harder to intercept intelligible messages. Packet switching, as Baran understood it, made perfect sense in the cold war context of his proposed system.

The US Air Force agreed with Baran's assessment and was eager to build a network based on his design. Internal Defense Department politics thwarted this plan, but Baran's ideas were widely disseminated among American military and civilian research institutes.

▶ FORGING PACKET SWITCHING IN THE WHITE HEAT: NETWORKS AND NATIONALISM IN THE UNITED KINGDOM

While the United States was caught up in the cold war in the early 1960s, the United Kingdom was experiencing political upheaval of a different type. Just as the Americans were worried about a 'science gap' between their country and the Soviet Union, so there were widespread fears in the United Kingdom of a 'technology gap' with the United States. In 1963 Harold Wilson was elected leader of the Labour Party, at a time when that party, and much of the general population, felt that the country was facing an economic crisis. Politicians on all sides warned that the nation was falling behind the other industrial powers in its exploitation of new technologies, that there was a 'brain drain' of British scientists to other countries, and that the country's technological backwardness was at least partly responsible for its economic malaise.[21]

Wilson addressed the technology issue head on in his address to the Labour Party's annual conference at Scarborough on 2 October 1963. Calling on labour and management to join in revitalizing British industry, Wilson stressed the importance of keeping up with the ongoing scientific and technological revolution, and he invoked a stirring vision of a new Britain 'forged in the white heat of this revolution'.[22] When Labour came to power in the 1964 general election, Wilson was eager to act on his vision by implementing a new economic and technological regime for Britain. To oversee national technological development, Wilson created the Ministry of Technology, a major new department that assumed control of the Atomic Energy Authority, the Ministry of Aviation, the National Research Development Corporation (NRDC), and a number of national laboratories. Mintech, as it came to be called, had two main aims: to transfer the results of scientific research to industrial development, and to intervene in industry so as to make private enterprise more efficient and competitive. Mintech was to have, in Wilson's words, 'a very direct responsibility for increasing productivity and efficiency, particularly within those industries in urgent need of restructuring or modernisation'.[23] These industries included machines tools, aviation, electronics, shipbuilding, and – above all – computing.

One of the British scientists who took the lead in computing research was Donald W. Davies of the National Physical Laboratory (NPL) in Teddington, a suburb of London. The National Physical Laboratory had been established in 1899 to determine values for physical constants, standardize instruments for physical measurements, and perform similar activities involving standards and materials testing.[24] NPL first became involved in computing in 1946, when a team at the laboratory, following a proposal by Alan Turing, built an early stored-program digital computer called the Pilot ACE. Davies had joined NPL in 1947 and had worked on the Pilot ACE; in 1960 he became superintendent of the division in charge of computing science, and in 1965 he was also named technical manager of the advanced computer techniques project.[25] Davies was thus in touch with both the latest advances in computing technology and the government's plans to use that technology to aid the British economy.

If the watchword for Baran was survivability, the priority for Davies was interactive computing. He was particularly interested in a technique called time sharing, which allows many people to interact with a computer at the same time. During discussions with British and American colleagues, Davies became aware of a widely perceived obstacle to interactive computing: inadequate data communications. In early time-sharing systems, the terminals had been directly connected to the computer and were located in an adjacent terminal room. As time went on, people began locating terminals at some distance from the computer itself, either for the user's own convenience or, in the case of commercial time-sharing services, to offer access to customers over a wide geographic area. Distant terminals could be connected to the computer using dial-up telephone links and modems, but long-distance telephone connections were very expensive, and for data transmission they were also inefficient. Computer messages, as noted earlier, tend to come in bursts with long pauses in between, so computer users paid dearly for telephone connections that were idle much of the time. The high cost of communications put pressure on users to work quickly, sacrificing the user-friendly quality for which time sharing had been invented in the first place.

Davies was perhaps even more aware of the cost issue than his American colleagues. In Britain, unlike in America, there was no flat rate for local telephone calls. Also, while American researchers tended to think in terms of academic computing, where users normally accessed the machine from a relatively short distance, Britain had a larger percentage of users whose computer access came from distant commercial systems.[26] Davies had a long-standing interest in switching techniques, and as he thought about the data communications problem, it occurred to him that a new approach to switching might offer a solution.[27] He knew that message switching was used in the telegraph system to make efficient use of lines, and he believed that by adapting this technique to computer communications he could achieve similar economies. Like Baran, Davies came from a background in computing, rather than communications, so he felt free to suggest a technique that departed from traditional communications techniques but took advantage of advances in computer technology. Davies proposed dividing messages into standard-sized 'packets', and having a network of computerized switching nodes that would use information carried in packet headers to route the packets to and from time-sharing computers. He called this technique 'packet switching'.

Packet switching in Davies's system

Like Baran, Davies saw that packet switching would allow many users to share a communications link efficiently. But he wanted that efficiency for a different purpose. Packet switching, in his view, would be the communications equivalent of time sharing: it would maximize access to a scarce resource in order to provide affordable interactive computing.[28]

Davies presented his network ideas publicly for the first time at a talk in March 1966, which was enthusiastically received by an audience of people active in computing, telecommunications, and the military. It was one of the latter, an attendee from the British Ministry of Defence, who gave Davies the surprising news that packet switching had already been invented a few years earlier by an American, none other than Paul Baran. The fact that the military man knew about this earlier development when Davies himself did not underscores the very different contexts in which packet switching evolved in the two countries. Baran's foremost concern had been survivability, which was underlined by his use of terms like 'raid', 'salvos', 'target', 'attack level' and 'probability of kill' in describing the hostile conditions under which his system was expected to operate.[29] Davies, on the other hand, did not view packet switching as a way to make the network survivable; after reading Baran he commented that 'the highly-connected networks there considered are not needed in a civil environment'.[30] Instead, he thought the pressing need was for a network that could serve the users of commercial time-sharing services. The Labour government specifically wanted to redirect research and development efforts away from military projects and toward civilian industry. Davies shared this concern, which is evident in his plan to survey businesses to find out their data communication requirements.[31] It also shows up in Davies's efforts to make the system easy to use. In his proposal for a national network he stressed that 'A further aim requirement we must keep in mind constantly is to make the use of the system simple for simple jobs. Even though there is a communication system and a computer operating system the user must be able to ignore the complexities'.[32]

The main benefit of packet switching for this type of system was that it would bring down the cost of data communications. Davies found further meanings in packet switching, however, that derived from his vision of a commercial network service. For instance, one of the merits Davies saw in packet switching was that it helped achieve fairness in access to the network. In an ordinary message switching system, each message had to be sent in its entirety before the next message could begin. In a packet switching system, users would take turns transmitting portions of their messages using time division multiplexing. If a user had a short message, such as a single command for a time-sharing system, the whole message could be sent in the first packet, while longer messages would take several time slots to transmit. The user with the short message would not have to wait behind users with long messages.[33] This kind of fairness was appropriate for a system where computers were serving ordinary people's everyday needs, rather than transmitting life-or-death messages through a command hierarchy.

Davies also believed that packet switching technology could itself become a commercial product, and thus contribute directly to Wilson's plan to revitalize the British economy. In a 1965 proposal to have the General Post Office (GPO) build a prototype for a national packet switching network, Davies emphasized that:

> Such an experiment at an early stage is needed to develop the knowledge of these systems in the GPO and the British computer and communications industry . . . It is very important not to find ourselves forced to buy computers and software for these systems from [the] USA. We could, by starting early enough, develop export markets.[34]

Davies's concern with issues of economics and user-friendliness underscores the national context in which he conceived the idea of a packet switching network. Davies did not envision a world in which his proposed network would be the only surviving communications system. Rather, he saw a world in which packet switching networks would need to compete with other communications systems to attract and serve the business user, and a world in which Britain would need to compete with the United States and other countries to offer innovative computer products.

In December 1965 Davies proposed the idea of a national packet switching network that would provide low-cost data communications across Britain. He envisioned the network as providing a number of services to business and recreational users, including remote data processing, point-of-sale transactions, database queries, remote control of machines, and even on-line betting.[35] In his scheme, a backbone of high capacity telephone lines would link major cities in the United Kingdom; the proposed network had multiple connections to most nodes, although it was not nearly as redundant as Baran's system. Like Baran, however, the NPL group designed the network with a dynamic, distributed routeing system, with each node making routeing decisions independently, according to current conditions in the network. The nodes would be connected by high-speed telephone lines so as to provide fast response times for interactive computing. Users would attach their computers, terminals, and printers to the nodes through dedicated interface computers at local sites.

Davies was convinced that the type of data communications infrastructure he was proposing would be necessary to keep Britain competitive in the information age. However, NPL did not have the resources or authority to build such a large network on its own. This authority belonged to the GPO, which ran the national postal and telephone networks, but the managers there had little knowledge of or interest in data communications. In consequence, Davies was only able to build a small prototype network, called the Mark I, rather than the nationwide system he had proposed.

One unusual characteristic of the Mark I that derived from the emphasis on user-friendliness was that all terminals, printers, and other peripheral devices were connected directly to the network. The network was actually interposed between a computer and its own peripherals, so that the network became, in a sense, internal to the computer. Using the network as a common communications channel for all components would make it possible for any pair of machines to interact. Normally, if a terminal user wanted to print a file, they would have to log-in to a host computer and send a command to a printer attached to that host. With the Mark I, however, the user could send a command directly from their terminal to the printer, without ever having to go through the host computer. Remote resources would be as easy to use as local ones, since the access procedures were identical. This was a radical concept in user interface design, one that would not become a commonplace feature of networked systems for another two decades.

PUTTING IT ALL TOGETHER: PACKET SWITCHING AND THE ARPANET

Baran and Davies had both envisioned nationwide networks that would use the new technique of packet switching, but neither one had been able to fully realize this goal. Instead, the first large-scale packet switching network would be built by the Advanced Research Projects Agency (ARPA). The design of this network would draw on the work of both Baran and Davies, but the network's builders had their own vision of what packet switching could achieve.

ARPA was one of many new American science and technology ventures that had been prompted by the cold war. Founded in 1958 in response to the Soviet Sputnik launch, ARPA's mission is to keep the US ahead of its military rivals by pursuing research projects that promise a significant advance in defence-related fields. ARPA is a small agency with no laboratories of its own; ARPA staff initiate and manage projects, but the actual research and development is done by academic and industry contractors. The agency is recognized even by critics for its good management and rapid development of new technologies, and has had some notable successes in transferring its technologies to the armed services and the private sector.[36]

ARPA has several project offices that fund research in different areas, which are created or disbanded as defence priorities change. The first ARPA offices directed research in behavioural sciences, materials sciences, and ballistic missile defence, and in 1962 ARPA added the Information Processing Techniques Office (IPTO). With the founding of IPTO, ARPA became a major funder of computer science in the United States, often exceeding university funding by significant amounts. IPTO has been the driving force behind several important areas of computing research in the United States, including graphics, artificial intelligence, time-sharing operating systems, and networking.[37]

ARPA's funding of basic research fitted in with the philosophy of Lyndon Johnson's administration. President Johnson advocated the use of agency funds to support basic research in universities in a September 1965 memo to his cabinet. Noting that funding by various federal agencies made up about two-thirds of total university research spending, he said that this money should be used to establish 'creative centers of excellence' throughout the nation. He urged each government agency engaged in research to take 'all practical measures . . . to strengthen the institutions where research now goes on, and to help additional institutions to become more effective centers for teaching and research'. Johnson specifically did not want to limit research at these centres to mission-oriented projects: 'Under this policy more support will be provided under terms which give the university and the investigator wider scope for inquiry, as contrasted with highly specific, narrowly defined projects'.[38]

A few months later the Department of Defense responded to Johnson's call with a plan to create 'centers of excellence' in defence-related research. According to this plan, 'Each new university program should present a stimulating challenge to faculty and students and, at the same time, contribute to basic knowledge needed for solving problems in national defense'.[39] IPTO created several computing research centres, giving large grants to universities such as MIT, Carnegie Mellon, and UCLA. By the end of the decade ARPA had funded a variety of time-sharing computers located at universities and other computing research sites across the United States. The purpose of its proposed network – the ARPANET – was to connect these scattered computing sites.

The ARPANET project was managed by Lawrence Roberts, a computer scientist who had conducted networking experiments at MIT's Lincoln Laboratories before joining ARPA in 1966. Roberts had a mandate to build a large-scale, multi-computer network, but he did not initially have a firm idea of how this should be done. He considered having pairs of computers establish a connection using ordinary telephone calls whenever they needed to exchange data, a method he had employed in his earlier experiments. But the high cost of long-distance telephone connections made this option seem prohibitively expensive. Roberts also worried that the ordinary phone service would be unacceptably prone to transmission errors and line failures. Although he was aware of the concept of packet switching, Roberts was not sure how to implement it in a large network.

With these issues still unresolved, Roberts attended a computing symposium in Gatlinburg, Tennessee, in October 1967, where he was to present

ARPA's tentative networking plans. Roger Scantlebury from NPL also presented a paper, and Roberts heard for the first time about Davies's packet switching ideas and the ongoing work on the Mark I. Afterwards, a number of conference attendees gathered to discuss network design informally, and Scantlebury and his colleagues advocated packet switching as a solution to Roberts's concerns about line efficiency. The NPL group influenced a number of American computer scientists in favour of the new technique, and they adopted Davies's term 'packet switching' to refer to this type of network. Roberts also adopted some specific aspects of the NPL design. For instance, Roberts had planned to use relative low-speed telephone lines to connect the network nodes. He later recalled that after the NPL representatives 'spent all night with me arguing about the thing back and forth . . . I concluded from those arguments that wider bandwidths would be useful'.[40] The ARPANET also used a packet format similar to NPL's. After the ARPANET project was underway, the firm of Bolt Beranek and Newman, which was awarded the main contract to build the network nodes, continued to interact with the NPL group.

Baran also became directly involved in the early stages of planning the ARPANET. Scantlebury had referred Roberts to Baran's earlier work, and soon after his return from Gatlinburg Roberts read Baran's *On Distributed Communications*. He would later describe this as a kind of revelation: 'suddenly I learned how to route packets'.[41] Some of the ARPANET contractors were already aware of Baran's work and had used it in their own research. Roberts recruited Baran in 1967 to advise the ARPANET planning group about distributed communications and packet switching.[42]

Through these various encounters, Roberts and other members of the ARPANET group were exposed to the ideas and techniques of both Baran and Davies, and became convinced that packet switching and distributed networking would be both feasible and desirable for the ARPANET. The new technique promised to make more efficient use of the network's long-distance communications links as well as enhance the system's ability to recover from the equipment failures that an experimental network would surely encounter. At the same time, however, packet switching was still an unproven technique that would be difficult to implement successfully. The decision to employ packet switching on such a large scale reflected ARPA's commitment to high-risk research: if it worked, the pay-off would be not only greater efficiency and ruggedness in the ARPANET itself, but also a significant advance in computer scientists' understanding of network properties and techniques. ARPA managers could afford – indeed, had a mandate – to think extravagantly, to aim for the highest pay-off rather than the safest investment.

► THE SOCIAL CONSTRUCTION OF PACKET SWITCHING

The three projects sponsored by Rand, NPL, and ARPA had much in common in their approach to packet switching, but also some crucial differences that helped the ARPANET play a more enduring and influential role than the others. Davies, Baran and Roberts each made technical choices based on specific local concerns, and the extent to which their systems were influential depended in part on whether others shared those concerns. For instance, Baran's system had many elements that were specifically adapted to the cold war threat, including the very high levels of redundancy, the location of nodes away from population centres, and the integration of cryptographic capabilities and priority/precedence features into the system design. None of these features was adopted by Davies or Roberts, neither of whom was concerned with survivability.[43] On the other hand, aspects of Baran's system that would be useful in a variety of situations – such as high-speed transmission, adaptive routeing and efficient packet switching – did find a home in later systems. As another example, Roberts shared Davies's commitment to supporting interactive computing, and therefore adopted Davies's higher line speeds, which were meant to provide users with a faster response from the computer.

One thing that Baran, Davies and Roberts had in common was the insight that the capabilities of a new generation of small but fast computers could be harnessed to transcend the limitations of previous communications systems. Telephone systems did not use computerized switches during this period, and message switching systems used large, expensive computers to handle messages at low speeds. When presented with the idea that a network could employ dozens of computers to operate its switches, people from the communications industry tended to be sceptical that contemporary computers would be both fast enough and cheap enough to make this idea feasible.[44] Indeed, the first of these 'minicomputers' did not appear until 1965, when the Digital Equipment Corporation introduced its PDP-8. The fact that packet switching relied on an innovative computer product helps explain why the technique was consistently explored by people in computing, rather than communications, even though it combined aspects of both fields.

In both the United States and the United Kingdom, computing technologies became policy instruments in the 1960s. In Britain, intervention in the computer industry was seen as a symbol of the Labour Party's commitment to modernization as well as an engine of economic growth, and the government made efforts to fund research and coordinate industrial production. In the United States, technological prowess was seen as a weapon in the cold war, and defence-related research was generously funded through organizations like Rand and ARPA. In both countries, individuals and organizations interested in pursuing computer networking often found it necessary to join government-sponsored projects or to present their work as responsive to contemporary political agendas.

But while computer networking had a political role in both countries, there were striking differences in the levels of government funding, in policy makers' interpretation of networks as a military or civilian technology, and in their inclination to intervene in private enterprise. These differences can be seen in the contrasting outcomes of the attempts by NPL and ARPA to build large-scale networks. The United States poured much more money into basic computing research than the United Kingdom, and most of that money was channelled through the Defense Department. This meant that Roberts not only had a generous budget for his project, but also was able to call on computer experts from around the country to help build the network. Davies had a much smaller budget at NPL. In the midst of a perceived economic crisis and a need to compete with the United States and other high-tech exporters, the British government tried to rationalize the computing industry and to encourage commercial spin-offs of government research. Eventually much of the research at places like NPL was directly focused on short-term commercial applications, and the Labour government's industrial policy limited the choice of computers available to Davies. The American government was less inclined to try to manage the domestic computer industry. Overall, Roberts had much more support and much less interference from the government than Davies.

Davies had been one of the earliest and most articulate advocates of packet switching, formulating a detailed plan for a national network at a time when the ARPANET was still just an idea. Yet by the middle of 1968 Davies was already lamenting that his project had been eclipsed by the American effort: 'As a force in this discussion NPL is too remote and our own demonstration as planned now is small-scale and likely to be delayed by the reductions in staff and administrative difficulties in purchasing computers'.[45] Despite their technological vision, neither Baran nor Davies could find the backing to build a national packet switching network, while Roberts was able to make the ARPANET an internationally recognized symbol of the feasibility of packet switching only a few years after he himself had learned of the technique.

The fact that packet switching had to be integrated into local practices and concerns led to very different outcomes in the three network projects. Some visions of packet networking were easier to implement, some turned out to be a better match for evolving computer technology, and some were more attractive to organizations in a position to sponsor network projects. Making packet switching work was not just a matter of having the right technical idea: it also required the right environment. Only after the ARPANET presented a highly visible example of a successful packet switching system did it come to be seen as a self-evidently superior technique. The success of the ARPANET may have depended on packet switching, but it could equally be argued that the success of packet switching depended on the ARPANET.

▶ **NOTES**

1 P. Baran (1990) Interview by Judy O'Neill, p. 11, 5 March, Menlo Park, CA. Transcript archived and owned by the Charles Babbage Institute.
2 Ibid., 12, 16.
3 Ibid., 14–15.

4 P. Baran (1960) *Reliable Digital Communications Systems Using Unreliable Network Repeater Nodes*, p. 3. Santa Monica, CA: Rand Corporation.
5 P. Baran (1964a) *On Distributed Communications*, RAND Report Series VIII, Part V, August. Santa Monica, CA: Rand Corporation.
6 M. Campbell-Kelly (1988) Data communications at the National Physical Laboratory (1965–1975). *Annals of the History of Computing* 9 (3/4): 221–47.
7 P. Baran (1964b) On distributed communications networks, p. 8.
8 P. Baran (1964a) op. cit. V, Part I.
9 G. E. Schindler (ed.) (1982) *A History of Engineering and Science in the Bell System: Switching Technology (1925–1975)*. New York: AT&T Bell Laboratories; *Long Lines* (1965) AUTOVON. 45 (3): 1–7. In 1966 the system was upgraded with AT&T's new electronic switches, and by 1971 the network had grown to 45 switches in the United States and Canada.
10 G. E. Schindler, ibid., p. 267; J. W. Gorgas (1968) The polygrid network for AUTOVON. *Bell Laboratories Record*, 46 (4): 223–7; *Long Lines* (1969) AUTOVON. 48 (9): 18–23.
11 P. Baran (1960) op. cit.
12 P. Baran (1964a) op. cit., V, Part II.
13 P. Baran (1964b) op. cit., p. 6.
14 E. F. O'Neill (ed.) (1985) *A History of Engineering and Science in the Bell System: Transmission Technology (1925–1975)*. New York: AT&T Bell Laboratories.
15 P. Baran (1990) op. cit., p. 18. See also AT&T Bell Laboratories, ibid. pp. 531–2.
16 P. Baran (1964b) op. cit., p. 4–5
17 The term 'packet switching' was coined by Donald Davies, as discussed below, but the word 'packet' seems to have been in common use in the data communications field and was used informally by Baran himself.
18 P. Baran (1964a) op. cit., XI, Part I.
19 P. Baran (1964b) op. cit., p. 6.
20 Ibid.
21 R. Coopey and D. Clarke (1995) *3i: Fifty Years Investing in Industry*. New York: Oxford University Press. D. Edgerton (1996) The 'White Heat' revisited: the British government and technology in the 1960s. *20th Century British History*, 7 (1): 53–90.
22 Quoted in Edgerton, ibid., p. 56.
23 H. Wilson (1971) *The Labour Government, 1964–1970: A Personal Record*, p. 8. London: Weidenfeld & Nicolson.
24 E. Pyatt (1983) *The National Physical Laboratory: A History*. Bristol: Adam Hilger.
25 M. Campbell-Kelly, op. cit., pp. 222–23.
26 Ibid., p. 225.
27 D. W. Davies (1986) Interview with Martin Campbell-Kelly, 17 March. Unpublished. Archived at NAHC (National Archive for the History of Computing), Manchester.
28 See D. W. Davies (1966b) 'Proposal for a digital communication network', NPL unpublished memorandum, June. Archived at NAHC.
29 P. Baran (1964b) op. cit., p. 2.
30 D. W. Davies (1966b) op. cit., p. 21.
31 D. W. Davies (1968a) 'Communication requirements of a Minitech computer network', p. 2, NPL unpublished memorandum, November. Archived at NAHC.
32 D. W. Davies (1966a) 'A computer network for NPL', unpublished memorandum, 28 July. Archived at NAHC.
33 M. Campbell-Kelly, op. cit., p. 26.

34 D. W. Davies (1965) 'Proposal for the development of a national communication service for on-line data processing', p. 8, unpublished memorandum, 15 December. Archived at NAHC.

35 Ibid.

36 A. Pollack (1989) America's answer to Japan's MITI. *New York Times*, 5 May, pp. 1, 8.

37 A. L. Norberg and J. E. O'Neill (1996) *Transforming Computer Technology: Information Processing for the Pentagon, 1962–1986*. Baltimore, MD: Johns Hopkins University Press; K. Flamm (1988) *Creating the Computer: Government, Industry, and High Technology*. Washington, DC: Brookings Institution; P. N. Edwards (1996) *The Closed World: Computers and the Politics of Discourse in Cold War America*. Cambridge, MA: MIT Press, have all documented how computing developments in the USA were largely driven by military considerations.

38 L. B. Johnson (1972) Statement of the president to the cabinet on strengthening the academic capability for science throughout the nation, in J. L. Penick Jr., C. W. Pursell Jr., M. B. Sherwood and D. C. Swain (eds) *The Politics of American Science, 1939 to the Present*, pp. 335–36. Cambridge, MA: MIT Press.

39 Department of Defense (1972) Project THEMIS, in J. L. Penick Jr., C. W. Pursell Jr., M. B. Sherwood and D. C. Swain (eds) *The Politics of American Science, 1939 to the Present, p. 337*. Cambridge, MA: MIT Press.

40 L. G. Roberts (1989) Interview by Arthur L. Norberg, April 4, San Mateo, CA. Unpublished transcript archived and owned by the Charles Babbage Institute.

41 A. L. Norberg and J. E. O'Neill, op. cit., p. 166.

42 Ibid.

43 Robert Taylor has emphasized that surviving an attack was *not* the point behind the ARPANET design. See K. Hafner and M. Lyon (1996) *Where Wizards Stay Up Late: The Origins of the Internet*, p. 10. New York, Simon & Schuster. Some of the ARPANET spin-offs built for the US Defense Department did try to incorporate the type of survivability and security features proposed by Baran (see H. B. Heiden and H. C. Duffield (1982) Defense Data Network, IEEE Electronics and Aerospace Systems Conference (EASCON). New York: Institute of Electrical and Electronics Engineers, pp. 61–75).

44 P. Baran (1990) op. cit. pp. 19–21; L. Roberts (1978) The evolution of packet switching. *Proceedings of the IEEE*, 66 (11): 1307–13; L. Roberts (1988) The ARPANET and computer networks, in A. Goldberg (ed.) *A History of Personal Workstations*, p. 150. New York: ACM Press; M. Campbell-Kelly, op. cit., p. 8.

45 D. W. Davies (1968b) 'Report on a visit to the United States of America, 20 April–10 May', NPL unpublished memorandum, May. Archived at NAHC.

▶ **BIBLIOGRAPHY**

Abbate, J. (1994) *From ARPANET to Internet: A History of ARPA-Sponsored Computer Networks, 1966–1988*. PhD thesis, University of Pennsylvania.

Baran, Paul, *et al.* (1964a) *On Distributed Communications*. RAND Report Series. Santa Monica, CA: RAND Corporation.

Baran, Paul (1964b) On distributed communications networks. *IEEE Transactions on Communications*, 1–9.

Barber, D. L. A. (1969) 'Visit to Canada and the USA, 22 September–17 October, 1969', NPL unpublished memorandum, October. Archived at NAHC.

Barber, D. L. A. (1979) 'A small step for Britain; a giant step for the rest?' NPL unpublished memorandum, April. Archived at NAHC.

Bolt Beranek and Newman Inc. (1973) *Interface Message Processors for the ARPA Computer Network: Quarterly Technical Report No. 4*, BBN Report No. 2717. Cambridge, MA: Bolt Beranek and Newman, Inc.

Burdick, E. and Wheeler, H. (1962) *Fail-Safe*. New York: McGraw-Hill.

Coopey, R. (1993) 'Industrial policy in the white heat of the scientific revolution', in R. Coopey, S. Fielding and N. Tiratsoo (eds) *The Wilson Government, 1964–1970*. New York: St. Martin's Press.

Data Processing (1969) Low-cost automatic message switching. November–December: 580–1.

Foy, N. (1986) Britain's real-time club. *Annals of the History of Computing*, 8 (4): 370–1.

Frank, H. (1990) *Interview by Judy O'Neill*, 30 March, Fairfax, VA. Unpublished. Transcript archived and owned by the Charles Babbage Institute.

Hadley, E. E. and Sexton, B. R. (1978) The British Post Office Experimental Packet Switched Service (EPSS) – A review of development and operational experience. *Conference Proceedings*, Fourth International Conference on Computer Communication: Evolutions in Computer Communications, Kyoto, pp. 97–102.

Hendry, J. (1990) *Innovating for Failure: Government Policy and the Early British Computer Industry*. Cambridge, MA: MIT Press.

Horner, D. (1993) The road to Scarborough: Wilson, Labour and the scientific revolution, in R. Coopey, S. Fielding and N. Tiratsoo (eds) *The Wilson Government, 1964–1970*. New York: St. Martin's Press.

IEEE Spectrum (1964) Scanning the issues, August: 114.

Jennings, D. M., Landweber, L. H., Fuchs, I. H., Farber, D. J. and Richards, A. W. (1986) Computer networking for scientists. *Science*, 231 (4741): 943–50.

Kahn, R. E. (1990) *Interview by Judy O'Neill*, 24 April. Reston, VA. Unpublished. Transcript archived and owned by the Charles Babbage Institute.

Kleinrock, L. (1990) *Interview by Judy O'Neill*, 3 April. Los Angeles. Unpublished. Transcript archived and owned by the Charles Babbage Institute.

Long Lines (1965) AUTOVON. 45 (3): 1–7.

Malik, R. (1989) The Real Time Club. *Annals of the History of Computing*, 11 (1): 51–2.

Morgan, A. (1992) *Harold Wilson*. London: Pluto Press.

Quarterman, J. S. (1990) *The Matrix: Computer Networks and Conferencing Systems Worldwide*. Burlington, MA: Digital Press.

Shute, N. (1957) *On the Beach*. New York: W. Morrow.

Tanenbaum, A. S. (1989) *Computer Networks*, 2nd edn. Englewood Cliffs, NJ: Prentice Hall.

Wichmann, B. A. (1967) 'An electronic typewriter', NPL unpublished memorandum, 26 May. Archived at NAHC.

Wilkinson, P. T. (1968) 'The main features of a proposed national data communication network', NPL unpublished memorandum, February. Archived at NAHC.

26 ▶ Manufacturing gender in military cockpit design

Rachel N. Weber

Recent work in both science and technology studies and feminist theory has focused on the military as an institution which has both guided technological development and has had a historic claim to masculinity (Cooke and Woolacott 1993; MacKenzie 1990; Law and Callon 1988; Roe Smith 1985; Enloe 1983). In attempting to dissect the historic link between militarism and male power, however, many feminists have accepted the biologically determinist notion that military technologies – the instruments of war – are extensions of the phallus and inextricably linked to the inherently male drive to dominate (Wheelwright 1992; Brownmiller 1975). If we were to apply a less deterministic framework to understanding military technologies, we might find that the 'inherent' masculinity of such technologies is socially constructed. For example, Pentagon officials and engineers have traditionally built a bias against women's bodies into the military technologies through the construction of engineering specifications and design guidelines.[1]

Many scholars of gender and technology have questioned women's access to particular technologies (Wajcman 1991). In the context of military aviation, one would ask questions regarding women's upward mobility in the profession; for example, are women limited because they are not trained, socialized, or permitted to fly certain aircraft? Solutions to these problems would lie in eroding barriers to these boundary markers, such as easing women-in-combat exclusions or other operational requirements.

A second approach – and the one which informs the subject of this article – asks questions about the technology itself. How are cockpits designed to accommodate women's bodies? When is a particular flight deck 'gender neutral,' and when is male bias embodied in the actual design, in the engineering specifications? How can biased technologies be altered to become more 'women friendly'?

Design bias is not restricted to the military; commercial technologies such as aircraft, automobiles, and architecture are also built to accommodate male anthropometry. Civilian and military contractors, however, have exhibited different degrees of commitment to the task of accommodating female operators into the design phase. Ironically, in the field of airframe manufacture, civilian contractors are lagging behind their military colleagues in attempting to rectify the problem of design bias against women; the Pentagon has led the movement to alter cockpit design to accommodate women and smaller-statured men . . .

▶ **TECHNOLOGICAL BIAS IN EXISTING AIRCRAFT**

Civilian and defense aircraft have traditionally been built to male specifications (Binkin 1993). Since women tend to be shorter, have smaller limbs and less upper-body strength, some may not be accommodated by such systems and may experience difficulty in reaching controls and operating certain types of equipment (McDaniel 1994). To understand how women's bodies become excluded by design and how difference becomes technologically embodied, it is necessary to examine how current military systems are designed with regard to the physical differences of their human operators.

To integrate the user into current design practices, engineers rely on the concepts of ergonomics and anthropometrics (McCormick and Sanders 1982). Ergonomics, also called 'human factors,' addresses the human characteristics, expectations, and behaviors in the design of items which people use. During World War II, ergonomics became a distinct discipline, practiced predominantly by the U.S. military. Ergonomic theories were first implemented when it became obvious that new and more complicated types of military equipment could not be operated safely or effectively or maintained adequately even by well-trained personnel. The term 'human engineering' was coined and efforts were made to design equipment that would be more suitable for human use.

Anthropometrics refers to the measurement of dimensions and physical characteristics of the body as it occupies space, moves, and applies energy to physical objects as a function of age, sex, occupation, and ethnic origin and other demographic variables. Engineers at the Pentagon and at commercial airframe manufacturers rely on the U.S. Army Natick Research Development and Engineering Center's '1988 Anthropometric Survey of Army Personnel,' in which multiple body dimensions are measured and categorized to standardize the design of systems. The Natick survey contains data on more than 180 body and head dimension measurements of a population of more than 9,000 soldiers. Age and race distributions match those of the June 1988 active duty Army, but minority groups were intentionally oversampled to accommodate anticipated demographic shifts in Army population (Richman-Loo and Weber 1996).

Technological bias within defense aircraft

Department of Defense acquisition policy mandates that human considerations be integrated into design efforts to improve total system performance
by focusing attention on the capabilities and limitations of the human
operator. In other words, the Defense Department recognizes that the best
defense technology is useless if it is incompatible with the capabilities and
limitations of its users. In the application of anthropometric data, systems
designers commonly rely on Military Standard 1472, 'Human Engineering
Design Criteria for Military Systems, Equipment and Facilities.' Like the use
of military specifications in the procurement process, these guidelines are
critical in developing standards; they embody decisions made which reflect
the military's needs and goals and are ultimately embodied in the technology (Roe Smith 1985).

These guidelines suggest the use of 95th and 5th percentile male dimensions in designing weapons systems. Use of this standard implies that
only 10 per cent of men in the population will not be accommodated by a
given design feature. If the feature in question is sitting height, the 5 per
cent of men who are very short and the 5 per cent who are very tall will not
be accommodated.

Accommodation becomes more difficult when more than one physical
dimension is involved, and several dimensions need to be considered in
combination. The various dimensions often have low correlations with
each other (e.g., sitting height and arm length). For example, approximately 52 per cent of Naval aviators would not be accommodated by a
particular cockpit specification if both the 5th and 95th percentiles were
used for each of the thirteen dimensions.

Because women are often smaller in all physical dimensions than men,
the gap between a 5th percentile woman and a 95th percentile man can be
very large (Richman-Loo and Weber 1996). Women who do not meet
requirements are deemed ineligible to use a variety of military systems.

The case of the Joint Primary Aircraft Training System (JPATS), used by
both the US Navy and the Air Force to train its pilot candidates, has been the
most publicized case of military design bias against women.[1] Engineers and
human factors specialists considered minimum anthropometric requirements needed by an individual to operate the JPATS effectively and wrote
specifications to reflect such requirements. For example, 'the ability to
reach and operate leg and hand controls, see cockpit gauges and displays,
and acquire external vision required for safe operation' was considered
critical to the safe and efficient operation of the system. Navy and Air
Force engineers determined the five critical anthropometry design 'drivers'
to be sitting height, functional arm reach, leg length, buttock-knee length,
and weight (Department of Defense 1993: 2).

Original JPATS specifications included a 34-inch minimum sitting height
requirement in order to safely operate cockpit controls and eject. This
specification is based on sitting height minimums in the current aircraft
fleet and reflects a 5th percentile male standard. However, at 34 inches,
anywhere from 50 to 65 per cent of the American female population is

excluded because female sitting heights are generally smaller than male. Therefore, JPATS, as originally intended, accommodated the 5th through 95th percentile male, but only approximately the 65th through 95th percentile female.

After successful completion of mandatory JPATS training, student pilots advance to intermediate trainers and then to aircraft-specific training. Therefore, if women cannot 'fit' into the JPATS cockpit or if the cockpit does not 'fit' women pilots, they will be unable to pursue aviation careers in the Navy or Air Force. In other words, design bias has far-reaching implications for gender equity in the military.

Technological bias within commercial aircraft

Despite a similar technological base, the cockpit technology encountered in civilian aviation differs from that found in the military. The role of the human being and the control processes available to him or her also will differ. For example, the extreme rates of acceleration experienced in military cockpits require elaborate restraining devices. Such restraints must be designed to fit the anthropometric characteristics of the intended users. Ejection is also an issue limited to military cockpit design. Much of the JPATS controversy centers on ejection seats and the need to provide safe ejection to lighter individuals.

In contrast, commercial aircraft do not reach the same high speeds as military planes, nor do they contain ejection seats. The seats in a commercial cockpit are adjustable to meet the varied comfort and safety requirements of the users. Thus certain anthropometrics such as height, weight, and strength do not have the same valence in commercial aviation as they do in the military. Many argue that commercial aircraft can accommodate a more variable population because the operating requirements are not as stringent as in the military.

However, the location of various controls on the commercial flight deck has been found to disadvantage women and smaller-statured men (Sexton 1988). Although the seats are more adjustable, individuals with smaller functional arm reach and less upper-body strength may still experience difficulties manipulating controls and reaching pedals. When smaller women are sitting in the co-pilot seat, some complain that they are not able to reach controls on the right side of the control panel. Reach concerns become increasingly important during manual reversion (when the system reverts to manual operation) even though electrical and hydraulic systems require smaller forces to actuate.

Cockpit design specifications have protected what has traditionally been a male occupation. Because both commercial and defense aircraft have been built for use by male pilots, the physical differences between men and women serve as very tangible rationales for gender-based exclusion. Although technology certainly is not the only 'cause' of exclusion and segregation, biased aircraft act as symbolic markers, used to delineate the boundaries between men's and women's social space. Reppy (1993: 6) notes that it is not that women are not physically capable of flying these particular aircraft or that they are not equally exposed to danger in other aircraft; rather denying women access to combat aircraft is a way of protecting a distinctly male arena. The technical artifact . . . has functioned to delineate the 'other.'

▶ **REGULATING ACCOMMODATION IN DEFENSE AIRCRAFT**

The decision to standardize any technology is often contested, occurring within a space where social, economic, and political factors vie for position. In this case, standardization involved altering technologies in order to adjust to a changed sociopolitical environment. In the military, cockpit technology had to be adjusted to the entry of women into the armed forces and their new roles within the services. The process of design accommodation in the military became a process of negotiation between various social groups who held different stakes in and interpretations of the technology in question (Pinch and Bijker 1984).

One could argue that negotiations over accommodation arose as a result of changes made in policies regarding women in combat. Former Secretary of Defense Les Aspin publicly recognized that women should play a greater role in the military when he issued a directive in April 1993 on the assignment of women in the armed forces. The directive states that

> the services shall permit women to compete for assignments in aircraft, including aircraft engaged in combat missions.
>
> The Army and Marine Corps shall study opportunities for women to serve in additional assignments, including, but not limited to, field artillery and air defense artillery.
>
> (Aspin 1993: 1)

Although the new policy gave women a greater combat aviation role and was intended to allow for their entry into many new assignments, the aircraft associated with these assignments precluded the directive from being implemented. The realization that existing systems could contain a technological bias against women's bodies despite the Congressional mandate for accessibility alarmed policy specialists at the Pentagon. This contradiction would potentially embarrass a new administration which was reeling from its handling of the gays in the military debacle and desperately trying to define a working relationship with an antagonistic Pentagon.

In May 1993 the Under Secretary of Defense (Acquisition) directed the Assistant Secretary of Defense (Personnel and Readiness) to develop a new JPATS sitting height threshold which would accommodate at least 80 per cent of eligible women. He delayed release of the JPATS draft Request for Proposal until a new threshold could be documented. This move led to the establishment of the JPATS Cockpit Accommodation Working Group which included representatives from the Air Force and Navy JPATS Program

Offices as well as from service acquisition, personnel, human factors, and flight surgeon organizations. After months of deliberation, the Working Group determined that a reduction of the sitting height requirement by 3 inches would accommodate approximately 82 per cent of the eligible female population (Department of Defense 1993).

Reducing the operational requirements would entail modifying existing cockpit specifications. Significant modifications were needed because the requirement for an ejection seat restricts the possibility of making the seat adjustable. In addition, the aircraft nose, rudder, and other flight controls would also need to be substantially modified to accommodate a smaller person. Further, since ejections at smaller statures and corresponding body weights had yet to be certified for safety, test articles and demonstrators had to be developed to ensure safe ejection (Dorn 1993).

After the May 1993 directive, many procurement specialists at the Pentagon were perplexed: a design which would accommodate the 5th percentile female through the 95th percentile male would have to incorporate a very wide variability of human dimensions. Some senior defense officials opposed such a change because they believed that such alterations would delay the development of the JPATS, would raise the price of training, and would be prohibitively expensive.

In opposition to these officials, pragmatists within the Pentagon – including most members of the Working Group – argued that it was both efficient and economical to integrate human factors into acquisition. Pragmatists felt that the technologies built for the military, as opposed to civilian markets, tended to privilege capability over maintenance and operability and hardware over personnel. They argued that with decreasing budgets, this could no longer be the case. Design changes, they claimed, would not only benefit women assigned to weapons systems originally designed for male operators, but would benefit smaller men as well. Studies have shown that smaller men also have difficulty operating hatches, damage control equipment, and scuttles on ships (Key, Fleischer and Gauthier 1993). Shrinking personnel resources and a changing demographic pool from which the military recruits also mandated that defense technologies be more closely matched to human capabilities. The pragmatists were quick to emphasize that the inclusion and accommodation of smaller men would be necessary given changes in the ethnic and racial make-up of the nation (Stiehm 1985).

Pragmatists also pointed to the prospect of foreign military sales to countries with smaller-sized populations, which would make design accommodation an important economic consideration as well. Edwin Dorn (1993), the Assistant Secretary of Defense, in a memorandum to the Under Secretary of Defense (Acquisition), stressed that

> a reduced JPATS sitting height threshold will also expand the accommodation of shorter males who may have previously been excluded from pilot training. For potential foreign military sales, this enhances its marketability in countries where pilot populations are of smaller average stature.

The pragmatists emphasized that cockpit accommodation would benefit all soldiers because it required the acquisition process to consider differences concerning capabilities and limitations. In pursuing this line of argument, they essentially neutered the discourse, erasing the specificities of women's bodies. By refusing to engage in a gendered discourse and instead emphasizing economic benefits, they hoped to appeal to a broader segment of the population and to a Pentagon traditionally hostile to women's issues.

In contrast to the Pentagon pragmatists, women's groups both within the military and outside supported the decision to alter the JPATS sitting height requirement on more ideological grounds. The fact that women were being excluded by the operational requirements and by the technology was central to their decision to support the changes. In general, feminism in the contemporary military environment is organized around ideals of parity and equal opportunity regarding career opportunities (Katzenstein 1993). Insisting that career advancement be based on qualifications, not biology, many argued that physical restrictions which disqualified women would unfairly limit women's mobility in the services.

Through informal networks and more formal associations such as the Defense Advisory Committee on Women in the Service (DACOWITS), new groups of activists set about to influence policy decisions about career opportunities for women. Women aviators organized around the issue of female accommodation and found a receptive audience in some of the new Clinton appointees, such as Edwin Dorn, Assistant Secretary of Defense. Unlike other changes imposed from the top, the decision to alter JPATS was part of a low-level process that began with limited intervention from high-ranking administrators (Brundage 1993).

Although the media spectacle of the Tailhook scandal[2] provided the necessary momentum for feminist groups in the military and brought gender issues to the forefront of national debates, the decision to accommodate more women in the JPATS cockpit was not without dissension. Some women officers – many of whom also considered themselves feminists – believed that, as one of the people I have interviewed told me, 'shrill cries for accommodation could be used against women politically.' They insisted that demanding special treatment would single women out in an institution which, on the surface, seeks to eradicate differences between the sexes. In a sense, they were asking women to ignore their difference and prove themselves on gender-neutral terms.

A few women pilots questioned the construction of the operational requirements and thresholds but insisted that the existing cockpits were not biased. Is it really necessary, some asked, to possess a sitting height of 34 inches to fly defense aircraft? Women with smaller sitting heights had flown during wartime, and many believed that pilots at shorter sitting heights were no less capable of flying safely. One woman claimed that 'the whole issue of height in aircraft is overstated, and just ignorance on the part of the Navy.'

As debates raged in the press and within the Working Group during 1993, the possibilities for technological variety began to close down. The Pentagon pragmatists attempted to stabilize the debate, but the public spectacle of the issue facilitated closure by broadening the deliberative arena. With the JPATS case, 'administrative' closure was achieved when the 1994 Defense Authorization Bill was passed. The bill included a provision which prevented the Air Force, the lead agency in the purchase of the JPATS, from spending $40 million of its $41.6 million trainer budget unless the Pentagon altered the cockpit design. John Deutsch (1992), then the Under Secretary of Defense, wrote a memo legitimizing the problem of accommodation of women in defense aircraft, stating:

> I believe the Office of the Secretary of Defense (OSD) should continue to take the lead in addressing this problem. Other platforms in addition to aircraft should be considered as well. We must determine what changes are practical and cost effective in support of Secretary of Defense policy to expand combat roles for females. I request that you take the lead in determining specification needs. Further, you should determine the impact of defense platforms already in production and inventory.
>
> (Deutsch 1992: 1)

After Working Group deliberations, the Air Force issued a revised JPATS Draft Request for Proposal that included a 32.8-inch sitting height threshold. The RFP identified crew accommodation as a key source selection criterion so that during the selection process, prospective contractors would be required to submit cockpit mock-ups which would be evaluated for their adherence to the revised JPATS anthropometric requirements. Candidates who adhered to and even exceeded these requirements stood the best chance of winning the contract.

Author's note: I wish to thank Judith Reppy, Susan Christopherson, Nina Richman-Loo, Sheila Jasanoff, Mary Katzenstein, Trevor Pinch, Olga Amsterdamska, and three anonymous referees for their comments and guidance. Financial support from the Transportation Research Board and the Peace Studies Program at Cornell University made this project possible. An earlier version of this paper appeared in the *Transportation Research Record* (Weber, 1995).

▶ NOTES

1 As there is a pronounced dearth of research in the area of gender and cockpit design, this project relies heavily on interviews conducted with human factors specialists at major airframe manufacturers (primarily Boeing and McDonnell Douglas), public sector research laboratories, and regulatory agencies. Because interviews were conducted during the very competitive source selection phase of procurement, interview subjects were reluctant to discuss this subject unless they were guaranteed strict anonymity. Due to these constraints, I have chosen to paraphrase interviews rather than use direct quotations.
2 The Tailhook scandal refers to the annual Tailhookers' (Navy carrier pilots) convention of 1991 where several women were sexually harassed by servicemen and

later went public with their charges. As a result, three admirals were disciplined, although none of the servicemen were officially charged.

▶ **REFERENCES**

Aspin, L. (1993) *Policy on the assignment of women in the armed forces.* 28 April. Washington, D.C.: Department of Defense.

Binkin, M. (1993) *Who will fight the next war? The changing face of the American military.* Washington, D.C.: Brookings Institute.

Brownmiller, S. (1975) *Against our will: Men, women and rape.* New York: Simon & Schuster.

Brundage, W. (1993) The changing self-definitions of the military and women's occupational specializations. Paper presented at the workshop on 'Institutional change and the U.S. military: The changing role of women,' October, Cornell University.

Cooke, M. and A. Woolacott (eds) (1993) *Collateral damage: Gender and war.* Princeton, NJ: Princeton University Press.

Department of Defense (1993) JPATS cockpit accommodation working group report. Unpublished report, 3 May.

Deutsch, J. (1992) Memorandum on JPATS cockpit accommodation working group report. Unpublished report, 2 December.

Dorn, E. (1993) Memorandum on JPATS cockpit accommodation working group report. Unpublished report, 19 October.

Enloe, C. (1983) *Does khaki become you? The militarization of women's lives.* Boston: South End Press.

Katzenstein, M. (1993) The formation of feminism in the military environment. Paper presented at the workshop on 'Institutional change and the U.S. military: The changing role of women,' October, Cornell University.

Key, E., E. Fleischer and E. Gauthier (1993) Women at sea: Design considerations. Paper presented at the Association of Scientists and Engineers, 30th Annual Technological Symposium, March, Houston, TX.

Law, J. and M. Callon (1988) Engineering and sociology in the Military Aircraft Project: A network analysis of technological change. *Social Problems* 35: 284–97.

MacKenzie, D. (1990) *Inventing accuracy.* Cambridge: MIT Press.

McCormick, E. and M. Sanders (1982) *Human factors in engineering and design.* New York: McGraw-Hill.

McDaniel, J. (1994) Strength capability for operating aircraft controls. In *Advances in industrial ergonomics and safety,* vol. 6, edited by E. Aghazadeh, 58–73. Bristol, PA: Taylor and Francis.

Pinch, T. and W. Bijker (1984) The social construction of facts and artifacts. *Social Studies of Science* 14: 399–441.

Reppy, J. (1993) New technologies in the gendered workplace. Paper presented at the workshop on 'Institutional change and the U.S. military: The changing role of women,' October, Cornell University.

Richman-Loo, N. and R. Weber (1996) Gender and weapons design: Are military technologies biased against women? In *It's our military too! Women and the U.S. military,* edited by J. Stiehm, 136–155. Philadelphia: Temple University Press.

Roe Smith, M. (ed.) (1985) *Military enterprise and technological change.* Cambridge: MIT Press.

Sexton, G. (1988) Cockpit-crew systems design and integration. In *Human factors in aviation*, edited by E. Weiner and T. Nagle, 495–526. San Diego: Academic Press.

Stiehm, J. (1985) Women's biology and the U.S. military. In *Women, biology, and public policy*, edited by J. Stiehm, 205–34. Beverly Hills, CA: Sage.

Wajcman, J. (1991) *Feminism confronts technology*. University Park, MD: Pennsylvania State University Press.

Weber, R. (1995) Accommodating difference: Gender and cockpit design in military and civilian aviation. *Transportation Research Record* 1480: 51–6.

Wheelwright, J. (1992) 'A brother in arms, a sister in peace': Contemporary issues of gender and military technology. In *Inventing women: Science, technology and gender*, edited by G. Kirkup and L. Smith Keller, 213–23. Milton Keynes: Open University.

27 ▶ The American Army and the M-16 rifle

James Fallows

Between 1965 and 1969, more than one million American soldiers served in combat in Vietnam. One can argue that they should never have been sent there, or that the strategy under which they were commanded ensured their demoralization and defeat. No one can argue that, while there, the soldiers should have been given inferior equipment. Yet that is what happened. During those years, in which more than 40,000 American soldiers were killed by hostile fire and more than 250,000 wounded, American troops in Vietnam were equipped with a rifle their superiors knew would fail when put to the test.

The rifle was known as the M-16; it was a replacement for the M-14, a heavier weapon that was the previous standard. The original version of the M-16, a commercial model developed by the Armalite company and known as the AR-15, was the most reliable, and the most lethal, infantry rifle ever invented. But within months of its introduction in combat, it was known among soldiers as a weapon that might jam and misfire, and could pose as great a danger to them as to their enemy. These problems, which loomed so large on the battlefield, were entirely the result of modifications made to the rifle's original design by the Army's own ordnance bureaucracy. The Army's modifications had very little to do with observation of warfare, but quite a lot to do with settling organizational scores. By the middle of 1967, when the M-16 had been in combat for about a year and a half, a sufficient number of soldiers had written to their parents about their pathetically unreliable equipment, and a sufficient number of parents had sent those letters to their congressmen, to attract the attention of a congressional investigating committee. The committee, headed by Representative Richard Ichord, a Democrat from Missouri, conducted an exhaustive inquiry into the origins of the M-16 problem. Much of the credit for the hearings belongs to the committee's counsel, Earl J. Morgan. The hearing record, nearly 600 pages long,[1] is a forgotten document, which received modest press attention at the time and calls up only dim recollections now. Yet it is the purest portrayal of the banality of evil in the records of modern American defense.

Nearly a century before American troops were ordered into Vietnam, weapons designers, especially in Europe, had made a discovery in the science of 'wound ballistics.' The discovery was that a small, fast-traveling bullet often did a great deal more damage than a larger round when fired into human or (for the experiments) animal flesh. The explanation lay in physics: when the bullet passed from a medium of one density, such as air, into a medium of different density, the bullet became unstable and began to tumble. This was true for bullets fired through air into water, and it was equally true for bullets as they entered human flesh. What impeded the bullet from tumbling was its own weight and momentum; the lighter the bullet, the more rapidly and wildly it would tumble end-for-end in flesh. A large artillery round might pass straight through a human body, but a small bullet could act like a gouge . . . [As rifle designer Eugene Stoner put it:]

> What it amounts to is that bullets are stabilized to fly through air, and not through water or a body, which is approximately the same density as the water. And they are stable as long as they are in the air. When they hit something, they immediately go unstable . . . If you are talking about a .30 caliber bullet [like that used in the M-14], that might remain stable through a human body . . . While a little bullet, being it has a low mass, it senses an instability situation faster and reacts much faster. This is what makes a little bullet pay off so much in wound ballistics.[2]

A far-sighted troop commander, General Wyman, had asked Stoner to design his rifle precisely to take advantage of the 'payoff' of smaller bullets. The AR-15, the precursor of the M-16, used .22-caliber bullets instead of the .30 caliber that had long been standard for the Army.

A second discovery about weaponry also lay behind the design of Eugene Stoner's AR-15. In his studies of combat units during World War II, S. L. A. Marshall found that nearly four fifths of combat soldiers never fired their weapons during battles. This finding prompted a closer look at the weapons the soldiers used. It turned out that one group of soldiers was an exception to this rule: those who carried the Browning automatic rifles (BAR). These were essentially portable machine guns, which could spray out bursts of continuous fire. The M-1's that the other soldiers carried were 'semiautomatic,' and required a separate trigger squeeze for each round. Within a combat group, firing would begin with the BAR man and spread out from him. The nearer a soldier with an M-1 stood to the BAR man, the more likely he was to fire. The explanation most often suggested was that the infantryman carrying a normal rifle felt that his actions were ultimately futile. As John Keegan said in *The Face of Battle*, 'Infantrymen, however well-trained and well-armed, however resolute, however ready to kill, remain erratic agents of death. Unless centrally directed, they will choose, perhaps badly, their own targets, will open and cease fire individually, will be put off their aim by the enemy's return of fire, will be distracted by the wounding of those near them, will yield to fear or excitement, will fire high, low, or wide.'[3] The normal infantryman could not see the enemy clearly or have any sense of whether he had hit. The BAR man, by contrast, had the sense that he could dominate a certain area – 'hose it down,' in the military slang – and destroy anyone who happened to be there.

From the end of World War II, there was a demand from some Army officers for a new infantry weapon that would be light, reasonably accurate, and capable of fully automatic fire. The response of the Army's ordnance organizations was to build the M-14. This was basically an automatic-firing, less solidly made version of the Army's previous standard, the M-1. Like the M-1, it used a large .30-caliber round. Its disadvantage was that it was virtually uncontrollable when in fully automatic firing. The explosive charge needed to propel the heavy bullets was so great, and the rifle itself so flimsily built in an effort to make it lightweight that the kick was ferocious. A soldier who used it on automatic fire was likely to get a nosebleed, in addition to being unable to control the weapon's aim. It was with this rifle that American troops trained in the early and middle sixties, and with it they went to Vietnam.

The M-14 was a product of the Army's own arsenal system, an informal congeries of weapons laboratories, private contractors, and the Army Materiel Command that is often generically known as the 'ordnance corps.' The ordnance corps had been in charge of small-arms design for the Army for more than a hundred years. In questions of technology, it emphasized the outlook of the 'gravel-bellies,' the sharpshooters and marksmen who measured a weapon by how well it helped them hit a target four hundred, five hundred, six hundred yards away in peacetime rifle competition. 'The M-14 had been developed on the premise that aimed-fire, the fire of the marksman, was of the utmost importance in combat,' a Rand employee named Thomas McNaugher wrote in a study of the M-16.[4] 'To the U.S. Army, it was more than a premise, it was a creed that had evolved over nearly a century since the service adopted its first rifle in 1855.'[5] Giving generous credit to the element of rationality in the ordnance corps' practices, McNaugher says that the marksman's philosophy was appealing because the 'Ordnance Department, the agency that developed and produced the service's rifles and ammunition, preferred tactics that stressed slow and deliberate fire because it meant less waste of ammunition and hence less strain on the Department's supply lines and production facilities.'[6]

For the marksman's purposes, a large, heavy round was ideal, since it remained steady in flight and was less sensitive to wind. Hand in hand with this mentality went an insistence on rigid technical specifications. If a round didn't leave the muzzle at 3,250 feet per second, it was no good; if it couldn't be fired in the Arctic and the Sahara and perform just as well in each place, it was not fit for army duty. These emphases had little to do with the experience of modern combat, in which most fire fights took place at a range of thirty to fifty yards or less, and in which speed and surprise were so important that it might often cost a soldier his life to take the time to aim his rifle, as opposed to simply pointing it in the right direction and opening up on automatic.

In its sociology, the ordnance corps was small-time, insular, old-fashioned. Its first instinct, when presented with a new technical possibility, was to reject it and stick to its own, traditional solutions. Twice since the Civil War, American Presidents have had to force the ordnance corps to adopt new rifles that had come from outside its own shop.

There was also an air of coziness in relations between the ordnance corps and the rifle and ammunition makers who supplied it. 'Sole source' contracts, which gave one company a monopoly on the Army's business, were not unusual. One of the most important of these, which would prove to have an especially crucial effect on the development of the M-16, was with the Olin-Mathieson Corporation, which since the end of World War II had been the Army's supplier of a kind of gunpowder known as 'ball powder.'

The ordnance corps had every reason to dislike the AR-15. It came from an outside inventor and threatened to replace a product of their own arsenal system, the M-14. It was not a gravel-belly's or a technician's rifle. And it proposed using what was, by the standards of the corps, a laughably small round – a .22-caliber bullet, the size kids used to shoot at squirrels. A popgun was all it was. In the early fifties the U.S. ordnance corps had fought a grueling battle against European governments in NATO, who wanted to have a small bullet adopted as the NATO standard. The ordnance corps' struggle to impose the .30-caliber bullet as NATO standard had been successful, but it had left much ill will in its wake. Having won that bitter struggle, the Army was not likely to surrender meekly on the same point in its own home territory.[7]

The M-14 was adopted as the Army's standard in 1957. At the same time, Eugene Stoner was completing the design of the rifle then known as the AR-15. By that time Stoner was known as one of the great figures in this special calling. Like some of the other outstanding American rifle designers – including John Browning, inventor of the Browning automatic rifle, who had to sell his weapon to foreign governments after rejection by the American ordnance corps – Stoner had never seen his models win easy acceptance from the Army. Stoner was working for the Armalite Corporation when he finished developing the AR-15.

The rifle combined several advantages. One was the lethal 'payoff' that came with its .22-caliber bullets. The smaller, lighter ammunition meant that the rifle could be controlled on automatic fire by the average soldier because its kick was so much less than the M-14's. The rifle itself was also lighter than the M-14. Together, these savings in weight meant that a soldier using the AR-15 could carry almost three times as many rounds as a man with the M-14. This promised to eliminate one of the soldier's fundamental problems in combat: running out of ammunition during a fire fight. The rifle had two other technical advantages. One was the marvelous reliability of its moving parts, which could feed, fire, extract, and eject 600 or 700 cartridges a minute and practically never jam. The other was a manufacturing innovation that drastically cut the cost of the weapon. The parts were stamped out, not hand-machined as in previous rifles, and they could be truly mass-produced. The stock was made of plastic, which further cut the cost. To traditionalists, this was one more indication that the AR-15 was not a real weapon. They said that you couldn't use a plastic rifle as a club. Stoner's reply, in effect, was that with the AR-15's reliability and its destructive power, you wouldn't need to . . .

Through 1962 and 1963, there followed a series of tests, evaluations, and counterevaluations by the American military, the repeated theme of which was the lightness, 'lethality,' and reliability of the AR-15. The results of one test, conducted by the Defense Advanced Research Projects Agency, were summed up in September 1962 by the Comptroller of the Defense Department:

> Taking into account the greater lethality of the AR-15 rifle and improvements in accuracy and rate of fire in this weapon since 1959, in overall squad kill potential the AR-15 is up to 5 times as effective as the M-14 rifle . . .
>
> The AR-15 can be produced with less difficulty, to a higher quality, and at a lower cost than the M-14 rifle.
>
> In reliability, durability, ruggedness, performance under adverse circumstances, and ease of maintenance, the AR-15 is a significant improvement over any of the standard weapons including the M-14 rifle. The M-14 rifle is weak in the sum of these characteristics . . .
>
> It is significantly easier to train the soldier with the AR-15 than with the M-14 rifle.
>
> Three times as much ammunition can be carried on the individual soldier within the standard weapon and ammunition load.[8]

Meanwhile the Army Materiel Command, home of the ordnance corps, was conducting its own evaluations of the AR-15. In these, too, there was consistency. The corps found little to admire in the AR-15, and many technical objections to it. It had poor 'pointing and night firing characteristics'; its penetration at long distance was also poor. The ordnance corps' recommendation was to stick with the M-14 until a 'radically' better model, based on advanced technology, emerged from research programs the ordnance labs had recently begun.

Early in 1963, with strong support from President Kennedy and Secretary of Defense McNamara, the Special Forces (Green Berets) asked for and got approval to use the AR-15 as their standard issue because they needed lightweight gear for mobility and stealth. The Army's Airborne units in Vietnam also got it, as did some operatives from the CIA. As the AR-15 attracted a greater and greater following among units actually operating in Vietnam, Secretary of the Army Cyrus Vance asked the Army's Inspector General to look again at the reasoning and evidence that had led the Army Materiel Command to reject the AR-15. His investigation found that the tests had been blatantly rigged. The M-14s used in the tests were all hand-picked, hand-made, 'matchgrade' weapons (suitable for marksmen's competitions), while the AR-15s were taken straight from the box. The ammunition for the M-14 had also come from a special, coddled lot. The inspector found that various organizations of the ordnance corps had met beforehand to discuss how to fix the tests. They agreed to take a dry run through the tests, and then (according to the printed minutes of their meeting) include in the final tests 'only those tests that will reflect adversely on the AR-15 rifle . . .'[9] The lines became more clearly drawn within the Pentagon, with the Air Force and the civilian leadership of the Defense Department (especially McNamara and his Secretary of the Army, Cyrus Vance) in favor of the AR-15 and the Army ordnance establishment opposed.

As the fighting in Vietnam grew more intense, procurement of the rifle began in late 1963, with 19,000 rifles for the Air Force and another 85,000 for the special Army units. Robert McNamara, in the interests of efficiency, designated the Army as the central procurement agency for all the services. It was at this point that the Army ordnance corps got hold of Eugene Stoner's AR-15, declared it to be inadequately 'developed,' and 'militarized' it into the M-16.

The first of several modifications was the addition of a 'manual bolt closure,' a handle that would permit the soldier to ram a cartridge in manually after it had refused to seat properly by itself. The Air Force, which was to buy the rifle, and the Marine Corps, which had tested it, objected vehemently to this change. An Air Force document said, 'During three years of testing and operation of the AR-15 rifle under all types of conditions the Air Force has no record of malfunctions that could have been corrected by a manual bolt closing device.'[10] Worse, they said, the device would add cost, weight, and complexity to the weapon, thereby reducing the reliability that had been its greatest asset. Years later, during the congressional testimony, Colonel Harold Yount, who had been a project manager at the Rock Island arsenal in 1963, was asked how this change could have been justified. Not on the basis of complaints, or of prior tests, Colonel Yount said. It was justified 'on the basis of direction.'[11] Direction from where? a congressman asked. Direction from his superiors on the Army staff, was all he would say. The fact was that General Earle Wheeler, the Army's Chief of Staff, had personally ordered the useless handle.

The next modification was to increase the 'twist' of the rifle's barrel, from a one-in-14-inches twist rate to one-in-12. More twist made the bullet spin faster as it flew, and therefore made it hold a more stable path; but by exactly the same process, it made the bullet more stable as it entered flesh, and thereby greatly reduced the shocking 'lethality' that had so distinguished the AR-15. In the face of the logic that led the Army to this decision, it is difficult to avoid the conclusion that reducing the M-16's 'lethality,' along with its other advantages over the beloved M-14, was precisely the intention of the change. The Army's explanation for the increased 'twist' of the barrel was that otherwise the rifle could not meet its all-environments test. To qualify as 'Army standard,' a rifle and its ammunition had to show that they would perform equally well at 65 degrees below zero and 125 above. On the basis of skimpy test evidence, an Arctic testing team concluded that the AR-15 did not do so well on the cold-weather portions of its test. Supposedly, the rounds wobbled in flight at 65 below. The Army's reaction was to increase the 'twist' and thereby decrease the 'lethality,' even though the rifle was due for shipment to the steaming jungles along the Mekong.

The final change was the most important. Like the others, it was publicly justified by a letter-of-the-law application of technical specification, but it seems to have been motivated by a desire to discredit the AR-15 as a competitor to the Army's own M-14.

Weapons designers speak of automatic rifles as 'resonant mechanisms,' in which several different cycles must all work in harmony. One of the determining factors for synchronizing these cycles is the explosive characteristic of the ammunition. Some powders explode very quickly, others build up pressure more slowly. Depending on that pattern, certain other decisions follow – for example, the location of the 'gas port,' or the proper cycling rate for inserting and extracting the bullets. Eugene Stoner had designed his AR-15 around a powder known as IMR (for 'improved military rifle'). It was produced by Du Pont, which sold it to Remington to fill the cartridges. It is made of nitrocellulose, sometimes known as guncotton, which is extruded like toothpaste and cut into little granules. All of the early tests of the AR-15 had involved IMR ammunition; it was the ammunition that the Air Force had accepted and that had proven so reliable in all field trials.

In June 1963 the Army Materiel Command conducted tests at Frankford Arsenal which showed that IMR powder would not do. Once again, it seems obvious that the test was designed to produce exactly this result. The problem, as with the barrel twist, was failure to meet a technical specification. For reasons that no members of the ordnance corps could ever satisfactorily explain to congressional investigators, the Army specified that the muzzle velocity for the rifle must average 3,250 feet per second plus or minus 40. In all its previous tests, and in its successful performance in combat in Vietnam, the AR-15 had never attained that velocity. The Army had tested the weapon thoroughly enough to know that when it was fired with the gunpowder it had originally been designed to use, its velocity averaged about 100 feet per second less. No testing panel had complained about the lower velocity. No problems had shown up in combat – quite the contrary. But when the Army's 'technical data package' for the M-16 was issued in 1963, it required the 3,250 fps muzzle velocity, and also specified that the pressure within the firing chamber could not exceed 52,000 pounds per square inch.

After a good deal of negotiation and haggling that lasted several months, the outcome of the Frankford Arsenal tests was that IMR ammunition could not meet the newly devised standards. To get the velocity up to 3,250 fps, it had to bring chamber pressure too close to the limit. In February 1964 the Army sent out a request to the manufacturers to come up with substitute powders. A few months later Du Pont said it would stop producing IMR, and Remington switched to the Army's 'sole-source' supplier of 'ball powder,' Olin-Mathieson. By the end of 1964 Remington was loading only ball powder in its cartridges for the rifle, which by now had been renamed the M-16.

Ball powder was first adopted by the Army early in World War II, for use in certain artillery rounds. It differs from IMR in being 'double-based' (made of nitrocellulose and nitroglycerine) and in certain other ways. Its most important difference is its explosive characteristics, for it burns longer and slower than IMR. Olin-Mathieson has long enjoyed a comfortable relationship with the ordnance corps as the 'sole-source' supplier of ball powder for many ammunition jobs.[12] Olin-Mathieson received contracts for some 89 million cartridges in 1964 alone, and far more as the war went on. More than 90 per cent of the cartridges used in Vietnam were loaded with ball powder.

After the Army had made the decision to switch to ball powder, it sent a representative, Frank Vee of the Comptroller's office, to try to get Eugene Stoner to endorse the change. Stoner had not been consulted on any of the modifications to his rifle, not the bolt closure nor the barrel twist nor the ball powder, and he thought that all were bad ideas. He recalled for the congressional committee his meeting with Vee:

> He asked me my opinion [about the specs requiring ball powder] after the fact. In other words, this was rather an odd meeting . . . I looked at the technical data package and he said, 'what is your opinion?' I said, 'I would advise against it . . .'
>
> I asked, 'so what is going to happen?' And he said, 'well, they have already decided this is the way they are going to go.' I said, 'so why are you asking me now,' and he said, 'I would have felt better if you had approved of the package.'
>
> And I said, 'well, we both now don't feel so good.'[13]

The reason for Stoner's concern was that the change of powders destroyed most of the qualities he had built into his rifle. With ball powder, the M-16 looked better on the Army's new specification sheets but worse in operation. There were two problems. One was 'fouling' – a powder residue on the inside of the gas tube and chamber that eventually made the rifle jam. The AR-15 had been designed so that its gas port stayed closed through the combustion of the powder, but that was for a different powder. The new ball powder was inherently dirtier; in addition, it burned longer, and was still burning when the gas port opened and let it burn into the gas tube. The other effect of ball powder was to increase the rifle's 'cyclic rate.' The AR-15, with all its interlocking mechanical cycles, had been designed to fire between 750 and 800 rounds per minute. When cartridges loaded with ball powder were used, the rate went up to 1,000 or more. 'When the Army said, 'No, we are going to use our ammunition,' the cyclic rate of the weapons went up at least 200 rounds per minute,' Stoner told the congressional committee.[14] 'That gun would jump from 750 to about 1000 rounds a minute, with no change other than changing the ammunition.'[15]

The consequences of a higher cyclic rate were immediate and grave. What had been a supremely reliable rifle was now given to chronic breakdowns and jams. In November 1965, engineers from Colt fired a number of rifles, some with the original IMR powder and some with ball. They reported: 'For weapons such as those used in this experiment, none are likely to fail with ammunition such as [IMR], whereas half are likely to fail with ammunition such as [ball powder].'[16] In December the Frankford Arsenal conducted another test for malfunctions. When M-16s were loaded with IMR cartridges, there were 3.2 malfunctions per 1,000 rounds, and .75 stoppages. When the same rifles were fired with ball powder, the failure rates were about six times higher (18.5 and 5.2, respectively). Under the central procurement policy, the Army's decision also forced the Air Force to switch to ball powder. The Air Force protested, pointing out that the rifles had been extremely reliable when loaded with IMR. One Air Force representative described a test in which 27 rifles fired 6,000 rounds apiece. The malfunction rate was one per 3,000 rounds, and the parts replacement rate one per 6,200 rounds. The rifle and its original cartridge worked fine, the Air Force insisted, even though they didn't happen to meet the specifications of 3,250 feet per second from the muzzle.[17]

In May 1966 there was one more report, this one the result of an extensive and unusually realistic series of tests held by the Army's CEDEC field test organization at Fort Ord. In these tests, the soldiers fired as squads, not as individuals; the targets resembled real battlefield targets; in that they were hard to see, and obscured by brush and other cover, there was simulated fire from the targets themselves, done in a pattern resembling that of combat; soldiers were run through the course only once, to avoid any familiarity with it. The conclusion was that the M-16 was more effective than the M-14 or the Soviet AK-47 (which was also tested), but that it was an unreliable weapon. The reason for the fouling, the jamming, and the breakdowns, the testers said, was the switch to ball powder.[18] By that time the Army was ordering ball powder in greater quantities than ever and shipping it to Vietnam.

In 1965, after the years of the advisers and the Special Forces, American troops began full-fledged ground combat in Vietnam. The regular Army and Marine units carried the old M-14. On arrival, they discovered several things about their weapon. One was that in jungle warfare the inaccurate, uncontrollable M-14 was no match for the AK-47, made in the Soviet Union, which their enemies used. Both rifles fired a .30-caliber bullet, but the AK-47's cartridges had a lighter bullet and were packed with less powder, which reduced the recoil to an endurable range. They also saw that the old AR-15s that had been used by the Special Forces had been a big hit in Vietnam. On the black market the weapon was going for $600 (the original price was around $100), to soldiers who were willing to sacrifice several months' pay to get hold of one.

One of those who noticed these patterns was William Westmoreland, then the commander of American forces in Vietnam. He saw that his men were doing very badly in the fire fights against the AK-47 and that the casualties were heavy. He also saw how the AR-15 performed, and near the end of December 1965 he sent an urgent, personal request for the M-16, immediately, as standard equipment for units in Vietnam.

The ordnance corps met this request with grudging compliance. The rifle would be sent to Vietnam, but only as a special, limited purchase. It would not be issued to American troops in Europe or in the United States; it would not replace the M-14 as the Army's standard weapon. Nor would it go to Vietnam under circumstances likely to show off its merits, because there was no backing off the requirement that its cartridges be filled with ball powder.

The climactic struggle over ball powder had occurred one year before Westmoreland's request, in 1964. As test after test showed that ball powder made the rifle fire too fast and then jam, the manufacturing company finally threw up its hands. Colt said that it could no longer be responsible for the M-16's passing the Army's acceptance test. It could not guarantee performance with the ball powder. One of the provisions of test was that the rifle's cyclic rate not exceed 850 rounds per minute, and six out of ten rifles were far above that when using ball powder. Don't worry, the Army said in an official letter; *you can use whatever ammunition you want for the tests*. But we'll keep sending our ball powder to Vietnam. Beginning in 1964, Colt used IMR powder so that its rifles would pass the acceptance tests, the Army promptly equipped those rifles with ball-powder cartridges and sent them to soldiers who needed them to stay alive. The Army's official reasoning on the matter was that since it did not recognize the theories that ball powder was the cause of the problems, why should it care which powder Colt used? Colt delivered at least 330,000 rifles under this agreement. After uncovering the arrangement, the Ichord committee concluded in its report:

> Undoubtedly, many thousands of these were shipped or carried to Vietnam, *with the Army on notice that the rifles failed to meet design and performance specifications and might experience excessive malfunctions when firing ammunition loaded with ball propellant* [emphasis in original] . . . The rifle project manager, the administrative contracting officer, the members of the Technical Coordinating Committee, and others as high in authority as the Assistant Secretary of Defense for Installations and Logistics knowingly accepted M-16 rifles that would not pass the approved acceptance test . . . Colt was allowed to test using only IMR propellant at a time when the vast majority of ammunition in the field, including Vietnam, was loaded with ball propellant. The failure on the part of officials with authority in the Army to cause action to be taken to correct the deficiencies of the 5.56-mm ammunition borders on criminal negligence.[19]

The denouement was predictable and tragic. In the field, the rifle fouled and jammed. More American soldiers survived in combat than would have with the M-14, but the M-16's failures were spectacular and entirely unnecessary. When they heard the complaints, ordnance officials said it only proved what they'd said all along, that it was a lousy rifle, anyway. The official Army hierarchy took the view that it was a question of improper maintenance. Officials from the Pentagon would go on inspection tours to Vietnam and scold the soldiers for not keeping the rifles clean enough, but there never seemed to be enough cleaning supplies for the M-16. The instruction leaflets put out by the Army told them that 'This rifle will fire longer without cleaning or oiling than any other known rifle,' and 'an occasional cleaning will keep the weapon functioning indefinitely.'[20]

At last the soldiers began writing letters – to their parents, to their girl friends, and to the commercial manufacturer of a rifle lubricant called Dri-Slide. The Dri-Slide company received letters like the following:

December 24, 1966

Dear Sir:

On the morning of December 22nd our company . . . ran into a reinforced platoon of hard core Viet Cong. They were well dug in and boy! Was it hell getting them out. During this fight and previous ones, I lost some of my best buddies. I personally checked their weapons. Close to 70 per cent had a round stuck in the chamber, and take my word it was not their fault.

Sir, if you will send three hundred and sixty cans along with the bill, I'll 'gladly' pay it out of my own pocket. This will be enough for every man in our company to have a can.

————, Spec. Fourth Class

March 9, 1967

Dear Sir:

I'm very much interested in your product, Dri-Slide. Being stationed here in Vietnam with the rain seasons coming, myself and other GI's I'm with, need something to keep our weapons from jamming up. The regular type oil that we are using collects too much dust, and the dust here is quite terrible.

————, Spec. Fourth Class[21]

Parents in Idaho received this letter from their son, a Marine:

Our M-16s aren't worth much. If there's dust in them, they will jam. Half of us don't have cleaning rods to unjam them. Out of 40 rounds I've fired, my rifle jammed about 10 times. I pack as many grenades as I can plus bayonet and K bar (jungle knife) so I'll have something to fight with. If you can, please send me a bore rod and a $1\frac{1}{4}$ inch or so paint brush. I need it for my rifle. These rifles are getting a lot of guys killed because they jam so easy.[22]

One man wrote to a member of the Armed Services Committee staff, recounting what his brother had told him about his experience in Vietnam:

He went on to tell me how, in battles there in Vietnam, the only things that were left by the enemy after they had stripped the dead of our side were the rifles, which they considered worthless. That when battles were over the dead would have the rifles beside them, torn down to attempt a repair because of some malfunction when the enemy attacked . . . This man speaking has been shooting since he was 15 . . . He said, 'part of me dies when I have to stand by and see people killed, and yet my hands are tied'.[23]

A letter that ended up in the office of Representative Charles W. Whalen, Jr., of Ohio:

I was walking point a few weeks back and that piece of you know what jammed 3 times in a row on me. I'm lucky I wasn't doing anything but reconning by fire or I wouldn't be writing this letter now. When I brought the matter up to the Captain, he let me test fire the weapon – well in 50 rounds it double fed and jammed 14 times. I guess I'll just have to wait till someone gets shot and take his rifle because the Captain couldn't get me a new one.[24]

Another, referred to Senator Gaylord Nelson of Wisconsin:

The weapon has failed us at crucial moments when we needed fire power most. In each case, it left Marines naked against their enemy. Often, and this is no exaggeration, we take counts after each fight, as many as 50% of the rifles fail to work. I know of at least two marines who died within 10 feet of the enemy with jammed rifles. No telling how many have been wounded on that account and it is difficult to count the NVA who should be dead but live because the M-16 failed. Of course, the political ramifications of this border on national scandal. I suppose that is why the Commandant and all the bigwigs are anxious to tell all that it is a wonderful weapon.

My loyalty has to be with these 18-year old Marines. Too many times (yesterday most recently) I've been on TF's awaiting medical evacuation and listened to bandaged and bleeding troopers cuss the M-16. Yesterday, we got in a big one . . . The day found one Marine beating an NVA with his helmet and a hunting knife because his rifle failed – this can't continue – 32 of about 80 rifles failed yesterday.[25]

When investigators from the congressional committee went to Vietnam, they confirmed another report: that one Marine had been killed as he ran up and down the line in his squad, unjamming rifles, because he had the only cleaning rod in the squad.[26]

▶ **NOTES**

1 Hearings, Special Subcommittee on the M-16 Rifle Program, Committee on Armed Services, U.S. House of Representatives, 90th Cong., 1st sess. (Referred to as 'Hearings.')
2 Ibid., pp. 4563–64.
3 Keegan, *Face of Battle*, p. 229.
4 Thomas McNaugher, 'Marksmanship, McNamara, and the M-16 Rifle: Innovation in Military Organizations,' *Public Policy* (Winter 1980), pp. 1–37.
5 Ibid., p. 4.
6 Ibid., p. 7.
7 A useful history of the rifle programs is laid out in the 'Report of the Special Subcommittee on the M-16 Rifle Program,' Report Number 26, Committee on

Armed Services, U.S. House of Representatives, 90th Cong., 1st sess., pp. 5321–72. (Referred to as 'Report.')

8 Ibid., p. 5327.
9 Ibid., p. 5330.
10 Ibid., p. 5333.
11 'Hearings,' p . 4701.
12 See GAO report B-146977, issued March 31, 1965.
13 'Hearings,' p. 4559.
14 Ibid., p. 4549.
15 Ibid., p. 4571
16 'Report,' p. 5356.
17 Ibid., p. 5355.
18 Ibid., p. 5357.
19 Ibid., pp. 5354 and 5370.
20 Ibid., p. 5263.
21 'Hearings,' pp. 4509–10.
22 Ibid., p. 4584.
23 Ibid., pp. 4582–83.
24 Ibid., p. 4583.
25 *Loc. cit.*
26 Ibid., p. 4873.

28 ▶ The Thor–Jupiter controversy

Michael H. Armacost

The purpose of this study is to contribute to an understanding of how the content of weapons policies is influenced by the character of the political process through which those decisions are made. Specifically, this study presents an analysis of the ways in which interservice competition affected the development, production, and deployment of the novel weapon system: the intermediate-range ballistic missile [IRBM].

The Thor–Jupiter* controversy between the US Air Force and Army provides the case study material for this analysis. Both services coveted the responsibility for developing and producing an operational IRBM system. In 1955, both advanced plausible and promising technical proposals – the Army for the Jupiter Missile System, the Air Force for its Thor system. During 1956, each sought to design a weapon system which would advance its qualifications for the deployment assignment. After the Air Force successfully obtained sole responsibility for deployment of an IRBM in November 1956, the focus of service rivalry shifted in anticipation of future struggles, to the organizational arrangements which would govern subsequent research and development efforts in the space field. Detailed study of the rival Army and Air Force efforts to design, develop, produce, and deploy an IRBM system should shed some light on the political dimensions of the choices between competing weapons systems . . .

Some features of the Thor–Jupiter controversy were undoubtedly unique. Army and Air Force differences were projected into the public realm with an extraordinary virulence. The blurring of established divisions of functions

* Thor was the name of the intermediate-range ballistic missile developed by the US Air Force, and Jupiter that of the missile developed by the US Army. Thor missiles were deployed in the United Kingdom between 1959 and 1963; Jupiter missiles were deployed in Italy and Turkey between 1960 and 1965. Variants of each were also used extensively in the American space programme.

by the development of missile delivery capabilities was unusual. The guided missile was not a direct derivative of either the airplane or of field artillery, Air Force and Army rhetoric to the contrary notwithstanding. It had characteristics similar to both. If the intrinsic technical characteristics of the IRBMs implied no obvious jurisdictional assignment, the stakes which the services invested in a favorable decision were such as to compound the difficulties of choice. Since long-range missiles could be efficiently employed only with nuclear warheads, moreover, the proprietary issue of control over the nuclear stockpile was also reopened.

More than simply the jurisdictional question of who would design, produce, and deploy the IRBM was at issue. Competing patterns of contractor relations contributed to the dynamics of contention. The controversy set not only the Army and Air Force against one another, but the airframe industry against other industrial aspirants for a major role in missile development as well. It also involved a spirited contest between sponsors and supporters of the 'arsenal system' and those who advocated reliance upon the 'weapon system manager' concept of development. In the former, government scientists, engineers, and technicians were at times engaged in the in-house fabrication and assembly of military weaponry, as well as being participants in the research, design, and component development stages of weapons innovation. In the latter, a military service provided managerial surveillance while contracting out to industry the responsibility for systems engineering, technical direction, and component production and assembly on new weapon systems. In the convergence of jurisdictional conflicts among the services with these industrial and administrative rivalries, the case is perhaps unique.

An element of generality is imparted to the conclusions, however, since the Thor–Jupiter controversy revealed the variable reactions of the services to novel strategic delivery capabilities made possible by rocket technology. While each technological revolution will be differently appreciated and exploited by the services, the differential impact of technical advance upon their fortunes, the forceful disruption of the prevailing distribution of power and responsibility among them by technological change, and the expectation that political-technological efforts may be undertaken on behalf of proprietary as well as strategic interests are predictable aspects of every revolutionary advance in military weaponry. . . .

Understanding of the process by which strategic decisions are formulated has been enriched by applying analogies from the fields of both international politics and legislative bargaining. The relevance of those analogies is amply confirmed in the case of the Thor–Jupiter controversy. The relations among the services, as among sovereign states, are marked by the existence of both conflict and accommodation. The formulation and execution of alliance policies, the application of limited reprisals, the persistent conduct of negotiations among quasi-sovereign entities, and efforts to arrange compromises through the activities of third party mediators are all to be found in the IRBM dispute. Such analogies call attention to the relative autonomy of the participants and their consequent need voluntarily to coordinate their interests, moderate their ambitions, and mitigate their differences in the search for a mutually acceptable policy.[1]

Limits to such analogies inhere in the fact that the bureaus, agencies, and departments engaged in the politics of weapons innovation are only 'quasi-sovereign.' While bargaining among them is pervasive, the potential for authoritative decision is greater in administrative than in international politics. Consequently, the 'strain toward agreement' – enhanced by the existence of shared objectives and the natural desire of bureaucrats to get on with the job – is strengthened by the knowledge that policy may be imposed if it cannot be freely negotiated.[2]

The techniques of legislative politics are also much in evidence in the Thor–Jupiter controversy. Difficult choices are invariably postponed. Responsibility for them is regularly devolved upon committees. Agreements are often facilitated by compromise and logrolling, and policies frequently reflect the lowest common denominator of consensus – occasionally achieved upon the basis of unrealistic assumptions.[3]

One would not want to overemphasize the lateral relationships in defense politics, however, at the expense of an awareness of hierarchical elements. As Paul Y. Hammond has noted, even the highest ranking committees in the US Government, such as the Joint Chiefs of Staff and the National Security Council, though they are bedeviled by the familiar failings of committees, contain intrinsically hierarchical elements. This is due to the fact that they serve the President. His sense of priorities, his idiosyncrasies, and his administrative style are thus 'reflected in the problems they take up, the staff work they do, the deliberative acts they perform, and the recommendations they offer.'[4]

In the search for an appropriate model of service politics, one must therefore account for both the pervasiveness of bargaining among semi-autonomous agencies and departments, and the residual elements of hierarchy to be found in the bureaucratic politics of the Pentagon. Defense politics in the 1950s certainly was marked by the aggressive maneuvering of rather autonomous service departments eager to maximize their share in the distribution of budgets, roles and missions, and research and development assignments. At the same time, those vested with formal authority for the management of the Department of Defense struggled to discipline the efforts of the services to a single strategic policy. In the Thor–Jupiter controversy the service departments performed prodigious feats of lobbying. The activities of the Secretary of Defense and his assistants more nearly resembled the interest aggregation functions of political party leaders. Thus one may hazard the hypothesis that in the realm of weapons innovation the services act as powerful *institutional* interest groups.[5]

The services, in this context, articulate and aggregate demands and present them as a program to those legally endowed and politically capable of authorizing action. They attempt to mobilize support for their programs through persuasion and bargaining. They seek to transform their recommendations into policy through the various channels of influence in the policy making process. Like interest group activity in any political system, interservice politics is likely to be conditioned by the substance of existing policy, by the prevailing procedures for policy making, and by those cultural norms which constitute the ground rules for politics in a particular environment.[6]

The pressure group model appears especially attractive, since it emphasizes the interaction and interdependence between a relatively homogeneous group of public officials endowed with the legal responsibility to formulate policy and a network of lobby groups, each seeking to implement its own 'partial view of the public interest.' This implies that weapons policies are not merely discrete contractual bargains negotiated by freewheeling and selfish partisans under the surveillance of Presidential or Congressional politicians who serve as 'referees.' On the contrary, the administrative politics of weapons innovation contain a 'mixture of authoritarian *and* equalitarian elements, the lateral relationships, and the hierarchy of authority.'[7] In short, the emphasis is neither upon an undisciplined pluralism nor upon transcendent central direction, but upon the dialectic between the services as pressure groups and the politically accountable leaders of the Pentagon as managers of a diffuse defense establishment. It is thus appropriate to inquire: in what ways, and to what extent do strategic ideas, the distribution of power and responsibility implicit in Defense Department organizational arrangements, and the climate of opinion regarding both technical innovations and interservice politics influence the pattern of bargaining and its results?

During 1955 the services had campaigned for approval of their respective developmental projects; in 1956 their interests turned to the even more contentious issues of operational control over intermediate-range ballistic missiles. Throughout the year the ambiguity of [Secretary of Defense] Wilson's public and private statements, the style of management in the Pentagon, and the competitive ambitions of the services, abetted by their zealous industrial suppliers, sparked a vigorous competition for control over the IRBM.[8] . . .

The basic issue was whether or not the Army would be permitted to deploy the weapon it was in the process of developing. Air Force responsibilities in the strategic retaliatory mission had long been acknowledged. Army theories justifying their use of intermediate-range ballistic missiles in tactical nuclear war situations enjoyed no such broad support.[9] . . .

Among the Joint Chiefs of Staff, the Army dared to hope for the benevolent neutrality of the Navy; they anticipated vehement opposition from the Air Force. In the event of split decisions from the Joint Chiefs of Staff, Chairman Admiral Radford would assume the role of arbitrator. In view of his skepticism toward Army efforts to justify operational control of their own IRBM, a favorable decision was unlikely.

Nor were the Army's prospects significantly improved by broadening the base of politics to include 'back stop'[10] organizations or industrial suppliers. Whereas the Navy and Air Force were consistently able to rely on the concentrated industrial support of the shipbuilding and airframe industries respectively, the Army had a much more diffuse group of suppliers. The

Army Association was neither as well established as the Navy League nor as well financed as the Air Force Association.

Particularly significant was the opposition the Army encountered from the Aircraft Industries Association.[11] As manned bombers and fighter aircraft approached obsolescence, the airframe industry faced an uncertain and ominous future, unless they were successful in capturing a sizable segment of the military missile field in both its development and production aspects.[12] Unlike airplanes, missiles would be produced in limited numbers, would require relatively little airframe, and would be exceedingly expensive to develop. Thus the industry covetously eyed major responsibilities in the research, design, development, test, and assembly phases of future missiles programs. . . .

The Air Force had already been acknowledged by [Secretary of Defense] Wilson as legitimate user of the IRBM. Thus, it was to their advantage to prevent a diffusion of authority and control over the new weapon. Their prospects appeared bright. It was widely assumed that either the [Air Force's] Thor or the [Army's] Jupiter project would be eventually cancelled, and the Air Force missile had been designated IRBM No. 1. Their industrial allies were numerous and influential. That this support was substantial is indicated by the fact that in 1955 the list of top defense contractors was topped by Boeing, North American, General Dynamics, and United Aircraft. All were, of course, airframe manufacturers; together they accounted for 19.3 per cent of the total defense procurement expenditures. Seven of the top ten corporations, eleven of the top fifteen, and thirteen of the top twenty were aircraft firms. Chrysler [the Army's contractor for the Jupiter project], on the other hand, had been the sixth largest defense supplier during the Korean build-up; by 1955, it had fallen to ninety-fourth place.[13] . . .

▶ STRATEGIES OF CONTROL: ARMY

The pursuit of objectives in a political environment involves the 'cultivation of an active clientele, the development of confidence among other governmental officials, and skill in following strategies that exploit one's opportunities to the maximum.'[14] All of these elements were apparent in the Army strategy, which was based primarily on the assumption that should they win the developmental competition with the Air Force, they stood a fair chance of deploying the Jupiter themselves. Their strategy was directed toward (1) facilitating the development of the Jupiter with all deliberate speed; (2) deferring a decision on roles and missions as long as possible; and (3) designing a missile compatible with their tactical doctrine, able to exploit the unique skills of the artilleryman, and capable of maximizing the importance of established Army performance of supporting missions in logistics, geodesy, and survey. . . .

Army spokesmen naturally emphasized the tactical usefulness of an extended-range missile. They sought to identify the IRBM with traditional Army functions. Extensions in range of artillery support weapons was simply a response to deeper enemy tactical targets and the need to deploy from less vulnerable rear positions. The implications of the Jupiter as a strategic weapon, should it be deployed near the front, were conveniently disregarded. Army officials did, however, emphasize the fact that missiles of even strategic range were deployed and operated by the Soviet Army as an extension of artillery rather than a sophisticated variant of the manned bomber.[15]

As he designed ground-launch equipment for the Jupiter, General Gavin, Assistant Chief of Staff for Research and Development, clearly thought in terms of a mobile field weapon. Such a concept was consonant with Army traditions, and it served to underscore their peculiar capabilities for handling such a weapon. As Air Force General Clarence Irvine commented: 'Gavin is thinking in terms of equipment, quite properly, to chaperone an Army in the field, as an extension of artillery. And with this concept there is nothing wrong with what he proposes to do.'[16]

Army partisans clearly hoped it would clinch the case for operational control, for they thought of themselves as the 'one service fully equipped and competent to provide all the supporting elements necessary to missile warfare.'[17] Their deployment concept, sharply differentiated from that of the Air Force, was serviceable for both deterrence strategy and their own tactical needs. As a mobile missile, it would enjoy immunity from surprise or pre-emptive attacks. Invulnerability could be achieved through dispersion and concealment, as well as mobility. Significantly, the Army alone could handle the weapon under field conditions. Their competence in camouflage and field defense, in logistic support of dispersed units, in the acquisition of target information from reconnaissance drones, in survey and geodesy, and in providing transportation to launching sites was vastly superior, they contended, to Air Force capabilities.[18]

The Jupiter might also provide them with a support system unaffected by weather, visibility, or enemy air defenses.[19] No new skills would be required to man the system; personnel already trained to launch other Army missiles could be easily adapted to the peculiarities of the IRBM. . . .

▶ STRATEGIES OF CONTROL: AIR FORCE

The Air Force had no objection to Army development and use of short range tactical missiles. The Army Jupiter program, however, not only appeared as an indication of growing interest in the strategic field, but threatened the classic mission of their medium-range bombers; namely 'sealing off the battlefield by striking at enemy railheads, troop concentrations, and supply areas far behind the front.'[20]

Since Air Force rights to an IRBM had been established, their problem was to retain exclusive control over the new weapon. For this purpose an appropriate strategy called for an early decision on roles and missions, a gradual detachment of the Navy from support of the Jupiter land-based missile project, a vigorous development program, and an acknowledged capacity to produce enough missiles to cover all strategic needs. . . .

Of [particular] concern to the Air Force was the growing interest displayed by the Army in rockets with strategic capabilities. Air Force doctrine, for quite cogent and plausible reasons, counseled against any dilution in the unified command and control arrangements governing the use of strategic delivery systems. The basic premise in Air Force thinking was that strategic airpower was 'something that can be employed as a *single instrument* . . . all of it, or any part of it, can be directed, controlled, if need be, from a single source.'[21]

The requirement of instantaneous, centralized control of strategic weapons was based upon the speeds of delivery systems, the tempo of modern warfare, the capacity for a high rate of fire, the relative invulnerability of ballistic missiles once in the air, and the consequent necessity for split-second decisions – unhindered by committees, coordinating mechanisms, or negotiated agreements among service commanders. Central direction would provide insurance that all essential targets were covered and that inadvertent escalation would not occur.

Moreover, in view of the anticipated mix of manned bombers and ballistic missiles to be employed on strategic missions, planning and operations required extensive coordination and intimate cooperation. 'There is nothing,' one writer emphasized, 'which should preclude the use of the existing command and control structure of the Strategic Air Command. The principle of central control of all strategic systems remains valid. I think it is absolutely necessary that all uses of large nuclear weapons, except in the missions of air defense and close support of the surface battle, be coordinated by a central agency.'[22]

Assuming, however, the desirability of centralized control over the new means of strategic delivery – and the logic of this position was compelling – why should the Air Force be endowed with such responsibilities? Army officials had readily conceded that the characteristics of the bomber required certain peculiar Air Force skills for its effective operation. With respect to ballistic missiles, however, they drew analogies with artillery, and contended that Air Force officers were suited neither by temperament nor training to field them.

To Air Force leaders it seemed self-evident that the problems of missile deployment and use were related to strategic bombing. In their view, the similarity of mission rather than the characteristics of weapons should be the vital determinant. Considering the IRBM as a strategic weapon, they argued that their experience with quick reaction time weapons and with coordinated strikes from widely dispersed bases, their capabilities in pre-hostilities reconnaissance, postattack damage assessment, and data processing conferred a clear advantage upon them as the operational agent for the Department of Defense over all strategic missiles.[23]

[Eventually, Armacost writes, the Air Force won control over both the Thor missile *and* the Army-developed Jupiter. The effects of conflict were still seen as the missiles entered the final phase of development and testing. There were disputes over turbopump design and over how to protect warheads from the heat generated by their re-entry into the atmosphere. But the most significant dispute continued to centre on whether to design the ground-support equipment to be relatively mobile or relatively fixed.]

Although the Air Force had been named the user of the missile, Army partisans were outspokenly critical of operational concepts preferred by the Air Force. Above all they were convinced of the essentiality of mobility or movability for the IRBM. The significance of mobility for survival was, they contended, the unalterable lesson of World War II. Certainly the experience of the German V-2 scientists and engineers confirmed this perspective.[24] Army officials were also sensitive to political considerations. By creating movable weapon systems that could be easily moved into or out of unstable regions, crises could be met without long-term commitments. But while political insight and doctrinal preference influenced their position, so, too, did their proprietary interests. Some officers still nurtured hopes of seeing the Wilson Memorandum [which, by limiting the Army's jurisdiction to missiles with ranges of 200 miles or less, had given the Air Force control over the Jupiter missile] revised. [Army Chief of Staff] General Taylor later explained Air Force distaste for mobile concepts as follows:

Although the Jupiter was specifically designed for field mobility, in November 1958, the Air staff directed the Army to remove this feature completely as if it were something unholy. The reason for this attitude is hard to determine. Perhaps it is also the fact that a mobile missile needs Army-type troops to move, emplace, protect, and fire it. Such troops include transportation units for mobility on the road and site construction, signal troops for field communications, infantry for close defense, and ordnance units for repair and maintenance. All these would be needed if the Jupiter were used in its mobile configuration. Thus, a decision to organize mobile ballistic missile units would in logic have led to transferring the operational use of the weapon back to the Army – where it should have been all the time.[25]

In designing ground handling equipment the Air Force proceeded from quite a different philosophy. Secretary of the Air Force, James Douglas, later explained Air Force instincts:

Their first inclination was to say, 'Here is another airbase problem. We will create launchers.' At that time there was more interest in a short reaction time than there was in hardening or protecting the missile system.[26]

Air Force officials harbored doubts about the feasibility of [making large missiles mobile]. General Irvine, for example, suggested that 'some people ought to take a look at some World War II pictures of what you could do to a bunch of trucks on the road with a bunch of fighters, and the Russians have lots of fighters.'[27] In reply to General Gavin's contention that their Jupiter and ancillary equipment could be moved to prearranged sites in two to four hours, he declared:

The laws of nature are the same for everybody; whether it is an Army truck or an Air Force truck, it sinks in the sand just the same. If you set up a hundred-thousand pound missile, we have the idea that maybe you ought to lay out a little piece of concrete about twenty feet square to put it on.[28]

Their different approaches reflected different conceptions of future contingencies, as well as past combat experiences. What was clear was that the Air Force view was more persuasive in the Office of the Secretary of Defense. Deputy Secretary Quarles elucidated the Department view to the Johnson Committee:

If you assume that we are fighting a war over a period of months in which we are maneuvering our missiles around and they are trying to destroy them, you might very well come out with von Braun's conclusion about his V-2s, that the mobile ones would survive and the fixed ones would not.

But what we are talking about instead is maintaining an instant retaliatory position with these missiles and one that can respond to tactical warning within fifteen minutes and actually launch the missiles in such a time.

Now to do that, you need arrangements associated with the missile's very rapid fueling and tanking up with liquid oxygen and conditioning the electronics equipment and warhead equipment and all the rest, which means that you must have arrangements, I won't call them fixed; they are not fixed; they are movable, as you say, but in any concept that it is mobile and you will get into action in any such length of time is just not realistic with these liquid oxygen systems.[29]

 NOTES

1 Comparisons between international politics and the domestic policy making process have long been noted by Professor William T. R. Fox in his lectures at Columbia University. They were also discussed in Roger Hilsman's early explorations of consensus-building in the making of policy. See his 'The Foreign Policy Consensus: An Interim Report,' *Conflict Resolution*, III (December, 1959), 367–71.

2 The suggestive phrases 'quasi-sovereign' and 'strain toward agreement' were coined by Professors William T. R. Fox and Warner R. Schilling respectively. While a consensus among the services will generally be easier to arrange than among nation states, Secretary of Defence Charles Wilson's plaintive exhortation to the services to 'at least . . . treat each other like allies' contains a useful reminder that this is not inevitably the case. Cited by Harry Howe Ransom, *Can American Democracy Survive Cold War?* (New York, Doubleday & Co., Inc., 1964), p. 80.

3 The similarities between legislative politics and strategy making were most clearly developed by Samuel P. Huntington, 'Strategic Planning and the Political

Process,' *Foreign Affairs*, XXXVIII (January, 1960), 291-92. See also Charles E. Lindblom, *The Intelligence of Democracy* (New York, The Free Press, 1965).

4 Paul Y. Hammond, 'Foreign Policymaking: Pluralistic Politics or Unitary Analysis?' *RAND P-2961-1* (February, 1965), 20.

5 Institutional interest groups are to be distinguished from anomic, associational, and non-associational groups which also engage in interest articulation. Institutional interest groups are formal organizations with designated political functions or legal responsibilities other than simply the articulation of interests, which nevertheless lobby as corporate groups on behalf of their own interests or those of other groups in society. See Gabriel Almond and Bingham Powell, Jr., *Comparative Politics: A Developmental Approach* (Boston, Little, Brown and Co., 1966), p. 77.

6 Harry Eckstein has analyzed the influence these variables exerted on British pressure group politics in his study *Pressure Group Politics* (Stanford, Stanford University Press, 1960), esp. pp. 15–39.

7 Hammond, 'Foreign Policymaking: Pluralistic Politics or Unitary Analysis?' *RAND P-2961-1*, 21.

8 Needless to say, neither the Army nor the Air Force was free to operate as a monolithic entity as each sought to fashion a strategy for obtaining operational control over the IRBM. No unanimity of either purpose or political tactics is to be anticipated in large multipurpose organizations like the military services. A careful reading of the Congressional testimony of leading officers and civilian officials of the services, as well as their later recollections of events recounted here, suggests that reasonably close coordination prevailed between the respective service secretaries, Chiefs of Staff, and technical organizations of the Army and Air Force with respect to service objectives in the Thor–Jupiter controversy. It is plausible to suppose that the degree of coordination is closely related to the intensity of this particular dispute.

9 U.S. Congress, Senate, Committee on Appropriations, *Hearings, Department of Defense Appropriations*, 1958, 85th Cong., 2nd Sess., p. 42.

10 For a discussion of the 'back stop' organizations supporting the services, see Samuel P. Huntington, 'Inter-service Competition and the Political Roles of the Armed Services,' *American Political Science Review*, LV (March, 1961), 40–52.

11 To be sure, the Army prime missile contractor, Chrysler Corporation, was free to join the AIA. Needless to say, they would not have been welcomed with open arms, and from the standpoint of the corporation, the $75,000 fee charged to members may have diminished their interest in membership.

12 See Charles J. V. Murphy, 'The Plane Makers Under Stress,' *Fortune*, LXI (June, 1960), 134 ff.; Charles J. V. Murphy, 'The Plane Makers Under Stress,' II, *Fortune*, LXII (July, 1960). pp. 111 ff.

13 Aircraft Electronics Rise to Top, *Business Week* (September 1, 1956), 33.

14 Aaron Wildavsky, *The Politics of the Budgetary Process* (Boston, Little, Brown and Co., 1964), pp. 64–65.

15 Senate Committee on Armed Services, *Satellite and Missile Program Hearings*, 1957–1958, p. 490.

16 Ibid., p. 964.

17 Claude Witze, 'Army Expands Scope: Challenges U.S. Air Force,' *Aviation Week* (November 5, 1956), 32.

18 Senate Committee on Armed Services, *Satellite and Missile Program Hearings*, 1957–1958, p. 1509.

19 Senate Committee on Armed Services, *Airpower Hearings*, 1956, p. 722.

20 Martin, *Saturday Evening Post*, CCXXX (November 9, 1957), 116.

21 Colonel Jerry D. Paige and Colonel Rayal H. Roussel, 'What Is Air Power?' *Air University Quarterly Review*, VIII (Summer, 1955), 5.

22 Major Gen. Charles M. McCorkle, 'Command and Control of Ballistic Missiles,' in Lt. Col. Kenneth F. Gantz (ed.) *Air Force Report on the Ballistic Missile: Its Technology, Logistics, and Strategy.*

23 Air Force philosophy was set forth clearly by Maj. Gen. Clarence Irvine, who explained to the Senate Special Preparedness Investigation Subcommittee that the objective of the Air Force on the Thor program was 'to create another weapons systems which would fit into the SAC [Strategic Air Command] war plan; and therefore, among other things we wanted fast reaction, the ability to shoot quickly, the same as in SAC we want an alert system to get the airplanes in the air quickly, while the people who fly them are still alive; and in the case of missiles, while there is still a man there to push the button.' Senate Committee on Armed Services, *Satellite and Missile Program Hearings*, 1957–1958, p. 964.

24 James M. Gavin, *War and Peace in the Space Age* (New York, Harper & Bros., 1958), pp. 76–77.

25 General Maxwell Taylor, *The Uncertain Trumpet* (New York, Harper & Bros., 1959), p. 141.

26 House Committee on Government Operations, *Missile Program Hearings*, 1959, p. 32.

27 Senate Committee on Armed Services, *Satellite and Missile Program Hearings*, 1957–1958, p. 962.

28 Ibid., p. 963.

29 Ibid., p. 2053.

29 ▶ The weapons succession process

Mary Kaldor

This essay is an attempt to explore the process by which one weapons system succeeds another.* In other words, it is about the process of military-technological change. Among students of arms control, there is a debate about whether military-technological change is determined by demand factors (i.e., the external requirements for weapons), or supply factors (i.e., domestic institutional and systemic pressures). My approach, which I call the weapons succession process, focuses on the interaction between these different types of explanation. To develop this approach, I have combined classical economics with recent theories of bureaucratic politics.

Classical economics was concerned with the mechanism for reconciling demand and supply, the process by which resources are organized to satisfy a particular need. This analysis originally confined itself to the satisfaction of material needs and the production of material goods or commodities. Classical economics started from the assumption that the act of production and the act of consumption were separated, and that some kind of mechanism had to be found to reconcile resources with needs and demand with supply. In the market system, price is the mechanism that determines the allocation of appropriate quantities of resources required to satisfy a particular set of needs expressed in consumer preferences. In an economy where consumer preferences change and where technical change leads to changing resource requirements, the conditions under which a stable set of relative prices can be obtained are extremely restricted.[1]

In a centrally planned economy, price is determined through a process of bureaucratic bargaining; planners, however, on the basis of restricted

* Based on research supported by the Social Science Research Council, U.K., and collective thinking by the Military Technology and Arms Limitation Group at the Science Policy Research Unit, University of Sussex.

assumptions about the nature of production and utility functions, often try to behave as though a market did exist and it were possible to determine a set of 'equilibrium prices.' In this situation, unsold goods or shortages become a signal of consumer preferences, which may not necessarily be heeded since producers are not concerned with profit or loss but with directives from above; the planner decides whether to take these signals into account.[2]

The act of production of material goods is separated from the act of consumption, but producers and consumers are both directly involved in the act of exchange. Therefore consumer preferences, as expressed in the act of purchase, are assumed to reflect productive possibilities. Yet this is not necessarily the case for other kinds of output. Collective goods – for example, state-provided health care or education – are purchased by the state and consumed by individuals. In other words, the act of purchase is socially or institutionally separate from the act of consumption. Collective goods have no price, or they have a price that is politically determined, such as a standard fee for a visit to the doctor. In a democratic system, preference for collective goods can be expressed, albeit slowly, through the political process; in totalitarian societies, that is much more difficult. In the case of armaments, the state is both purchaser and consumer, but, because wars are discontinuous, the act of consumption is separated from the act of purchase over time. In other words, the act of purchase is temporally separated from the act of consumption. The price of armaments is determined on the basis of cost-plus mark-up; it does not vary according to the utility of the weapon because there is no test of utility short of war. The only limit to price is the overall military budget, which is at once the sum of individual decisions about armaments and the reflection of the state's priorities. Similarly, all orders are delivered; shortages or overproduction are mistakes rather than signals of consumer preference. Only war provides a mechanism for reconciling the appropriate quantities of resources to a set of needs. Battle, said Clausewitz, can be compared to the act of exchange. But war is discontinuous; therefore, some kind of proxy mechanism must be developed.

Before exploring this proxy mechanism, it is useful to specify two other general concepts. First, I shall use the concept of the *weapons system*, which was originally developed by the U.S. Air Force in the 1950s. The weapons system, at one level, is a piece of hardware. It consists of the delivery system – the delivery platform (an aircraft, missile launcher, tank, or ship) and the weapon (a munition, such as a bomb, warhead, or shell and its delivery means such as a gun, missile, torpedo, or rocket) – combined with the means of command and communication. It is also a technology, a body of interrelated knowledge, and a set of linked techniques, all needed to develop, produce, and operate the hardware. And it is a social organization of people and institutions who possess that knowledge and are responsible for the technology – scientists, engineers, workers, managers, soldiers, technicians, bureaucrats, etc. They possess their knowledge as members of the social organization and not as individuals.[3]

The second concept is that of *assimilation*. Military technology has to pass through the various phases of invention, innovation, integration, and eventually, obsolescence. At each stage, different constituencies are involved, such as laboratories, corporations, military units, or bureaucratic departments. The stages roughly correspond to appropriation categories in the military budget: invention is the stage of basic and applied research; innovation involves full-scale development and testing; and integration involves procurement, operation, and maintenance. Integration also involves the adaptation of prevailing military doctrine and contingent military organization toward the new technology. A military technology that has passed through the three stages of invention, innovation, and integration is fully assimilated. By and large, a weapons system is a fully assimilated military technology.[4]

In examining the process whereby one weapons system succeeds another, we thus look at the mechanism whereby demand and supply constituencies interact at each stage – the relations that characterize the social organization of the weapons system.

 DEMAND

Explanations for weapons acquisition that focus on the external requirement for armaments – for instance, the Soviet threat – can be described as demand theories. Supply theories or technology-push theories seek the main explanation for weapons acquisition in the web of domestic interests involved in the process of acquisition.[5] An important subset of supply theories consists of models of bureaucratic politics, in which the acquisition of weapons is explained in terms of the complex interplay of different armed services and government departments.[6]

Demand theories are primarily concerned with the *use* of armaments, as though there were no distinction between use and acquisition, as though requirements defined by some external situation could be immediately translated into resources through the agency of some 'rational' decision maker. If, however, the act of purchase is separated in time for the *use* of the weapons, then the peacetime demand for weapons has to be assessed in terms of the situation of purchase as well as the potential or imagined situation of use. For this reason, I include bureaucratic-politics theories as part of the demand side of the model. The potential requirement for weapons as defined by the international situation can be said to be the *systemic aspect* of demand. How this is mediated by the perceptions of the armed services, various bureaucratic departments, and politicians represents the *institutional aspect* of demand.

A more complete analysis of systemic aspects of demand would require an assessment of theories of international conflict and the current global order as well as an analysis of the state and the way in which competing priorities are established. Some introductory remarks are relevant, however. It is worth noting that the imagined contingencies in which weapons might be used in the period 1945–90 remained relatively static. Of particular interest is the pervasiveness of the European contingency (and, to a lesser extent, the Pacific contingency) in American defense planning despite the fact that nearly all wars between 1945 and 1990 took place outside of Europe. Was this merely an instance of bureaucratic conservatism, or did it reflect genuine concern about the situation of Europe?

My hypothesis would be that there exists a systemic interest in sustaining the significance of the European and Pacific contingencies. 'Wars,' says Raymond Aron, 'still have in our time the function of precipitating the crystallisation of political entities.' By keeping the memory of wars alive, the postwar political entities were preserved. In other words, the continued emphasis on the European and Pacific contingencies was a way of reminding enemies and allies of the political results of World War II. It suggests that the experience of World War II might be repeated, with a comparable set of victors. The persistence of naval doctrines based on the experience of Trafalgar in 19th-century British military planning might be explained in the same way. Of course, states that challenge the existing political entities might have an interest in alternative contingencies, and that is obviously the case in the third world. But the hegemonic military powers, who are the largest military spenders and who tend to dictate styles of arms procurement, have an inherent tendency to define and anticipate current contingencies based on past experience.[8] In other words, there may be systemic interests in a conservative set of military preferences.

Whether or not this hypothesis can be sustained, it is reinforced by bureaucratic politics. The overwhelming conservatism of military institutions, armed services, and bureaucratic departments is the theme of much of the bureaucratic-politics literature. Institutional resistance to major innovations such as the longbow, the breech-loading gun, the machine gun, the tank, the submarine, and the airplane has been extensively documented.[9] The old aphorism that generals always fight the last war now has a basis in organizations theory.

Conservatism can partly be explained by the uncertainties that exist in the absence of the test of war. Since it is only with hindsight that we can establish the significance of major innovations like those mentioned above, it is not surprising that military men cling to technologies that have been tested in war. Moreover, there is an organizational interest in conservatism. The technologies embodied in weapons systems entail particular forms of social organization. New technologies pose a risk for organizational survival.

Each military unit is associated with a certain military mission, and certain capabilities are required to carry out that mission. Coulam points out that each of the armed services tends to have a set of 'standard scenarios' in which these missions would be performed. One of these standard scenarios becomes the 'dominant scenario' of the organization. The persistence of the deep interdiction role of the U.S. Air Force, or the emphasis on aircraft speed, which was thought to be so important in World War II, can thus be explained in terms of the way organizations subjectively define their tasks. In his case study of the F-111 aircraft, Coulam shows that

McNamara, as Secretary of Defense, failed to question Air Force requirements that he had earlier opposed:

> When he assumed office anxious to expand limited war capabilities,
> McNamara was not presented with a wide range of options covering the
> whole variety of possible approaches to the tactical air mission. Instead,
> he received only two options; those of the Navy and of the Air Force.
> These options represented years of effort by each of the services and
> hundreds of agreements among interested parties within each service.
> They were not static, hypothetical abstractions. They were firmly on
> track, difficult to modify or to stop. They were powerful forces for
> persuasion, with a whole array of organizational resources behind
> them, generating enough information and heat to overwhelm the
> consideration of *imagined* alternatives lacking organizational sponsor-
> ship.[10]

The point here is that every alternative is an imagined one in a peacetime
situation. It is therefore not surprising if institutionalized past experience
plays a predominant role in shaping the imagination.

The dominant scenarios are, by and large, related to an imagined global
war, which is derived from the World War II experience. That war was the
last great test of military utility; it determines the present shape of military
organization. Thus, both systemic and institutional factors would suggest a
static demand function *between* wars. In other words, in contrast to economic consumer preferences, the weapons succession model in peacetime
would assume a static set of consumer preferences.

It could be argued that one of the lessons of World War II was the
importance of technical superiority. If the political results of that war are
to be preserved, then the West (as a proxy for the Allies) must display a
capability for winning a future equivalent of World War II at a more
advanced level of technology. Likewise, the military organization, having
learned the lesson of World War II, must incorporate technical change into
its organizational routines. To some extent, that is the case. Janowitz refers
to 'trend innovation,' whereby technical change is routinized within the
organization.[11] Coulam's notion of 'selective incrementalism' refers to the
linear improvement of parameters, like aircraft speed, that are defined as
important by the dominant scenario.

The emphasis on technical change was much less pronounced in the
Soviet Union than in the West. Institutionally, the same tendencies for
conservatism could be observed in the Soviet armed forces as in the Western
armed forces, and the same interests in preserving the political results of
World War II can be intuited. Many writers, however, have noted a much
greater doctrinal emphasis on simplicity and quantity in the Soviet armed
forces than in the West. This emphasis could be as much a lesson of World
War II as is the emphasis on new technology. The mass production capabilities of the United States and the Soviet Union were probably, in the last
analysis, the decisive factor in World War II. The fact that this emphasis was
retained in the Soviet Union but not in the West suggests that the difference

has to be explained in terms of differences in the nature of supply institutions.

Bureaucratic-politics models are descriptive rather than explanatory, and the only thing they can predict is inertia – that the future will look much like the past. They cannot explain technological change. If we presuppose a competitive model in which each institution competes for its share of the budget, and in which its competitive strength is proportionate to its size (i.e., its past share of the budget), then we can expect institutional stability. Indeed, this kind of bureaucratic model, sometimes known as input-output conservatism, has been well documented for the Soviet Union.[12] Technological change would tend to threaten such conservatism. Most theorists do, of course, include advocacy of technical change as a determinant of bureaucratic competition, but this is not endogenous to the model. In effect, what these theorists are explaining is how bureaucratic politics mediates an impetus for technological change. Such an impetus could come from the political leadership (for demand reasons), or it might be prompted by other domestic institutions.

In other words, even if limited 'evolutionary' technological change were to occur in the absence of other pressures for technical change, we could still conclude that major impulses for radical technical change must come *either* from supply-side factors *or* from wars. In other words, in peacetime, supply-side pressures explain the weapons succession process.

► SUPPLY INSTITUTIONS

There are two types of supply institutions: those associated with the *invention* stage of the weapons succession process and those associated with the *innovation* stage. The former are primarily government, university, or private nonprofit laboratories. From these emerge new military technologies, some of which may be 'revolutionary' in the sense that they challenge existing doctrine and organization. Such, for example, are the technologies that are now being developed from the sciences of the 1960s and 1970s, approximating nuclear weapons in their implications. Many of these technologies emerge from the defense laboratories, and sometimes acquire maverick constituencies within the armed forces. In general, however, they are likely to make slow progress through the institutions involved in innovation and integration; as in the case of earlier 'revolutionary' technologies, their assimilation tends to be resisted.

At the innovation stage, it is necessary to categorize different institutions for development and production.[13] In the West, prime contractors generally undertake responsibility for development and production of complete weapons systems. Such enterprises are 'sovereign': they are responsible for their own financial viability and are free to secure the necessary financial resources through obtaining contracts wherever possible. Sovereign enterprises are usually private companies, but some European nationalized companies, such as *Aerospatiale* in France, were expected to behave as though they were sovereign.

Typically, however, Western prime contractors are also 'dependent,' in the sense that they depend on the military sector for the bulk of their contracts. In cases where contractors are financially diversified, it is more correct to say that their defense divisions, which are expected to be financially viable, are dependent. Sovereign and dependent enterprises must obtain continuous contracts to ensure capacity employment. This is particularly important with respect to research and development. In practice, production capacity fluctuates substantially; indeed, sales and employment fluctuation tends to be greater in the military sector than in other sectors. Research and development, on the other hand, tend to utilize more specialized equipment and more highly skilled workers than does production. Companies that depend on their technological capabilities in order to obtain contracts cannot afford to disband design teams. Hence, whereas military units may be preoccupied with continued *operation*, the main supply enterprises are concerned with continuous *development* and, to a lesser extent, *production*.

In order to ensure continuous development, these enterprises must obtain new orders as soon as development of a particular weapons system is completed. Because the enterprise is dependent, the new orders are for new weapons systems. This is the process known as the follow-on imperative.[14] The effort to obtain follow-on orders involves intense competition between the main contractors. This competition takes on a technological rather than a price form: it is directed toward product rather than toward process improvement. Because the price is determined on the basis of cost-plus mark-up, a low price would reduce the total size of the market. Once the contract is awarded, the government has little choice but to go ahead with development regardless of cost; hence, perpetual cost overruns. That is why the companies compete by offering technological improvements that will appeal to their customers, the armed forces.

These improvements, however, are severely constrained by the organizational rigidities of the armed forces. New technologies can only get through the innovation and integration stages if they conform to the requirements of the dominant scenario. So long as they are directed toward improvement in the performance of missions that were established nearly 40 years ago and are currently defined through a set of performance characteristics or parameters (speed, payload, or protection) that have hardly changed, the technologies may be extremely radical in hardware terms – microprocessors, nuclear devices, directed energy, variable geometry aircraft, and so forth. I have termed this contradictory conservative but dynamic form of technological change 'baroque.'[15]

There are other types of supply institutions in the West. The arsenal system – the in-house army factories and navy shipyards – consists of enterprises that are dependent but not sovereign. The royal ordnance factories and royal dockyards in Britain and the *directions techniques* in France were similar types of enterprises. Their capacity was guaranteed by the state; excess capacity appeared as a charge on the military budget. In contrast to sovereign enterprises, the arsenal system appears, historically, to be resistant to change and responsive to the conservatism of its customers. The perturbations caused to the arsenal system by war are vividly described in James Fallows's story of the M-16 disaster during the Vietnam War, when the traditional resistance of the 'small-time, insular, old-fashioned' arsenal system to a new rifle led to the development of the M-16, which jammed continually, and consequently cost many lives.[16]

A final type of military supply enterprise to be found in the West is both sovereign and independent. It consists of private companies that are not dependent on government orders to secure their financial viability: certain electronics companies in the United States, as well as electronics and engineering companies in Japan and West Germany. Some revolutionary technologies do appear to have emerged from this type of enterprise – for example, small missiles for use against aircraft, tanks, and ships; remotely piloted vehicles (RPVs); and cluster munitions. There have also been process improvements leading to cheaper, simpler types of aircraft and armored vehicles. This type of enterprise is more usual in nonhegemonic states such as West Germany and Japan, for reasons that can easily be deduced.

In the former Soviet Union, research institutes, design bureaus, and production plants were organized as separate entities, although they all came under the responsibility of nine defense industry ministries, most of which were established in the late 1940s.[17] The research institutes that were attached to the central ministries compare with the various nonprofit laboratories in the West that are associated with the *invention* stage of the weapons succession process.

The Soviet design bureaus and production plants were associated with the *innovation* stage. The former enjoyed varying degrees of autonomy, depending on the level of technology. For low-technology items such as tanks, artillery, and ships, the design bureaus were attached to the production plants. In aviation, they were relatively independent. And in the Ministry of General Machinery, which makes ballistic missiles, the design bureaus appeared to be attached to research institutes.

Neither the design bureaus nor the production plants were sovereign enterprises; their budgets were guaranteed by the state. In the case of the design bureaus, about 25 per cent of capacity might fluctuate, depending on specific contracts. To some extent, therefore, competition between the design bureaus was significant. Moreover, successful designs which led to production contracts could buy prestige, state prizes, and even monetary rewards. However, the possibilities for revolutionary designs were severely constrained by the doctrinal emphasis of the armed forces on simplicity and quantity on the one hand, and by the conservatism of the production plants on the other.

In the production plants, civilian production was deliberately used to cushion military production. Excess capacity, if it existed, was a mistaken or a deliberate consequence of planning. The conservatism of both military and civilian production enterprises has been widely documented.[18] Technical change was resisted because it disrupted the fulfillment of quantitative

	Sovereign	Non-sovereign
Dependent	'Baroque'	Conservative
Independent	Revolutionary	?

Figure 29.1 Technical change and the characteristics of supply institutions

planning indicators on which future budgets might depend, and disrupted established supply lines.

The consequence was what some writers have described as 'evolutionary' or 'conservative' technological change – technological change that was incremental *both* with respect to performance characteristics *and* with respect to productive technology. Whereas 'baroque' technological change may disturb the social organization of producers and developers, conservative technological change preserves the overall social organization of the weapons systems.

One can explain the nature of peacetime technical change with respect to the character and possible changes in supply institutions as shown in the following matrix (see Figure 29.1). There were aspects of demand in the Soviet Union that might temper this conclusion, however. First, decisions about armaments were made at a much higher governmental level than in the West. Defense industry alumni have testified to the personal interest taken by Soviet leaders – especially Stalin and, later, Defense Minister Ustinov – in procurement decisions.[19] Second, because the state was ultimately the consumer, a much greater degree of consumer sovereignty existed in the military sector than in other sectors of Soviet society. Defense was a privileged sector, able to cut through red tape, commandeer machinery and parts, and so forth; military representatives, known as *voyenpreds*, were located at production plants to prevent bottlenecks and ensure standards of quality. Consumer preferences, even if they were static, were thus transmitted to producers. Finally, there were instances in which the political leadership used its role as a powerful consumer to introduce whole new technologies, representing entirely new social organizations – for example, the atomic bomb, the ICBM, and the air defense programs.

For this reason, defense was regarded as one of the few sectors in the Soviet Union in which radical technological change was possible. New technologies did not, by and large, disturb existing organizations; they were simply *additional* organisations. They derived neither from war nor – because Soviet institutions constrained indigenous innovation – from new internal discoveries. Rather, they were 'learned' from the West and tended to introduce Western social organizations into the Soviet Union. Since they were developed in the West through a process of baroque technical change, they introduced baroque organization into an otherwise conservative system.

▶ **DEMAND VERSUS SUPPLY**

Since, in peacetime, the act of purchasing weapons is separated from the act of consumption, there is no mechanism for assessing the appropriate quantity of resources required to fill potential military tasks. A static set of consumer preferences, established in previous wars, is combined with dynamic technological competition between suppliers of arms to produce 'baroque' technological change.

Baroque technological change represents 'improvements' to successive weapons systems which can pass through the phases of invention, innovation, and integration without disturbing the social organization of the users, although it can and does disturb the organization of developers and producers. The prime contractors remain stable, but the composition of subcontractors changes radically. Because baroque technical change is incremental with respect to performance, it preserves user organization, although it may be radical with respect to productive technology. This finding of 'step-ladder' technical change, combining the dynamics of the armorers with 'persistence of doctrines geared to earlier plateaux,' has been observed in previous peacetime arms races.[20]

The cost of marginal improvements to a given set of performance characteristics tends to increase sharply beyond a certain performance level.[21] 'Baroque' technology is characterized by soaring cost, increased technical complexity, and diminishing improvements in performance. Indeed, it is often argued – and apparently borne out by the experience of war – that the utility of 'baroque' technology has actually declined because of complexity, unreliability, lack of durability, logistical problems, and vulnerability.[22]

In the Soviet Union, the prevailing form of technical change was 'conservative.' Conservative technical change is incremental with respect to both performance and production. A static set of consumer preferences is combined with a linear evolution in supplier capabilities. Technological change preserves both user and supplier organizations. A form of reaction did appear to introduce 'baroque' tendencies into Soviet military technology, however.

'Revolutionary' technology challenges both 'baroque' and 'conservative' technology – the social organization of the users of weapons systems as well as the social organization of developers and producers. The wars in Vietnam and in the Middle East, as it were, impelled some 'revolutionary' technologies from the invention stage to the innovation stage, calling into question the utility of prevailing types of military technology. Examples are so-called 'smart' weapons and area destruction weapons such as cluster munitions.[23] This finding parallels the conclusion in Freeman *et al.* for civilian technology systems: 'Exogenous science and new technology tend to dominate in the early stages while demand tends to take over as the industry becomes established.'[24] The prevailing military-industrial institutions do not, however, adapt to the impact of war any more rapidly now than in the past. The Vietnam War was not considered relevant to the

European contingency. The Falklands War, although it seems to have demonstrated the vulnerability of modern warships, was used as a justification for aircraft carriers. The pace of assimilation is determined by the intensity and duration of war on the one hand, and by the entrenched rigidities of prevailing military-industrial institutions on the other. In broader, more systemic terms, we can say that the pace of assimilation is determined by the extent to which war can challenge the prevailing global order and industrial culture in which the military-industrial institutions are embedded.

Is there a fourth category, that of *non-sovereign, independent* institutions? Is it possible to conceive of enterprises whose budget and employment are guaranteed by the state, but which are responsive to non-governmental consumer preferences? Such institutions would be designed to be flexible and responsive to change without involving the brutality of the market mechanism.

► SUMMARY AND CONCLUSION

This analysis of the weapons succession process began with the concept of a weapons system that combines the technology of development, production, and use of particular families of hardware. Technology has been defined as a body of knowledge possessed by individuals by virtue of their membership in particular social organizations. The weapons succession process thus can be understood or explained in terms of a set of social organizations that are concerned with the supply of and demand for weapons, how they relate to each other, and how they compete and co-operate to produce technological change.

I have argued that, in the acquisition of weapons, the mechanism for reconciling demand with supply is war. In the absence of war, a proxy mechanism is developed in which consumer preferences are defined subjectively by the institutions of demand, while supply possibilities emerge from the operational requirements of the supply institutions. The interests of both demand and supply institutions are, moreover, shaped by their wider systemic roles.

Throughout most of the post-war period, demand institutions were extremely stable, leading to inherent conservatism in the acquisition of weapons. Supply institutions were not so stable, however; the requirements varied according to their nature. In describing three different types of supply institutions, I have shown that they generate different types of dynamic impulse for technological change. In the West, the predominant large private corporation, dependent on government orders, creates an 'industrial imperative' for rapid product improvement. In the East, the supply institutions were fundamentally bureaucratic organizations with the same inertial tendencies that are characteristic of demand institutions; therefore, there was not the same pressure for technological change.

The contradictory combination of stable consumer preferences and dynamic production imperatives may be characterized as 'baroque' technological change. Ultimately, such technological change challenges the stability of demand institutions and their role in the wider global system; it also challenges the continuity of supply institutions themselves. One set of implications includes the growing frequency of limited wars and the emergence of new types of supply institutions, especially in such countries as West Germany and Japan, which are capable of developing 'revolutionary' technology in response to radical changes in demand brought about through war. The countervailing pressures for arms limitation and appropriate institutional changes are an alternative set of implications. Future research needs to focus on the tension between various aspects of the weapons succession process and the ways in which these tensions could be relieved through arms limitation policies.

▶ NOTES

1 Luigi L. Passinetti, *Structural Change and Economic Growth: A Theoretical Essay on the Dynamics of the Wealth of the Nations* (Cambridge: Cambridge University Press, 1981).
2 Janos Kornai, *Growth Shortage and Efficiency* (Oxford: Basil Blackwell, 1982).
3 Mary Kaldor, *The Baroque Arsenal* (London: André Deutsch, 1982).
4 Julian Perry Robinson, 'Arms Control and the Assimilation of Chemical Weapons,' *International Journal* (Toronto, Canadian Institute of International Affairs), 36 (Summer 1981), 515–34.
5 For an elaboration of this distinction, see Ray Curnow and others, 'General and Complete Disarmament: A Systems Analytic Approach,' *Futures* (October 1976), 384–96.
6 Classic examples of bureaucratic politics theories include Graham T. Allison, *Essence of Decision: Explaining the Cuban Missile Crisis* (Boston: Little, Brown, 1971); Morton Halperin, *Bureaucratic Politics and Foreign Policy* (Washington: Brookings Institution, 1974); John D. Steinbruner, *The Cybernetic Theory of Decision: New Dimensions of Political Analysis* (Princeton: Princeton University Press, 1974).
7 Aron, 'War and Industrial Society: A Reappraisal,' *Millennium. Journal of International Studies* 7 (Winter 1978/79), 195–210.
8 I have explored this argument in 'The Concept of Common Security,' in Stockholm International Peace Research Institute (SIPRI), *Policies for Common Security* (London and Philadelphia: Taylor & Francis, 1985), 37–52.
9 See, for example, Graham T. Allison and Frederick Morris, 'Armaments and Arms Control: Exploring Determinants of Military Weapons,' *Daedalus* 104 (Summer 1975); Bernard Brodie, 'Technological Change, Strategic Doctrine, Political Outcomes,' in Klaus Knorr, ed., *Historical Dimensions of National Security Problems* (Lawrence: University Press of Kansas, 1976); John Ellis, *The Social History of the Machine Gun* (New York: Pantheon Books, 1973); Barton C. Hacker, 'Resistance to Innovation: The British Army and the Case Against Mechanization,' *Actes du XIIIième Congrès International d'Histoire des Sciences* 2 (1974); Irving

B. Holley, Jr., *Ideas and Weapons: Exploitation of the Aerial Weapon by the United States in World War II. A Study in the Relationship of Technical Advance, Military Doctrine and the Development of Weapons* (New Haven: Yale University Press, 1953).

10 Robert F. Coulam, 'The Importance of the Beginning: Defense Doctrine and the Development of the F-111 Fighter-Bomber,' *Public Policy* 23 (Winter 1975), 1–37, at 25 (emphasis added).

11 Morris Janowitz, *The Professional Soldier* (New York: Free Press, 1960).

12 Alec Nove, *The Soviet Economic System* (London: George Allen & Unwin, 1977).

13 See Mary Kaldor, 'Military R & D: Cause or Consequence of the Arms Race?' *International Social Science Journal* 35 (No. 1, 1983), 25–45.

14 The term was developed by James A. Kurth; see his 'Why We Buy the Weapons We Do,' *Foreign Policy* No. 11 (Summer 1973), 33–56.

15 Kaldor (fn. 5).

16 James Fallows, 'M-16: A Bureaucratic Horror Story,' *Atlantic Monthly* 247 (June 1981), 56–66. [See also pp. 382–94 of this volume.]

17 The ministries were Aviation, Defense Industry, Shipbuilding, General Machinery, Medium Machinery, Radio, Electronics, Communications, and Machine-Building. For a detailed description, see David Holloway, 'Innovation in the Defence Sector,' in Ronald Amann and Julian Cooper, *Industrial Innovation in the Soviet Union* (New Haven and London: Yale University Press, 1982), 276–366.

18 Nove (fn. 14); Arthur J. Alexander, 'Decision-Making in Soviet Weapons Procurement,' *Adelphi Papers No. 47 and 48* (London: International Institute for Strategic Studies, Winter 1978/79); Ronald Amann, Julian M. Cooper and R. W. Davies, eds, *The Technological Level of Soviet Industry* (New Haven and London: Yale University Press, 1977).

19 See the autobiography of one of the most famous aircraft designers, A. S. Yakovlov, *The Aim of a Lifetime* (Moscow: Progress Publishers, 1972). A similar point was made by a leading radar designer, Anatol Fedoseev, 'Design in Soviet Military R & D: The Case of Radar Research in Vacuum Electronics,' mimeo., seminar given at Center for Russian Studies, Harvard University, March 1983.

20 Christopher Harrie, 'Technical Change and Military Power in Historical Perspective,' a 'New Conventional Weapons and East-West Security,' Part 1, *Adelphi Paper No. 144.* (London: International Institute for Strategic Studies, Spring 1982), 5–13.

21 This tendency was originally noted by the German economist, Wolff, in 1912. See Christopher Freeman, John Clark, and Luc Soete, *Unemployment and Technical Innovation: A Study of Long Waves and Economic Development* (London: Frances Pinter, 1982).

22 See, for example, Kaldor (fn. 5); Chuck Spinney, *Defense Facts of Life* (Washington, DC: Program Analysis and Evaluation Division, Department of Defense, 1980); James Fallows, *National Defense* (New York: Random House, 1981).

23 Julian P. Perry Robinson, 'Qualitative Trends in Conventional Munitions: The Vietnam War and After,' in Mary Kaldor and Asbjörn Eide, eds, *The World Military Order: The Impact of Military Technology on the Third World* (London: Macmillan, 1979), 64–109.

24 Freeman (fn. 26).

30 ▶ Theories of technology and the abolition of nuclear weapons

Donald MacKenzie

In memory of Agatha C. Hughes

If theories are to be more than ways of speaking, they must help us understand concrete situations. In this chapter, I invite the reader to take part in a thought experiment designed to examine this dimension of theoretical perspectives in the social studies of technology.[1] Imagine that a prime minister or president is examining the feasibility of the permanent, worldwide abolition of nuclear weapons. She commissions you to tell her what the social studies of technology has to say about this. 'You talk much, using much jargon, about the emergence of new technologies,' she says. 'Can you not tell us, simply and clearly, what your field has to say about getting rid of them?'

I propose this as an exercise in thinking through the practical meaning of the theoretical perspectives with which our field abounds. I sketch my own answer below, focusing on four important sets of perspectives:

1 technological systems theory, as developed above all by Thomas P. Hughes;
2 cognitive perspectives, especially notions of tacit, local knowledge, as developed, above all, by Harry Collins;
3 'structural sociology', in which broad category I include Marxism, modernization theory and patriarchy theory;
4 actor-network theory, as developed by authors such as Michel Callon, Bruno Latour and John Law.

The need for brevity means that I will be oversimplifying these perspectives, and ignoring other approaches that could be drawn on: for example, 'cultural studies' perspectives; anthropological perspectives; Bijker and Pinch's 'relevant social groups' approach; and perspectives inspired by ethnomethodology or discourse analysis. I ask only this of those who are appalled by my theoretical crudity: contribute to the discussion by showing how a more faithful, or a more diverse, theoretical analysis adds to our understanding of the feasibility of nuclear disarmament.

The *desirability* of abolishing nuclear weapons is my focus in only one, restricted sense. I do not attempt to weigh the risks of abolition (might the probability of conventional war increase? might states be more likely to turn to chemical or biological weapons?) against the risks of their indefinite continued presence (nuclear weapons accidents? theft of nuclear weapons? nuclear war through miscalculation or irrationality? and so on). Instead, my concern is with what in most discussions of comprehensive nuclear disarmament is taken as the dominant risk of the endeavour: that an abolition agreement is either circumvented by a state hiding away some weapons or is subsequently broken by nuclear rearmament.

I make a number of assumptions about nuclear disarmament. First, I assume it would be implemented gradually, perhaps over a period of 20 to 30 years (see, for example, MccGwire 1994); it would, for example, be over that kind of period that a weapon would have to be hidden away. Second, I assume that any plausible nuclear disarmament agreement would involve more than just the dismantling of existing arsenals. I assume it would include acceptance of intrusive inspection and monitoring of suspected nuclear facilities, at least as intrusive as the mechanisms in place under current arms limitation treaties and the Nuclear Non-Proliferation Treaty. Third, I assume a prohibition on possessing or testing long-range ballistic missiles (the most important 'dedicated' nuclear weapons delivery systems), and possibly a ban on long-range cruise (air-breathing) missiles as well. Fourth, I assume the existence of a well-resourced intelligence effort, including use of both human intelligence and remote surveillance, to search for violations such as the construction of unmonitored nuclear facilities. Finally, I assume that 'breakout' from a treaty to eliminate nuclear weapons would have to be covert. In other words, I assume that if attempts at violation were detected, there would be both the capacity and the willingness to take effective, limited military action against them (akin to the 1981 Israeli air strike that rendered inoperable the nuclear reactor under construction at Osirak in Iraq).

▶ NUCLEAR WEAPONS FROM A 'SYSTEMS' PERSPECTIVE

Modern technical artifacts are typically parts of technological systems. Systems, Tom Hughes asserts, manifest growth patterns driven above all by the need to maximize load factor – the ratio of average usage (which determines revenue) to peak usage (which determines the necessary level of capital investment). Technical innovation focuses around the elimination of the obstacles to growth – 'reverse salients' as Hughes calls them (Hughes 1983).

The systems perspective implies that technologies cannot be understood in isolation, but only in their contexts, especially their systemic contexts.

The outside world, for example, might have detected Saddam Hussein's nuclear weapons programme earlier had this been kept in mind. What helped keep the Iraqi programme secret was that Iraq adopted an apparently outdated method of uranium enrichment: electromagnetic isotope separation.[2] This places enormous demands upon electricity supply: the uranium separation plant at Oak Ridge, Tennessee, consumed more electricity in 1945 than the whole of Canada produced in World War II (Albright and Hibbs 1991: 19). Both the International Atomic Energy Agency (responsible for Non-Proliferation Treaty inspections) and US intelligence seem implicitly to have taken for granted that no one would adopt this method of separation, so its main elements were no longer classified, there were no controls over its component technologies, and no one seems to have monitored purchases of items of these technologies (Davis and Kay 1992). Viewed in isolation, then, electromagnetic separation was indeed an inferior and superseded technology, but viewed in the Iraqi context, in which energy was plentiful but secrecy a priority, it was an eminently sensible choice.

Nuclear weapons as the products of technological systems

If artifacts are parts of a technological system, then they can often be rendered useless by disconnection from that system: a washing machine, say, does not work if unplugged from its electricity supply. If nuclear weapons are products of, and parts of, large technological systems, could they be rendered powerless by being unplugged from those systems?

To make even a single nuclear weapon requires quantities of the order of a kilogram or more of materials that either exist in nature only in minuscule amounts (plutonium) or are more common but intermingled with other materials (uranium-235, which has to be separated from the more common isotope, uranium-238). Although state-of-the-art plutonium production and uranium separation are no longer the massive industrial activities they were at the time of the Manhattan Project, the capacity to detect them is at the core of the Nuclear Non-Proliferation Treaty.

Unfortunately, control over the production of plutonium and enrichment of uranium, though clearly important, would on its own be a shaky underpinning of comprehensive nuclear disarmament. The chief problem is the existence of large, poorly documented, stockpiles of these materials. The US military stockpile of plutonium, for example, stands at 99.5 metric tons, but 111.4 tons have been produced or otherwise acquired by the USA over the half century of the nuclear age. The best estimate of the amount removed from the inventory by nuclear tests, radioactive decay, accidents, transfers and losses in production and handling (in the lining of pipes, ventilation ducts, glove boxes, and so on) is 9.1 tons. The remaining 2.8 tons cannot be accounted for (Fetter 1996). No one seriously believes that it has been stolen; the discrepancy is far more likely to be the result of errors in estimation and bookkeeping. Nevertheless, 2.8 tons of plutonium is the equivalent of more than 1000 nuclear weapons, and uncertainties of this

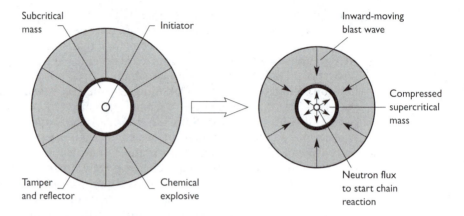

Figure 30.1 A simple *atomic* or *fission* bomb, of the standard 'implosion' design. Detonation of a shell of chemical high explosives suddenly compresses a core of plutonium and/or enriched uranium, making the core sufficiently dense to sustain a chain reaction in which neutrons split atoms, releasing large amounts of energy and generating more neutrons which split more atoms, and so on.

In a more sophisticated *boosted* weapon, the initiator is external, and two gases – tritium and deuterium – are injected into the heart of the core as it implodes. The neutrons generated by their fusion greatly increase the efficiency of the fission chain reaction. An atomic bomb can also be turned into a *hydrogen* bomb by adding what is called a 'secondary,' which contains thermonuclear 'fuel': typically lithium deuteride. Neutrons released by the initial fission explosion bombard the lithium, transforming it into tritium, and the extreme pressure and heat cause the tritium to fuse with the deuterium, releasing an enormous amount of energy.

magnitude (uncertainties which are almost certain to be far larger in the case of the former Soviet Union) are clearly very disturbing from the point of view of ensuring that existing stockpiles of fissile materials are adequately accounted for.

However, the 'system dependencies' of nuclear weapons are not restricted to plutonium and highly enriched uranium. Conventional designs of atomic bombs involve an 'initiator', which produces a sudden, intense burst of neutrons to start the chain reaction (see Figure 30.1). The most common design of initiator in unsophisticated nuclear weapons employs polonium. The initiator is crushed by the compression of the weapon's core, mixing the polonium with beryllium: the alpha particles that the former emits in large quantities strip neutrons from the latter. Polonium is made by irradiating bismuth in a nuclear reactor. It was, for example, being made in the pile at Windscale that caught fire in October 1957, in Britain's worst nuclear accident (Urquhart 1983).

Polonium-210 has a half-life[3] of only 138 days, so initiators containing it have to replaced frequently, and stocks of the substance are of limited value.

In part because of this, polonium is no longer used in the initiators of sophisticated nuclear weapons. However, these weapons are vulnerable (albeit more slowly) to the decay of tritium, a rare radioactive gas which is used both in their initiators and for 'boosting' (enhancement of a fission chain-reaction by injection of gaseous fusion elements).[4] Tritium's half-life is 12.3 years, so the tritium reservoirs (small pressure vessels containing tritium) of nuclear weapons need to be replaced after they have been in service for some time. Without replacement, there will come a time when boosting of a hydrogen bomb's fission 'primary' will be inadequate to trigger fusion in the 'secondary' that gives it most of its destructive force. At its most extreme, tritium decay in the primary's initiator might mean no significant nuclear explosion. Tritium decay is already a pressing issue for the USA. Since April 1988, it has had no means of producing tritium. The arms reduction process has prevented the situation from becoming a crisis by making available the contents of the tritium reservoirs of decommissioned warheads, but according to Department of Energy estimates that source will become inadequate by 2011.

Tritium, an isotope of hydrogen which exists in nature only in trace quantities, is again the product of large technical systems: a replacement tritium source could cost the USA $3000 million to $6000 million. Hitherto, tritium has been made by irradiating lithium-6 in specially-modified nuclear reactors, followed by separation in a specialized plant;[5] the cut-off of US tritium production was the result of safety problems with the reactors at Savannah River, South Carolina. The Department of Energy is considering an alternative route, but that is equally large-scale: constructing a large underground particle accelerator, powered by a 350 megawatt electricity supply, the equivalent of that of a small city (US Department of Energy 1995).

It is impossible, on the basis of the open literature, to be sure just how quickly and how completely a sophisticated nuclear arsenal, denied access to tritium, would succumb to the effects of tritium decay. Clearly, the vulnerability of different weapon designs will vary. The only overall estimate I know of is in a 1981 Los Alamos report, which says that a cut-off of access to tritium 'would mean the certain disablement, in approximately a decade, of one half of the [US] nuclear weapons stockpile that uses tritium' (Peabody 1981: 8).[6] However, provided the initiators of nuclear weapons (which seem to require much less tritium than boosting does) keep working, 'disablement' does not mean complete ineffectiveness: at most, it means failure of boosting and failure of the secondary (see Figure 30.1). Unaugmented primary explosions, however, would be enormously less destructive than full thermonuclear explosions. The explosive yield of a modern thermonuclear weapon is typically in excess of the equivalent of 100,000 tons of TNT. An unaugmented primary explosion, however, is likely to be of the order of 500–5000 tons (Kalinowski and Colschen 1995, Appendix A: 4; Collina 1996: 42). These figures correspond to massive explosions (between $\frac{1}{40}$ and $\frac{1}{4}$ of the Nagasaki bomb), but to a significantly less fearsome threat than full-scale thermonuclear blasts.

Tritium has civil uses that amount to a world market of around 400 grams per year (in terms of quantities, the most important usage is in runway landing lights for remote airfields) (Kalinowski and Colschen 1995: 140–1). The contents of the tritium reservoir of a single nuclear weapon – around 4 grams (Mark *et al.* 1988: Note 3) – therefore amount to around 1 per cent of the annual civil world market, and so to the kind of quantities whose diversion from civil uses ought to be detectable.

Nuclear weapons as components of technological systems

A nuclear device is not a weapon unless it can be delivered to its target. For terrorist use, transport in a truck or ship might be adequate (the proverbial suitcase bomb would require substantial skills in miniaturization), and for tactical use nuclear weapons have been fashioned into artillery shells, depth-charges and land mines. The main practical delivery means, however, are either aircraft or missiles.

Aircraft are the less demanding, in that a heavy bomber can carry and drop nearly all designs of nuclear weapon, at least if one is prepared to take off with the bomb fully assembled and live. However, aircraft can be intercepted, a particularly important drawback if a nation has only a small nuclear arsenal. Outside of certain restricted social contexts (most importantly the US air force), missiles, especially ballistic missiles, have nearly always been seen as the best means of delivering nuclear weapons to their targets.

Developing a long-range ballistic missile, and integrating a nuclear weapon into it, are demanding tasks of systems engineering. Indeed, the latter, as a distinct engineering discipline, largely began with the first American intercontinental ballistic missile (Atlas) and first submarine-launched missile (Polaris).[7] Non-proliferation policy latched only belatedly onto this 'system' aspect of nuclear weaponry: the current Missile Technology Control Regime was established, at the initiative of the USA, only in 1987. Ballistic missiles with relatively short ranges, of the order of 600 kilometres – most commonly the Soviet Scud, or derivatives of it – are relatively widely diffused. Nevertheless, moving from these to an adequate intercontinental ballistic missile, using only indigenous resources, is a demanding task: one, for example, that took China nearly three decades (Lumpe *et al.* 1992).

Missile delivery places important constraints on the designer of a nuclear weapon. There is a strict trade-off between the weight of a missile's payload and its range. The most immediate issue is to reduce the quantity of chemical explosives needed to detonate the weapon, since these are responsible for much of its bulk. This increases the demands for skill in the design and fabrication of an implosion system (which uses simultaneous detonation of high explosives, shaped into a lens structure, to produce an inward-moving blast wave and create a supercritical mass by compression: see Figure 30.1). This, for example, was a major problem faced by Iraqi nuclear weapons designers. Although it seems clear that their government wanted a nuclear warhead for its Scud missiles, Iraqi designers seem to have lacked confidence that they could achieve the requisite compression at the required weight. The design on which they were working when the 1991 Allied bombing interrupted the programme was still too heavy to be carried by a Scud, but the compression of the fissile material in that design was judged by them to be the best they could achieve (and, indeed, they were still some way from practical implementation of that design) (Albright and Kelley 1995: 63).

The need for warhead miniaturization typically grows as one moves from short-range to long-range missiles. The only way to avoid it is to build extremely large missiles (like the first Soviet ICBM, the SS-6), but this has drawbacks. It complicates silo basing, and largely rules out what is generally agreed to be the most secure deployment technique: submarine basing. So the miniaturization of nuclear warheads was a key goal in the USA, and, eventually, in the Soviet Union too.

Miniaturization places considerable demands on designers' skills, demands that even in the established nuclear weapons states have not always been met. For example, a warhead with a high yield-to-weight ratio was crucial to the first US submarine-launched ballistic missile programme, Polaris, and had been promised by bomb physicist Edward Teller. Despite imaginative systems engineering (instead of fitting a warhead inside a re-entry vehicle, as had been done in other programmes, the two were designed as an integral unit), designers at the Livermore nuclear weapons laboratory, established by Teller, could not meet the original yield specification. They also discovered in a 1958 test that their design was not one-point safe: a test detonation of the high explosive at a single point (as might occur in an accident) produced a nuclear explosion equivalent in energy to 100 tons of TNT. To make the weapon safer, they inserted a spool of wire into the plutonium 'pit' to prevent an accidental detonation, but found later that the mechanism to withdraw the wire often failed, so many of the first Polaris warheads would have been duds (Spinardi 1994: 55).

More generally, it is miniaturization that makes boosting, and thus tritium, crucial to a sophisticated arsenal. Although more sophisticated mechanical designs and more efficient implosion are important, it is, according to one US weapons designer, boosting that is 'mainly responsible for the remarkable 100-fold increase in the efficiency of fission weapons' since 1945 (Westervelt 1988: 56).

▶ NUCLEAR WEAPONS AND TACIT, LOCAL KNOWLEDGE

Technical knowledge is seldom fully explicit. Tacit knowledge – the kind of knowledge we deploy in being able to ride a bicycle without being able to say precisely how it is that we do so – is often important to the successful design and operation of technologies. Sociologists of science, notably Harry Collins, have suggested that even in the most advanced modern science tacit knowledge is largely a local phenomenon. Tacit skills are often not widely diffused, but are the properties of relatively small groups of people, and are transmitted hand-to-hand and face to face (see, for example, Collins 1974; Ferguson 1977, 1992).

An emphasis on tacit, local knowledge challenges one of the deepest assumptions about nuclear weapons, held by their opponents as well as by their supporters, that: 'the knowledge required to build nuclear . . . weapons can never be eliminated' (Blechman and Fisher 1994/5: 94). In a stimulating recent article, actor-network theorist Michel Callon suggests the following thought experiment, adapted from Herbert Simon: 'Imagine coloring theoretical statements in red, and all other inscriptions and skills incorporated in human beings and instruments, in green. A Martian contemplating our science from its planet would see a vast green ocean streaked with very occasional and fragile red filaments' (Callon 1994: 402).

For my current purpose, I want to alter Callon's vivid metaphor, and apply the same red colour both to theoretical statements *and* to skills incorporated in instruments, while leaving only skills incorporated in human beings as green. What would our Martian now see, particularly if looking not at science in general, but at the knowledge needed to design and build a nuclear weapon?

My reasons for adopting the different 'colouring scheme' are portability and mortality. Theoretical statements and instruments are portable and (within limits, as regards the latter) immortal. Skill incorporated in human beings – and tacit knowledge is, quite literally, incorporated, embodied knowledge, as the German expression *Fingerspitzengefühl* ('fingertip feeling'; intuition) reminds us – is portable only along with its human possessors, and shares their mortality. If humanly-incorporated skill is needed to design and build nuclear weapons, and if that skill is local (i.e., concentrated in nuclear weapons programmes, not widely diffused in other activities), then it could be lost as a consequence of a prolonged hiatus in which no new nuclear weapons were designed and built. While nuclear weapons clearly could be reinvented (if it was done once, it can be done again), 'reinvention' might eventually become the correct term: the necessary tacit knowledge would have to be created afresh.

I have explored elsewhere the balance of explicit and tacit knowledge needed to design and build nuclear weapons, and summarize only briefly here.[8] Three types of evidence are relevant: what the original invention of nuclear weapons reveals about the necessary knowledge; what their spread tells us; and what nuclear weapons designers tell us about the knowledge they deploy.

The history of the Manhattan Project suggests that explicit knowledge – at least explicit knowledge in physics – was not, on its own, adequate to permit an atomic bomb to be designed successfully. The physicists involved in the Manhattan Project thought at first that the difficulties lay in producing the requisite quantity of plutonium or enriched uranium, not in turning that fissile material into a bomb. Physicist Edward Teller recalls being advised by future Nobel laureate Eugene Wigner not to join the new laboratory at Los Alamos because its task – designing the atomic bomb –

would be too easy. Los Alamos was originally planned to be no bigger than the physics department of a large university. However, it ended up needing several thousand staff, many of whom were engaged in a multitude of apparently humdrum engineering design tasks that the physicists had underestimated. Technological skill turned out to be just as important as knowing nuclear physics.

The spread of nuclear weapons, on the other hand, seems to suggest that the knowledge needed to design an atomic bomb was then highly portable. Six nations, in addition to the USA, have successfully exploded nuclear devices, and two others are generally agreed to have – or, in the case of South Africa, to have had – the capacity to do so (see Table 30.1). Other than in the case of Britain (some of whose early designers had taken part in the Manhattan Project), this spread seems not to have taken place by the mechanisms by which tacit knowledge spreads: movement of people, or extended face-to-face, hand-to-hand instruction. However, the spread of nuclear weapons is not decisive evidence against the importance of tacit knowledge. The atomic bomb projects subsequent to the Manhattan Project have something of the character of reinvention, rather than copying based upon explicit knowledge. All took longer than the original process of invention (see Table 30.1); all for which details are known were large, labour-intensive projects involving hundreds or thousands of scientific staff; all the cases for which detailed information is available faced a multitude of practical problems; and the possession of explicit information from previous projects did not ease their tasks dramatically.

Certainly, the explicit knowledge generated by the Manhattan Project spread quickly. Soviet intelligence learnt of the West's atomic bomb research as early as September 1941, and a number of agents subsequently gave the Soviets considerable technical data. (Spying, it is worth noting, is an experiment in the nature of knowledge, because spies that remain in place cannot normally transmit tacit knowledge, but only such knowledge as can be spoken or written down.) In June 1945, the Los Alamos physicist Klaus Fuchs gave the Soviets a sketch, measurements and a detailed description of the plutonium implosion weapon to be tested at Trinity Site and later to devastate Nagasaki. As they now admit, Soviet bomb designers then set out simply to copy the Trinity bomb. Despite the fact that Soviet physicists were among the most sophisticated in Europe, and despite the top priority Stalin gave their work, it took them slightly longer than the Americans had taken to make the original. Producing plutonium in the war-devastated Soviet Union required a massive effort, but also important was the fact that explicit knowledge of the sort passed on by Fuchs and the other agents did not solve all the problems of bomb design. The requisite technological skills had to be created afresh.

Britain, too, began by trying to copy the Trinity/Nagasaki bomb. The task again took longer than the Manhattan Project, led to a design that differed significantly from the original, and involved a multitude of practical problems. For example, the high explosive lenses used to compress the bomb's fissile core had to have a very precise shape. Learning to cast and to machine high explosive with the requisite precision was far from easy, and lenses

Table 30.1 Approximate chronologies of successful nuclear weapons development programmes

	Start of nuclear weapons development programme	Date of first atomic test explosion (*) or weapon (†); type of first device	Date of first thermo-nuclear test explosion (*) or weapon (†)	Significant personal contact with previously successful weapons design team?	Began with attempt to copy previous design?
US	1942	1945*, Pu imp; U gun	1952*	no	no
USSR	1945	1949*, Pu imp	1953*	no	yes
UK	1947	1952*, Pu imp	1957*	yes	yes
France	1955	1960*, Pu imp	1968*	no	?
China	c. 1955	1964*, U imp	1967*	no	no
Israel	c. 1957 (?)	c. 1966†, Pu imp	?†	?	?
India	c. 1964	1974*, Pu imp		?	?
South Africa	1971	1979†, U gun		?	?
Pakistan	c. 1975	1998*, U imp		?	yes (?)

Pu = plutonium; U = uranium; imp = implosion
Note: At the time of writing, India and Pakistan have not yet tested, and seem not to have constructed, thermonuclear weapons; South Africa did not develop them; whether Israel has done so is unclear. Pakistan almost certainly possessed atomic weapons well before its 1998 tests.

always seemed to shrink. A humble remedy, PVC tape, was used to fill in spaces and reduce settlement in early British explosive lenses!

More recent nuclear states have had the advantage of sometimes being able to buy key equipment rather than having to make it. But even those whose purchases were most successful, notably Iraq, found that they also needed time to develop the skills to operate the equipment successfully: for example, Iraq was never able to get its centrifuges and electromagnetic separation plant to work satisfactorily.

Understanding of nuclear explosions is most fully explicit at the weapons laboratories of the nuclear powers, which have developed large computer programs, called 'codes', to assist nuclear weapons design. A modern American code can consist of up to a million lines of program, and runs on the world's most powerful supercomputers. Yet weapons designers at Los Alamos and Livermore, interviewed for this research, were adamant that the codes did not eliminate the need for human judgement. The demands of the miniaturization of nuclear weapons, and especially the consequent reliance on boosting, left their job still a difficult art.

None of the above evidence about tacit knowledge is conclusive. For example, even the most detailed historical treatments of the Manhattan Project leave unclear the balance of tacit and explicit knowledge involved in its technological rather than scientific aspects. The duration of projects is a less than decisive indicator of the hardness of the task, because it is clearly affected also by the financial and material resources available. Nuclear weapons designers may well have their own reasons (celebration of their unusual trade; desire to head off a comprehensive nuclear test ban) for emphasizing the continued need in their work for tacit as well as explicit knowledge.

Furthermore, there is some evidence that the tacit knowledge required to develop nuclear weapons is not entirely specific to existing nuclear weapons programmes. Civil nuclear power programmes also demand skills in the handling and machining of fissile materials and in neutronics (the study of the behaviour of neutrons in fissile materials), understanding of which is necessary for designing both nuclear reactors and nuclear weapons. Perhaps even more crucially (because it appears to be the hardest part of the design of implosion weapons, the type preferred by all post-1945 programmes apart from the South African: see Table 30.1), practical experience of sophisticated high explosive systems can be gained in other spheres, notably the designing of armour-piercing shaped charges for anti-tank warfare.

Finally, the demand for tacit knowledge appears to be at its highest in the design of an atomic bomb (or the fission primary of a hydrogen bomb). Given the availability of both a suitable fission bomb and of the unusual materials necessary to make a secondary (such as lithium-6 deuteride, the standard thermonuclear 'fuel'),[9] then the key to the design task appears to be a piece of explicit knowledge: the Teller-Ulam configuration.[10] The nuclear weapons designers interviewed for this research seemed more confident of the adequacy of explicit understanding of secondaries than of primaries (in particular, boosted primaries, where they saw the demand for tacit knowledge at its peak). In addition, the pattern of the spread of the hydrogen bomb is different from that of the atomic bomb, with three of the four states which made the transition from an atomic to a hydrogen bomb doing so quicker than did the USA. Fortunately, there is no known way of making a hydrogen bomb without first having an atomic bomb. So if designing and building the latter demands tacit knowledge as well as theoretical understanding and instruments (and, despite all the above provisos, I feel the balance of evidence suggests that it does), then this remains an important barrier.

▶ THE SOCIAL PRECONDITIONS OF NUCLEAR WEAPONS

Nuclear weapons are not self-evidently necessary or desirable. Only under certain social conditions do they appear so, and the contribution of what I am calling 'structural sociology' to understanding the abolition of nuclear weapons lies in its potential insights into these social conditions. Might one way of ensuring that the abolition of nuclear weapons was irreversible be to eliminate their social preconditions? After all, if these social preconditions were not present, there might be no incentive to seek nuclear rearmament.

For nuclear disarmament to be irreversible, the precondition that requires elimination would appear to be the risk of major war between industrialized nations. The 'system' and 'tacit knowledge' perspectives suggest that it may be slower and harder for industrialized nations to re-create nuclear weapons capabilities than is often assumed; nevertheless they clearly *can* do so. In a prolonged, major war between industrialized countries, or in a situation where such a war seems likely, the chances of an agreement to abolish nuclear weapons breaking down appear substantial.

Is it wholly utopian to imagine a world in which the probability of major war between industrialized nations is low? 'Realist' political scientists point out that these nations' pre-1945 record is not encouraging in that respect, and they attribute the rarity of such war since 1945 to the role of nuclear weapons in making the likely costs of such war outweigh any benefits. Yet it is important to note that this mutual vulnerability of industrialized nations is not the result of nuclear weapons alone. Imagine, for example, the consequences of a bombing raid (using conventional bombs) on a fuelled-up nuclear power station, or reprocessing plant. Such facilities certainly are plausible targets, if only because of their potential role in nuclear rearmament. Furthermore, the changes in war's cost-benefit equation are not simply on the cost side. For example, the mutual benefits of the European Union have been such that most observers would agree that war between those traditional foes, France and Germany, is now unlikely. If the pattern of mutually beneficial interconnectedness and interdependence were to become well entrenched worldwide, then that would be grounds for hope.

Here we stand on one of the classic divides of social theory. The optimistic point of view, for example advocated in the nineteenth century by Herbert Spencer, could be called 'modernization theory': that as societies modernize and industrialize, trade and democracy spread, and warlike influences diminish. The argument can even be given a feminist twist. If warlike tendencies and male power are associated – and the evidence of this doleful correlation is strong – and if modernization implies a weakening of patriarchy (a plausible, though of course not an entirely self-evident conclusion) then here is another reason, independent of the traditional ones, for optimism.

However, on the other side of the theoretical divide stand not only political-science realism (with its assumption that the behaviour of states is essentially independent of their internal structure) but also Marxism, which, at least in the twentieth century, has emphasized capitalism's role as a cause of war, rather than as a force for peace. There is simply insufficient empirical evidence to be certain whether the optimists or the pessimists are right. Michael Mann, for example, finds no overwhelming evidence that capitalist societies are more militaristic than non-capitalist ones, but equally no clear reason to argue that they are intrinsically more pacific (Mann 1984).

If capitalism's record is ambiguous, there is perhaps better reason to believe the argument, originating ultimately from Immanuel Kant, that democracy is a force for peace. 'Democracies don't attack each other', declared President Clinton in his 1994 State of the Union address. The matter is not empirically entirely clear-cut: for example, if Wilhelmine Germany counts as a democracy (which is certainly arguable), then the events of August 1914 are clearly an exception to Clinton's generalization; and a near-industry in recent political science has devoted itself to detailed debate about the claim.[11] Furthermore, electorates can sometimes be gripped by war fever, and some of the late twentieth century's minor wars, such as the British recapture of the Falklands, or the US invasion of Grenada, have been popular electorally. Nevertheless, the enormous costs of full-scale war – even conventional war – between advanced countries are very different from those of recent military interventions in the non-industrialized or less industrialized world. Democracy may not guarantee perpetual peace, but it is plausibly a force for peace.

▶ ACTOR-NETWORK THEORY

Actor-network theory, however, would regard all the theoretical perspectives discussed in the previous section – political-science realism, modernization theory, 'structural' feminism, Marxism and 'democratic peace' theory – as sharing, for all their diversity, the same theoretical flaw. Each seeks to identify a stable feature of social structure, whether it be the state system, industrialization, patriarchy, capitalism, or democracy, and to discover its effects on the propensity to war. Actor-network theory is deeply sceptical about the existence of stable social structure. There is simply no macro-structure constraining events, suggests Bruno Latour, but a constantly open-ended interaction between a multitude of actors. Actor-network theory inherits the interactionist or ethnomethodological insistence that social structure is a precarious outcome, rather than a firm given, and moreover it insists that the outcome is as much 'technical' as it is 'social' in its construction (see Chapter 8).

The bearing of actor-network theory on armament processes is straightforward, and has for example been explored by Steven Flank in a useful study of the Indian and South African nuclear weapons programmes (Flank 1993/4), and by this author in a study of the development of nuclear missile guidance (MacKenzie 1990). The development of weapons is always heterogeneous engineering, to borrow John Law's phrase: it is always simultaneous engineering of social phenomena as well as of physical ones (Law 1987). A successful weapons programme can indeed plausibly be seen as a network linking physical artifacts and human beings. Weapons system developers have often to spend as much time constructing and maintaining their relationships to human actors (politicians, industrialists, senior officers, the multifarious forms of 'bureaucratic politics') as they do forging physical artifacts.

The usefulness of an actor-network perspective from the point of view of opponents of the nuclear arms race was clear: it suggested a search for the varied and often surprising ways in which the network-building processes of armament could be disrupted (MacKenzie 1990). However, applying the approach to nuclear *dis*armament is a problem of a different order. Actor-network theory prohibits the 'structural sociology' approach of identifying, and then eliminating, the social preconditions of nuclear weapons, for it suggests that the 'social' and the 'technical' are too intertwined, and their interactions too open-ended and too unpredictable, for this to be possible. It would, for example, condemn Marxist pessimism and 'modernization' optimism alike.

This might suggest that the voice of the actor-network theorist in discussions of nuclear disarmament must always be one of extreme caution. If 'everything is uncertain and reversible', as John Law neatly sums up a central aspect of actor-network theory,[12] then there is always the possibility that the ingenious, heterogeneous engineering of a nuclear weapons developer can circumvent whatever technical or social underpinnings of disarmament we seek to create.

On the other hand, actor-network theory's message may not be quite as bleak as that formulation suggests. Unlike realist political science, which believes that states have definite, fixed propensities, actor-network theory asserts that the properties of all actors are not inherent but are relational, so it leaves open the possibility that the very act of nuclear disarmament might change the way states behave (a point I turn to below in discussing the iterative nature of nuclear disarmament). Actor-network theory could also be taken as suggesting the possibility of creating counter-networks (Mort 1995) that might give the abolition of nuclear weapons some stability. Perhaps actor-network theory's greatest virtue is its openness to the way in which 'technical' innovations can transform 'social' possibilities. It was, for example, just such an innovation, the reconnaissance satellite, that helped make possible the partial stabilization of the nuclear arms race in the 1970s and 1980s, a story whose technical dimensions are only beginning to be told (Wheelon 1997).

Of course, it would be naive to expect a 'technical fix' for the problems of making nuclear disarmament irreversible. However, the potential for the development of verification technologies should not be underestimated. The very radioactivity of the key components of nuclear weapons is here their weakness. Plutonium separation, for example, is already readily detectable: the krypton-85 plumes from the reprocessing plants at Sellafield and La Hague can be measured hundreds of kilometres downwind (Weiss *et al.* 1986). Uranium separation is harder to detect, but it is noteworthy that the USA reportedly gained evidence of Saddam Hussein's nuclear weapons programme from traces of radioactive isotopes on the clothing of released hostages (Moyland 1997: 18).

Furthermore, verification ought to be seen as a socio-technical rather than as a technical process. Even a modest nuclear weapons programme is likely to involve at least several hundred staff (MacKenzie and Spinardi 1995) and to involve activities that impinge on, and are therefore detectable by, a much larger number of people. This points to the importance of what Joseph Rotblat calls 'societal verification': systematic measures to encourage and facilitate reporting of violations by citizens of the countries concerned (Rotblat *et al.* 1993). If 'whistle blowers' are encouraged and protected, and perhaps even rewarded financially (though this last involves the possibility of financially-motivated false alarms), then the achievement of secrecy becomes much more difficult. Actor-network theory's critique of conventional notions of social structure rests on the premise that innovation can make previously large actors small; the converse of this is that small actors (peace movements, non-governmental organizations, even the safeguards branch of the International Atomic Energy Agency, whose annual budget of around $100 million (Moyland 1977: 24) is a tiny fraction of the defence budgets of the major powers) can become large. This leads us back to the question of democracy discussed in the previous section. If governments can punish their citizens arbitrarily, then human whistle blowing cannot be relied upon (though even North Korea and Iraq have had their defectors), and we must turn primarily to technical means of verification. As democracy and human rights spread, however, reliance upon societal verification becomes more plausible.

▶ **DISCUSSION**

The bad news

What have we learnt about the feasibility of permanent nuclear disarmament? No individual mechanism for achieving this goal seems likely to be watertight. The route via the control of fissile materials has a reassuring 'physicality', and remains the most important single barrier to further proliferation, but is daunting as a means of ensuring the permanence of the disarmament of the existing nuclear states. Control over new production is relatively straightforward; the major problem is the existence of large, badly documented, stockpiles. Control over polonium and tritium has the advantage that radioactive decay limits the usefulness of covert stockpiles of these materials – relatively quickly in the case of polonium, more slowly for tritium. To the extent that one can be sure that all existing nuclear weapons worldwide depend for their efficacy on tritium, polonium or other rapidly decaying substances, then one can begin to have some assurance that one barrier to disarmament – the risk of it being circumvented by a small number of nuclear weapons being hidden away – may be less alarming than it seems. If disconnected from sources of polonium and tritium, key components of such weapons would eventually waste away.

However, neutron generators not employing polonium exist: the most likely substitute would be a tritium-deuterium generator, which requires much less tritium than boosting does (conceivably as little as ten milligrams).[13] Furthermore, even the much larger quantities of tritium necessary to keep boosting in a small covert nuclear arsenal in working order could be obtained for a long period by secreting a tritium stockpile and purifying it periodically, or conceivably by producing it in a particle accelerator on a scale perhaps compatible with clandestine operation.[14] Furthermore, failure of boosting caused by lack of tritium means only a weaker nuclear explosion, not no nuclear explosion. Even an initiator may not be absolutely necessary. For example, former Los Alamos designer Theodore Taylor has suggested that if a nuclear weapon is built with reactor-grade rather than weapons-grade plutonium (which is difficult, but known to be possible), then the high level of spontaneous neutron emission in reactor-grade plutonium might make it possible to do without an initiator (McPhee 1974).

So there are limits to the extent to which the elimination of nuclear weapons can be made assuredly permanent by controls on the nuclear weapons production system. The difficulty with the requirement for tacit knowledge is that it only slows the reinvention of nuclear weapons: it does not make it impossible. It took roughly three years for the requisite knowledge to be created in the first place in the Manhattan Project. That project had the advantage that, unlike a future covert nuclear weapons programme, it was not in practice constrained by the need for secrecy: there were no German satellites or reconnaissance aircraft able to observe Los Alamos, Hanford or Oak Ridge. Against this, however, must be weighed the fact many of the technologies that had to be created afresh in the Manhattan Project are now available as machines and instruments, and have been much refined. We may perhaps be modestly confident that the need to acquire tacit knowledge afresh through trial-and-error learning means that future programmes (especially covert ones) should not be much faster than the Manhattan Project. There is, however, little reason to expect them to be much slower.

The 'structural sociology' approach to abolishing nuclear weapons by removing their social preconditions is also not watertight. The predictive power of social science is notoriously weak, so only time passing and trust building up can give us confidence that the risk of major war between industrialized countries is sufficiently low. Finally, the strength of actor-network theory – its emphasis on the open-endedness, unpredictability, and socio-technical nature of innovation – is here a source of weakness as well as strength. That small actors may grow, societal verification become possible, and the secrecy required for covert nuclear weapons programmes become unachievable, are all possibilities, not certainties. Actor-network theory is a descriptive, not a predictive, enterprise.

The good news

So none of the four perspectives that I have reviewed – 'system', 'tacit knowledge', 'social precondition', and 'actor-network' – offers a watertight route to permanent nuclear disarmament. The good news, in contrast, is in part simply the diversity of the mechanisms they suggest. To judge the feasibility of nuclear disarmament requires us to judge the efficacy of these mechanisms *as an entirety*. In contrast, the existing literature on the topic tends to consider individual facets of the issue in isolation (often along disciplinary lines), not as a totality.

The other aspect of the good news is that there is evidence that the *interactions* between the 'system', 'tacit knowledge', 'social preconditions', and 'actor-network' aspects of the problem are benign, from the viewpoint of the permanent elimination of nuclear weapons. I have already mentioned one such interaction: between the actor-network emphasis on undermining the socio-technical possibility of secrecy, and the 'social preconditions' emphasis on the spread of democracy. But perhaps particularly important are the interactions between the 'system' and the 'tacit knowledge' constraints on nuclear weapons development.

From the point of view of the demand for tacit knowledge, the simplest kind of atomic bomb is the gun design (in which the critical mass is formed by shooting one subcritical mass into another, using low-powered propellant explosive, rather than the high explosive used in implosion). The gun design avoids the complex detonics and electronics of implosion. The developers of a gun weapon could, furthermore, have reasonable confidence, without a full nuclear test, that it would work. The Hiroshima bomb was used without a full test of its design. South Africa, likewise, felt able to make do with only a laboratory test of its gun weapon.[15]

However, even for nuclear weapons laboratories with years of experience, building a satisfactory gun weapon from plutonium is a demanding task. In July 1944, in the most serious internal crisis faced by the Manhattan Project (its scientific director, Robert Oppenheimer, had to be persuaded not to resign), the plutonium gun was abandoned; and subsequent efforts to build one (pursued as late as the 1970s) have also not come to fruition (Hansen 1988: 21; Hoddeson *et al.* 1993). The problem is that even weapons-grade plutonium has a much higher rate of spontaneous neutron emission than uranium, and in a gun the critical mass is formed only relatively slowly, so the device tends to 'fizzle': to suffer a premature, partial detonation.

Building a gun would therefore in practice mean reliance on highly enriched uranium, rather than plutonium, and so if materials have to be produced afresh, uranium enrichment technology is needed, not just access to a suitable nuclear reactor. For all the progress in enrichment, it remains a demanding technology. Furthermore, greater quantities of fissile material are needed for a gun than for an implosion weapon (in a gun, there is little increase in the density of fissile material, and so critical mass is greater), and gun weapons are harder to miniaturize than implosion weapons. So the problems of delivery are greater for a gun weapon.

More generally, the more sophisticated the delivery system (missile versus bomber; ICBM (intercontinental ballistic missile) versus short-range missile), the greater the demands on warhead design. A uranium gun, to be carried by a bomber, can be and has been developed on the basis of explicit knowledge, plus the kind of trial-and-error learning and testing (short of a full nuclear test) that is hard to detect. A relatively simple plutonium implosion weapon to be carried by a bomber also seems feasible without nuclear testing (albeit much harder: the belief that it would be necessary to test an implosion weapon was one reason South Africa opted for a gun design) (De Villiers *et al.* 1993: 102). Moving without a full nuclear test from a weapon to be carried by a bomber to one to be carried by even a short-range missile is harder still (only in 1952, ten years after the start of the Manhattan Project, did the USA deploy a nuclear weapon weighing less than the 1000 kilograms of the Scud's payload).[16] Israel probably has done this, and possibly Pakistan too; Iraq would probably have succeeded eventually. In the absence of testing, however, there might be significant doubts that such a weapon would actually work with any reliability.

Making the currently paradigmatic strategic weapon – a miniaturized hydrogen bomb with a boosted-fission primary, to be carried by an ICBM or submarine-launched missile – almost certainly requires retracing many of the steps (including nuclear testing) already taken by the nuclear weapons states. One significance of tacit knowledge is that even those latter states might have to retrace some of their own steps were they, after a sufficiently long hiatus, to decide to rebuild their arsenals. Blueprints, documents, data and recollections from their original programmes would help, but might not suffice. In particular, as readers of Harry Collins might anticipate, even a detailed blueprint may not allow exact replication of the original weapon:

> Material batches are never quite the same, some materials become unavailable, and equivalent materials are never exactly equivalent; 'improved' parts often have new, unexpected failure modes; different people (not those who did the initial work) are involved in the remanufacturing; vendors go out of business or stop producing some products.
>
> (Miller *et al.* 1987: 3)

Some of these differences would matter, and some would not: knowing whether they would or not is precisely the kind of issue where experienced judgement may be necessary. Miniaturized nuclear weapons (especially those designed to maximize yield/weight or yield/diameter ratios, and to economize on special materials, as much of the recent US stockpile appears to have been[17]) are not a 'forgiving' technology. Designs can be uncomfortably close to what some of those involved call the 'cliff': the region where performance becomes very sensitive to internal and external conditions, with the result, for example, that the explosion of a 'primary' might fail to ignite the 'secondary'.

In addition, not just warheads but the missiles themselves would have to be tested, assuming existing missiles to be destroyed in a disarmament agreement or placed under international surveillance to be used as space boosters. A long range missile is a sophisticated piece of systems engineering.[18] The many test failures in early US missile work, and the occasional failure of even the most modern space boosters, such as Ariane-5, suggest that if this development has to be started from scratch, without its developers having previous experience,[19] the resultant missiles are un-likely to work with any reliability without reasonably extensive tests, and missile flight-tests, like nuclear explosions, are detectable.

Interrelated system and tacit knowledge considerations also reduce the risk of an agreement to eliminate nuclear weapons being circumvented effectively by secretly retaining some weapons. Nuclear weapons, like all complex technological structures, age – through radioactive decay, cor-rosion, chemical decomposition and materials creep (Rosengren 1983). These processes are generally slow, but – if a timescale of 20 to 30 years is being considered – important. To my knowledge, no existing design of nuclear weapon could be left unattended that long and then be expected to work with any reliability. Counteracting the effects of ageing requires maintenance of the weapon and replacement of aged parts. Access to the latter (especially to replenished tritium reservoirs) may be a problem; the process may be vulnerable to detection via societal verification; and main-tenance and replacement are processes requiring skill. For example, in some of the Soviet weapons inherited by the Ukraine, hydrogen built up danger-ously, and the Ukrainians found that, despite nationalistic sentiments, they could deal with this problem only by calling in the weapons' original Russian designers. This was a relatively straightforward problem; sophisti-cated, 'near the cliff' designs may well throw up subtler problems as the decades pass.

Finally, there is an evident trade-off between the size of an arsenal and the visibility of its development. Large-scale production of fissile materials involves isotopic 'signatures' that, as noted above, are vulnerable to remote monitoring techniques, and that might also be detectable via their impact on electric load curves (a 'system' suggestion I owe to Tom Hughes). Even assembly of weapons from covert stocks of fissile materials would risk detection via whistle-blowing and 'societal verification', so again there would be an incentive to keep such work to a small scale. However, the smaller the arsenal, the less likely is it that its possessors would be content with interceptable aircraft or unreliable missiles to carry it, and so the more important would be the system and knowledge constraints on integrating nuclear weapons and missile delivery systems.

▶ **CONCLUSION**

The risk that a nuclear weapons abolition agreement would be circum-vented by secret development of a large arsenal of sophisticated weapons thus appears to be limited. On the other hand, the risk of the covert development of a small arsenal of relatively crude nuclear weapons, carried by bombers or short-range missiles, is not negligible. Relatively effective defences against such an arsenal are feasible, and it is unlikely to be decisive militarily against a large, widely-dispersed conventional force, equipped and trained to continue to function in the face of limited nuclear attack – a force, in other words, of the sort that the USA and its NATO allies currently deploy. However, the potential of even a small, unsophisticated nuclear arsenal as a terror weapon would be considerable. Furthermore, the risks of another form of nuclear rearmament – by either or both parties to a prolonged, major war between industrialized nations – are also not negligible, at least until such a time as we can be confident that the probability of such war is very low.

So it seems to me that we have to approach the abolition of nuclear weapons as a *process*: we do not know, and should not pretend to know, now, the solution to all the problems it may throw up (Booth and Wheeler 1992). However, this is not a reason for delay in starting: actions taken now – for example, to control tritium production worldwide and begin the task of monitoring the tritium stockpile – could greatly ease later stages of the process.

Like almost everything else in our world, from the economic success of nations to the stability of empirical knowledge,[20] the abolition of nuclear weapons rests ultimately upon trust. As our everyday experience teaches us, trusting is a process: trust is built up through, and sometimes destroyed by, interactions. Trust is also, at least in the context of nuclear weapons, an intrinsically socio-technical matter. If we look only at 'the social', then, unless we are prepared to rely to an undue extent on modernization theory, we can, at best, proceed only very slowly indeed: reasonable certainty that there will be no more war between advanced industrial nations is hardly likely to be gained within the 20–30 year time scale envisaged here. Similarly, no informed observer can look at the current state of 'technical' verification and conclude that it is watertight: the avenues for circumventing it are just too evident.

If, however, we consider nuclear disarmament as a socio-technical process, then perhaps – just perhaps – it becomes feasible. The decision to embark on a worldwide process of nuclear disarmament would be, of course, a remarkable declaration of international trust, particularly among the eight overt or *de facto* nuclear weapons states (the USA, Russia, Britain, China, France, India, Pakistan and Israel). If technical verification efforts and an increasing scope for societal verification suggested that all the parties to the agreement were living up to the obligations they had undertaken at a first stage of its implementation, initial trust would then, most likely, be seen as warranted, and the parties involved might then proceed to the next stage of the implementation of disarmament. The very process might help bring about some of its social preconditions, such as the emergence of stable democratic governments in those nuclear weapons states that currently lack them, and an alleviation of the regional tensions (notably between India and Pakistan, and between Israel and its Arab neighbours) that have fostered nuclear armament. There could be no guarantee that an iterative process (Fetter 1996: 40) of this kind would actually converge to the goal of a permanent zero level of nuclear weaponry: a significant breach of trust by any of the parties to such an agreement would almost certainly lead to the others refusing to undertake the next step. Nevertheless, there seems to be a reasonable chance that zero might be obtained, and might be permanent.

Such a process might even find unexpected allies. Compared to the other goals currently being offered to those in the nuclear weapons laboratories, it might prove technically inspiring. Said one nuclear weapons designer: 'Zero in 2005. I can get enthused by that project. What a neat way to cap off your career' (quoted in Weisman 1994: 22). The elimination of all nuclear weapons by 2005 is unrealistic, but major steps in that direction may well be possible by then. Certainly, if the thought of the abolition of nuclear weapons can be pondered in a nuclear weapons laboratory, perhaps the rest of us need to take it seriously too.

▶ NOTES

1 This article also appears in Olivier Coutard (ed.) (1999) *The Governance of Large Technical Systems.* London: Routledge. The empirical research drawn on here was supported by the UK Economic and Social Research Council, mainly under the Science Policy Support Group Initiative on Science Policy Research on Defence Science and Technology (grant Y307253006), and the interviews were conducted mainly by Graham Spinardi, to whom I am very grateful. This paper is a much-elaborated version of one presented to the conference on Large Technical Systems and Networks, 27–30 September 1995. I am grateful for helpful comments by audiences there and at subsequent seminars at the University of Keele and MIT.

2 In electromagnetic separation, a gaseous uranium compound travels through a strong magnetic field. The paths of the lighter uranium-235 ions curve more sharply than those of uranium-238 ions, so creating two beams with different isotopic composition.

3 Half-life is the average time interval within which half of any quantity of a substance will undergo radioactive decay.

4 Public attention was first drawn to the significance of tritium for nuclear disarmament by Wilkie (1984).

5 For a description of the process, see Wilkie (1984).

6 All US nuclear weapons now use tritium. Substances other than tritium (such as helium-3, tritium's decay product) might be used in boosting, but 'use of these is considered not to be within reach of present weapons technology' (Mark *et al.* 1988: 116 fn.).

7 In his book on the history of systems engineering, Tom Hughes devotes a chapter to Atlas (Hughes 1998).

8 See MacKenzie and Spinardi (1995) for the sources of the points drawn on in the following paragraphs.

9 At normal temperatures, lithium deuteride is a solid, so this makes possible a 'dry' hydrogen bomb, in which interactions with neutrons generate tritium from lithium-6. See, for example, Rhodes (1995: 306).

10 There has been much discussion in the open literature (for example, Morland 1979; Hansen 1988; Rhodes 1995) of possible details of the Teller-Ulam configuration. Fortunately, there is no need here to add to this discussion.

11 For a useful introduction to this literature, see Owen (1994).

12 Comment made at Keele conference 'Actor-network Theory and Beyond', July 1997.

13 There are alternatives, such as a plutonium-238/beryllium generator, but the rapid decay of plutonium-238 raises the same problem as with polonium.

14 Fetter (1996: 24) calculates that a 50-gram stockpile of tritium, enough for around twelve weapons, could be maintained by an accelerator with a 200 kilowatt beam power.

15 The test involved making a fissile assembly supercritical, but only for a very short period of time, so that no explosion took place. Richard Feynman compared the original Manhattan Project version of this experiment to 'tickling the dragon's tail' (Hoddeson *et al.* 1993: 347).

16 See the weight data in Hansen (1988).

17 I do not know whether or not this last point is true of the other stockpiles.

18 This is true of cruise missiles as well as ballistic ones, hence the possible benefits of banning the former as well as the latter.

19 It should be noted that experience relevant to ballistic missile building can be gained from work on space boosters, as the two technologies are very similar.

20 On the latter, see Shapin (1994). On the former, see Fukuyama (1995).

▶ REFERENCES

Albright, D. and Hibbs, M. (1991) Iraq's nuclear hide-and-seek. *Bulletin of the Atomic Scientists*, September: 14–23.

Albright, D. and Kelley, R. (1995) Has Iraq come clean at last? *Bulletin of the Atomic Scientists*, November/December: 53–64.

Blechman, B. M. and Fisher, C. S. (1994/5) Phase out the bomb. *Foreign Policy*, 97: 79–95.

Booth, K. and Wheeler, N.J. (1992) Beyond nuclearism, in R. Cowen Karp (ed.) *Security without Nuclear Weapons? Different Perspectives on Non-Nuclear Security*, pp. 21–55. Oxford: Oxford University Press.

Callon, M. (1994) Is science a public good? *Science, Technology and Human Values*, 19: 395–424.

Collina, T. Z. (1996) Strike up the ban. *Bulletin of the Atomic Scientists*, January/February: 41–4.

Collins, H. M. (1974) The TEA set: tacit knowledge and scientific networks. *Science Studies*, 4: 165–86.

Davis, J. C. and Kay, D.A. (1992) Iraq's secret nuclear weapons program. *Physics Today*, 45 (7): 21–7.

De Villiers, J. W., Jardine, R. and Reiss, M. (1993) Why South Africa gave up the bomb. *Foreign Affairs*, 72 (5): 98–109.

Ferguson, E. S. (1977) The mind's eye: nonverbal thought in technology. *Science*, 197 (26 August): 827–36.

Ferguson, E. S. (1992) *Engineering and the Mind's Eye*. Cambridge, MA: MIT Press.

Fetter, S. (1996) *Verifying Nuclear Disarmament*. Washington, DC: Stimson Center, occasional paper 29.

Flank, S. (1993/4) Exploding the black box: the historical sociology of nuclear proliferation. *Security Studies*, 3: 259–94.

Fukuyama, F. (1995) *Trust: The Social Virtues and the Creation of Prosperity*. New York: Free Press.

Hansen, C. (1988) *US Nuclear Weapons: The Secret History*. Arlington, TX: Aerofax.

Hoddeson, L. Hendriksen, P. W., Meade, R. A. and Westfall, C. (1993) *Critical*

Assembly: A Technical History of Los Alamos during the Oppenheimer Years. Cambridge: Cambridge University Press.

Hughes, T. P. (1983) *Networks of Power: Electrification in Western Society, 1880–1930.* Baltimore, MD: Johns Hopkins University Press.

Hughes, T. P. (1998) *Rescuing Prometheus.* New York: Pantheon.

Kalinowski, M. B. and Coschen, L. C. (1995) International control of tritium to prevent horizontal proliferation and to foster nuclear disarmament. *Science & Global Security,* 5: 131–203.

Law, J. (1987) Technology and heterogeneous engineering: the case of the Portuguese expansion, in W. E. Bijker, T. P. Hughes and T. J. Pinch (eds) *The Social Construction of Technological Systems: New Directions in the Sociology and History of Technology,* pp. 111–34. Cambridge, MA: MIT Press.

Lumpe, L., Gronlund, L. and Wright, D. C. (1992) Third World missiles fall short. *Bulletin of the Atomic Scientists,* March: 31–7.

MccGwire, M. (1994) Is there a future for nuclear weapons? *International Affairs,* 70: 211–28.

MacKenzie, D. (1990) *Inventing Accuracy: A Historical Sociology of Nuclear Missile Guidance.* Cambridge, MA: MIT Press.

MacKenzie, D. and Spinardi, G. (1995) Tacit knowledge, weapons design, and the uninvention of nuclear weapons. *American Journal of Sociology,* 101: 44–99.

McPhee, J. (1974) *The Curve of Binding Energy.* New York: Farrar, Straus & Giroux.

Mann, M. (1984) Capitalism and militarism, in M. Shaw (ed.) *War, State and Society,* pp. 25–46. London: Macmillan.

Mark, J. C., Davies, T. D., Hoenig, M. M. and Leventhal, P. L. (1988) The tritium factor as a forcing function in nuclear arms reduction talks. *Science,* 241 (2 September): 1166–8.

Miller, G. H., Brown, P. S. and Alonso C. T. (1987) *Report to Congress on Stockpile Reliability, Weapon Remanufacture, and the Role of Nuclear Testing.* Livermore, CA: Lawrence Livermore National Laboratory, UCRL-53822.

Morland, H. (1979) The H-bomb secret. *The Progressive,* November: 14–23.

Mort, M. (1995) *Building the Trident Network.* Ph.D. thesis, University of Lancaster.

Moyland, S. van (1997) *Verification Matters: The IAEA's Programme '93 + 2'.* London: Verification Technology Information Centre.

Owen, J. M. (1994) How liberalism produces democratic peace. *International Security,* 19 (2): 87–125.

Peabody, A. T., Jr. (1981) *Some Political Issues related to future Special Nuclear Materials Production.* Los Alamos, NM: Los Alamos National Laboratory.

Rhodes, R. (1995) *Dark Sun: The Making of the Hydrogen Bomb.* New York: Simon & Schuster.

Rosengren, J. W. (1983) *Some Little-Publicized Difficulties with a Nuclear Freeze.* Marina del Rey, CA: R&D Associates, RDA-TR-112116-001.

Rotblat, J. (1993) Societal verification, in J. Rotblat, J. Steinberger and B. Udgaonkar (eds) *A Nuclear-Weapon-Free World: Desirable? Feasible?* pp. 103–18. Boulder, CO: Westview.

Shapin, S. (1994) *A Social History of Truth: Civility and Science in Seventeenth-Century England.* Chicago: University of Chicago Press.

Spinardi, G. (1994) *From Polaris to Trident: The Development of US Fleet Ballistic Missile Technology.* Cambridge: Cambridge University Press.

Urquhart, J. (1983) Polonium: Windscale's most lethal legacy. *New Scientist,* 97 (31 March): 873–5.

US Department of Energy, Office of Reconfiguration (1995) *Final Programmatic*

Environmental Impact Statement for Tritium Supply and Recycling. Washington, DC: Department of Energy, DOE/EIS-0161.

Weisman, J. (1994) Early retirement for weaponeers? *Bulletin of the Atomic Scientists,* July/August: 16–22.

Weiss, W., Stockburger, H., Sartorius, H. *et al.* (1986) Mesoscale transport of ^{85}Kr originating from European sources. *Nuclear Instruments and Methods of Physics Research,* B17: 571–4.

Westervelt, D. R. (1988) The role of laboratory tests, in J. Goldblat and D. Cox (eds) *Nuclear Weapon Tests: Prohibition or Limitation,* pp. 47–58. Oxford: Oxford University Press.

Wheelon, A. D. (1997) Corona: the first reconnaissance satellites. *Physics Today,* February: 24–30.

Wilkie, T. (1984) Old age can kill the bomb. *New Scientist,* 16 (February): 27–32.

▶ Bibliography

Abbate, J. (1998) *Inventing the Internet*. Cambridge, MA: MIT Press.

Adams, J. (1995) *Risk*. London: University College London Press.

Aitken, H. G. J. (1976) *Syntony and Spark: The Origins of Radio*. New York: Wiley.

Alder, K. (1997) *Engineering the Revolution: Arms and Enlightenment in France, 1763–1815*. Princeton, NJ: Princeton University Press.

Amann, R., Cooper, J, and Davies, R. W. (eds) (1977) *The Technological Level of Soviet Industry*. New Haven, CT, and London: Yale University Press.

Appleton, H. (ed.) (1995) *Do It Herself: Women and Technical Innovation*. London: Intermediate Technology Publications.

Armacost, M. H. (1969) *The Politics of Weapons Innovation: The Thor-Jupiter Controversy*. New York: Columbia University Press.

Arrow, K. (1962) The economic implications of learning by doing. *Review of Economic Studies*, 29: 155–73.

Arthur, W. B. (1994) *Increasing Returns and Path Dependence in the Economy*. Ann Arbor, MI: University of Michigan Press.

Ball, D. (1980) *Politics and Force Levels: The Strategic Missile Program of the Kennedy Administration*. Berkeley, CA: University of California Press.

Baker, J. P. (1996) *The Machine in the Nursery: Incubator Technology and the Origins of New-born Intensive Care*. Baltimore, MD: Johns Hopkins University Press.

Barker, J. and Downing, H. (1980) Word processing and the transformation of patriarchal relations of control in the office. *Capital and Class*, 10: 64–99.

Barnaby, F. (1980) The military tail wagging the political dog. *Guardian*, 9 October.

Barnaby, F. (1981) Social and economic reverberations of military research. *Impact of Science on Society*, 31, 73–83.

Barnes, B. (1982) *T. S. Kuhn and Social Science*. London and Basingstoke: Macmillan.

Barnes, B. and Edge, D. (eds) (1982) *Science in Context: Readings in the Sociology of Science*. Milton Keynes: Open University Press.

Barnes, B. and Shapin, S. (eds) (1979) *Natural Order: Historical Studies of Scientific Culture*, 143–63. Beverly Hills, CA: Sage.

Barnes, B., Bloor, D. and Henry, J. (1996) *Scientific Knowledge: A Sociological Analysis*. London: Athlone, and Chicago: Chicago University Press.

Beard, E. (1976) *Developing the ICBM. A Study in Bureaucratic Politics*. New York: Columbia University Press.

Beck, U. (1992) *Risk Society: Towards a New Modernity*. London: Sage.

Berman, R. and Baker, J. (1982) *Soviet Strategic Forces: Requirements and Responses*. Washington: Brookings.

Beynon, H. (1992) The end of the industrial worker, in N. Abercrombie and A. Warde (eds) *Social Change in Contemporary Britain*. Cambridge: Polity Press.

Bijker, W. E. (1995) *Of Bicycles, Bakelites, and Bulbs: Toward a Theory of Sociotechnical Change*. Cambridge, MA: MIT Press.

Bijker, W. E., Hughes, T. P. and Pinch, T. (eds) (1987) *The Social Construction of Technological Systems*. Cambridge, MA: MIT Press.

Blauner, R. (1964) *Alienation and Freedom: The Factory Worker and His Industry*. Chicago: University of Chicago Press.

Bloor, D. (1976) *Knowledge and Social Imagery*. London: Routledge and Kegan Paul.

Braun, E. and Macdonald, S. (1978) *Revolution in Miniature: The History and Impact of Semiconductor Electronics*. Cambridge: Cambridge University Press.

Braverman, H. (1974) *Labor and Monopoly Capital: The Degradation of Work in the Twentieth Century*. New York: Monthly Review Press.

Burns, T. (ed.) (1969) *Industrial Man*. Harmondsworth: Penguin.

Callon, M. (1986) Some elements of a sociology of translation: domestication of the scallops and the fisherman of St. Brieuc Bay, in J. Law (ed.) *Power Action and Belief: A New Sociology of Knowledge?* London: Routledge.

Campbell-Kelly, M. and Aspray, W. (1996) *Computer: A History of the Information Machine*. New York: Basic Books.

Cardwell, D. S. L. (1971) *From Watt to Clausius: The Rise of Thermodynamics in the Early Industrial Age*. London: Heinemann.

Cardwell, D. S. L. (1972) *Technology, Science and History*. London: Heinemann.

Chabaud-Rychter, D. (1994) Women users in the design process of a food robot: innovation in a French domestic appliance company, in C. Cockburn and R. Furst-Dilic (eds) *Bringing Technology Home: Gender and Technology in a Changing Europe*. Buckingham: Open University Press.

Cipolla, C. M . (1965) *Guns, Sails and Empires: Technological Innovation and the Early Phases of European Expansion, 1400–1700*. New York: Pantheon Books.

Cockburn, C. (1983) *Brothers: Male Dominance and Technological Change*. London: Pluto.

Cockburn, C, and Ormrod, S. (1993) *Gender and Technology in the Making*. London: Sage.

Collins, H. M. and Yearley, S. (1992) Epistemological chicken, in A. Pickering (ed.) *Science as Practice and Culture*, pp. 301–26. Chicago: University of Chicago Press.

Constant, E. W. II (1978) On the diversity and co-evolution of technological multiples: steam turbines and Pelton water wheels. *Social Studies of Science*, 8: 183–210.

Constant, E. W. II (1980) *The Origins of the Turbojet Revolution*. Baltimore, MD and London: Johns Hopkins University Press.

Coombs, R., Saviotti, P. and Welsh, V. (1987) *Economics and Technological Change*. London: Macmillan.

Cowan, R. S. (1979) From Virginia Dare to Virginia Slims: women and technology in American life. *Technology and Culture*, 20: 51–63.

Cowan, R. S. (1983) *More Work for Mother: The Ironies of Household Technology from the Open Hearth to the Microwave*. New York: Basic Books.

David, P. A. (1992) Heroes, herds and hysteresis in technological history: Thomas Edison and 'The Battle of the Systems' reconsidered. *Industrial and Corporate Change*, 1: 129–80.

David, P. A. (1997) *Path dependence and the Quest historical economics: One more chorus of the ballad of QWERTY*. University of Oxford, Discussion Papers in Economic and Social History, No. 20.

Dinneen, G. P. and Frick, F. C. (1977) Electronics and national defense: a case study. *Science*, 195, (18 March): 1151–5.

Dosi, G. (1982) Technological paradigms and technological trajectories: a suggested interpretation of the determinants of technical change. *Research Policy*, 11: 147–62.

Dosi, G. (ed.) (1990) *Technical Change and Economic Theory*. London: Pinter.

Douglas, M. and Wildavsky, A. (1982) *Risk and Culture: An Essay on the Selection of Technological and Environmental Dangers*. Berkeley, CA: University of California Press.

Duden, B. (1991) *The Woman Beneath the Skin: A Doctor's Patients in Eighteenth-Century Germany*. Cambridge, MA: Harvard University Press.

Easlea, B. (1983) *Fathering the Unthinkable. Masculinity, Scientists and the Nuclear Arms Race*. London: Pluto.

Edge, D. O. (1974–5) Technological metaphor and social control. *New Literary History*, 6: 135–47.

Edgerton, D. (1993) Essay review of MacKenzie (1990) *British Journal of the History of Science*, 26: 67–95.

Edwards, P. N. (1996) *The Closed World: Computers and the Politics of Discourse in Cold War America*. Cambridge, MA: MIT Press.

Ehrenreich, B. and English, D. (1975) The manufacture of housework. *Socialist Revolution*, 26: 5–40.

Ellul, J. (1964) *The Technological Society*. New York: Vintage.

Elster, J. (1983) *Explaining Technical Change: A Case Study in the Philosophy of Science*. Cambridge: Cambridge University Press.

Enloe, C. (1983) *Does Khaki Become You?* London: Pluto.

Fallows, J. (1981) *The National Defense*. New York: Random House.

Firestone, S. (1970) *The Dialectic of Sex*. New York: William Morrow & Co.

Flamm, K. (1988) *Creating the Computer: Government, Industry, and High Technology*. Washington, D.C.: Brookings.

Galison, P. (1987) *How Experiments End*. Chicago: University of Chicago Press.

Galison, P. (1997) *Image and Logic: A Material Culture of Microphysics*. Chicago: University of Chicago Press.

Gansler, J. S. (1982) *The Defense Industry*. Cambridge, MA, and London: MIT Press.

Gershuny, J. (1983) *Social Innovation and the Division of Labour*. Oxford: Oxford University Press.

Gilfillan, S. C. (1935a) *The Sociology of Invention*. Chicago: Follett.

Gilfillan, S. C. (1935b) *Inventing the Ship*. Chicago: Follett.

Goldstine, H. H. (1972) *The Computer from Pascal to von Neumann*. Princeton, NJ: Princeton University Press.

Gowing, M. (1982) Principalities and nuclear power: the origins of reactor systems. *The Nuclear Engineer*, 23: 70–6.

Greenwood, K. and King, L. (1981) Contraception and abortion, in Cambridge Women's Studies Group, *Women in Society*. London: Virago.

Greenwood, T. (1975) *Making the MIRV: A Study of Defense Decision Making*. Cambridge, MA: Ballinger.

Gross, P. R. and Levitt, N. (1994) *Higher Superstition: The Academic Left and its Quarrels with Science*. Baltimore, MD: Johns Hopkins University Press.

Gutting, G. (1984) Paradigms, revolution and technology, in R. Laudan (ed.) *The Nature of Technological Knowledge: Are Models of Scientific Change Relevant?* pp. 47–65. Dordrecht: Reidel.

Habakkuk, H. J. (1962) *American and British Technology in the Nineteenth Century: The Search for Labour-Saving Inventions*. Cambridge: Cambridge University Press.

Hafter, D. (1979) The programmed brocade loom and the 'Decline of the Drawgirl', in M. M. Trescott (ed.) *Dynamos and Virgins Revisited: Women and Technological Change in History*, pp. 49–66. Metuchen, NJ and London: Scarecrow Press.

Hales, M. (1993) Human-centred systems, gender and computer supported co-operative work, in B. Probert and B. W. Wilson (eds) *Pink Collar Blues: Work, Gender and Technology*. Melbourne: Melbourne University Press.

Haraway, D. J. (1985) A manifesto for cyborgs: science, technology, and socialist feminism in the 1980s. *Socialist Review*, 15 (2): 65–108.

Haraway, D. J. (1997) *Modest Witness: Second Millennium. Female Man Meets Onco Mouse*^{TM}. New York: Routledge.

Hayden, D. (1981) *The Grand Domestic Revolution: A History of Feminist Designs for American Homes, Neighborhoods and Cities*. Cambridge, MA: MIT Press.

Headrick, D. (1981) *The Tools of Empire: Technology and European Imperialism in the Nineteenth Century*. New York: Oxford University Press.

Hecht, G. (1998) *The Radiance of France: Nuclear Power and National Identity after World War II*. Cambridge, MA: MIT Press.

Heritage, J. (1984) *Garfinkel and Ethnomethodology*. London: Polity.

Hilton, R. H. and Sawyer, P. H. (1963) Technical determinism: the stirrup and the plough. *Past and Present*, 24: 90–100.

Hollowawy, D. (1977) Military technology, in R. Amann, J. Cooper and R. W. Davies (eds) *The Technological Level of Soviet Industry*. New Haven, CT and London: Yale University Press.

Holloway, D. (1982) Innovation in the defence sector: battle tanks and ICBMs, in R. Amann, J. Cooper and R. W. Davies (eds) (1977) *The Technological Level of Soviet Industry*. New Haven, CT and London: Yale University Press.

Hughes, T. P. (1969) Technological momentum in history: hydrogenation in Germany 1898–1933. *Past and Present*, 44: 106–32.

Hughes, T. P. (1971) *Elmer Sperry: Inventor and Engineer*. Baltimore, MD and London: Johns Hopkins University Press.

Hughes, T. P. (1976) The science-technology interaction: the case of high-voltage power transmission systems. *Technology and Culture*, 17: 646–62.

Hughes, T. P. (1983) *Networks of Power: Electrification in Western Society, 1880–1930*. Baltimore, MD and London: Johns Hopkins University Press.

Hughes, T. P. (1989) *American Genesis: A Century of Invention and Technological Enthusiasm, 1870–1970*. New York: Viking.

Illich, I. (1975) *Tools for Conviviality*. London: Fontana.

Joerges, B. (in press) Do politics have artifacts? *Social Studies of Science*.

Jordanova, L. (1989) *Sexual Visions: Images of Gender in Science and Medicine Between the Eighteenth and Twentieth Centuries*. London: Harvester Wheatsheaf.

Kaldor, M. (1982) *The Baroque Arsenal*. London: Deutsch.

Keller, E. F. (1984) *Reflections on Gender and Science*. New Haven, CT: Yale University Press.

Kranakis, E. (1997) *Constructing a Bridge: An Exploration of Engineering Culture, Design, and Research in Nineteenth-Century France and America*. Cambridge, MA: MIT Press.

Kuhn, T. S. (1970) *The Structure of Scientific Revolutions*, 2nd edn. Chicago: University of Chicago Press.

Kumar, K. (1995) *From Post-industrial to Post-modern Society: New Theories of the Contemporary World*. Oxford: Blackwell.

Landes, D. S. (1969) *The Unbound Prometheus: Technological Change and Industrial Development in Western Europe from 1750 to the Present*. Cambridge: Cambridge University Press.

Latour, B. (1987) *Science in Action*. Milton Keynes: Open University Press.

Latour, B. (1991) *The Pasteurization of France*. Cambridge, MA: Harvard University Press.

Latour, B. (1993) *We have never been Modern*. Hemel Hempstead: Harvester.

Latour, B, (1996) *Aramis*. Cambridge, MA: Harvard University Press.

Laudan, R. (ed.) (1984) *The Nature of Technological Knowledge: Are Models of Scientific Change Relevant?* Dordrecht: Reidel.

Law, J. (1987) Technology and heterogeneous engineering: the case of Portuguese expansion, in W. E. Bijker, T. P. Hughes and T. Pinch (eds) *The Social Construction of Technological Systems*. Cambridge, MA: MIT Press.

Lazonick, W. (1981) Factor costs and the diffusion of ring spinning in Britain prior to World War I. *Quarterly Journal of Economics*, 96: 89–109.

Leavitt, J. W. (1986) *Brought to Bed: Childbearing in America 1750 to 1950*. New York: Oxford University Press.

Liebowitz, S. J. and Margolis, S. E. (1990) The fable of the keys. *Journal of Law and Economics*, 33: 1–25.

Liebowitz, S. J. and Margolis, S. E. (1995a) Path dependence, lock-in, and history. *Journal of Law, Economics, and Organization*, 7: 205–26.

Liebowitz, S. J. and Margolis, S. E. (1995b) Policy and path-dependence: from QWERTY to Windows 95. *Regulation*, 3: 33–41.

Luhmann, N. (1993) *Risk: A Sociological Theory*. Berlin: de Gruyter.

McGaw, J. (1982) Women and the history of American technology. *Journal of Women in Culture and Society*, 7: 798–828.

MacKay, H. and Gillespie, G. (1992) Extending the social shaping of technology approach: ideology and appropriation. *Social Studies of Science*, 22: 685–716.

MacKenzie, D. (1984) Marx and the machine. *Technology and Culture*, 25: 473–502.

MacKenzie, D. (1990) *Inventing Accuracy: A Historical Sociology of Nuclear Missile Guidance*. Cambridge, MA: MIT Press.

MacKenzie, D. (1996a) *Knowing Machines: Essays on Technical Change*. Cambridge, MA: MIT Press.

MacKenzie, D. (1996b) How do we know the properties of artefacts? Applying the sociology of knowledge to technology, in R. Fox (ed.) *Technological Change: Methods and Themes in the History of Technology*, pp. 247–63. London: Harwood.

McKinlay, J. B. (1981) From 'Promising Report' to 'Standard Procedure' – Seven stages in the career of a medical innovation. *Milbank Memorial Fund Quarterly*, 59 (3): 374–411.

McKinlay, J. B. (ed.) (1982) *Technology and the Future of Health Care* (Milbank Reader 8) Cambridge, MA and London: MIT Press.

Marglin, S. A. (1976) What do bosses do? The origins and functions of hierarchy in capitalist production, in A. Gorz (ed.) *The Division of Labour: The Labour Process and Class Struggle in Modern Capitalism*, pp. 13–54. Brighton: Harvester.

Martin, E. (1987) *The Woman in the Body: A Cultural Analysis of Reproduction*. Boston, MA: Beacon.

Marx, Karl ([1867]1976) *Capital: A Critique of Political Economy*, vol. 1. Harmondsworth: Penguin.

Mathias, P. (1972) Who unbound Prometheus? Science and technical change 1600–1800, in A. E. Musson (ed.) *Science, Technology and Economic Growth in the Eighteenth Century*. London: Methuen.

Mayr, O. (1976) The science-technology relationship as a historiographic problem. *Technology and Culture*, 17: 663–72. (Extracts reprinted in B. Barnes and D. Edge (eds) (1982) *Science in Context: Readings in the Sociology of Science*. Milton Keynes: Open University Press.)

Melman, S. (1970) *Pentagon Capitalism. The Political Economy of War*. New York: McGraw-Hill.

Miles, I. (1988) *Home Informatics: Information Technology and the Transformation of Everyday Life*. London: Pinter.

Misa, T. J. (1988) How machines make history and how historians (and others) help them to do so. *Science, Technology, and Human Values*, 13: 308–31.

Moore, G. E. (1965) Cramming more components onto integrated circuits. *Electronics*, 38 (19 April): 114–17.

Mueller, W. F. (1964) Origins of DuPont's major innovations, 1920–1950, in J. R. Bright (ed.) *Research, Development and Technological Innovation*, pp. 383–401. Homewood, IL: Irwin.

Mumford, L. (1964) Authoritarian and democratic technic. *Technology and Culture*, 5: 1–8.

Musson, A. E. and Robinson, E. (1969) *Science and Technology in the Industrial Revolution*. Manchester: Manchester University Press.

Negroponte, N. (1995) *Being Digital*. Sydney: Hodder & Stoughton.

Nelson, R. R. and Winter, S. G. (1974) Neoclassical vs evolutionary theories of economic growth: critique and prospectus. *Economic Journal*, 84: 886–905.

Nelson, R. R. and Winter, S. G. (1982) *An Evolutionary Theory of Economic Change*. Cambridge, MA and London: Harvard University Press.

Nelson, R. R., Winter, S. G. and Schuette, H. L. (1976) Technical change in an evolutionary model. *Quarterly Journal of Economics*, 90: 90–118.

Noble, D. F. (1977) *America by Design. Science, Technology and the Rise of Corporate Capitalism*. Oxford: Oxford University Press.

Noble, D. F. (1979) Social choice in machine design: the case of automatically controlled machine tools, in A. Zimbalist (ed.) *Case Studies on the Labour Process*, pp. 18–50. New York and London: Monthly Review Press.

Noble, D. F (1984) *Forces of Production. A Social History of Industrial Automation*. New York: Knopf.

Norman, C. (1979) Global research: who spends what? *New Scientist*, 26 July: 279–81.

Nye, D. (1990) *Electrifying America: Social Meanings of a new Technology*. Cambridge, MA: MIT Press.

Ogburn, W. F. and Thomas, D. (1922) Are inventions inevitable? *Political Science Quarterly*, 34: 83–98.

Ordway, F. I. III and Sharpe, M. R. (1979) *The Rocket Team*. New York: Crowell.

Pacey, A. (1976) *The Maze of Ingenuity: Ideas and Idealism in the Development of Technology*. Cambridge, MA and London: MIT Press.

Pain, D., Owen, J., Franklin, I. and Green, E. (1993) Human-centred systems design: a review of trends within the broader systems development context, in E. Green, J. Owen and D. Pain (eds) *Gendered by Design: Information Technology and Office Systems*. London: Taylor & Francis.

Papanek, V. and Hennessey, J. (1977) *How Things Don't Work*. New York: Pantheon.

Parrott, B. (1983) *Politics and Technology in the Soviet Union*. Cambridge, MA and London: MIT Press.

Parsons, T. (1956) The American family: its relations to personality and the social structure, in T. Parsons and R. Bales (eds) *Family, Socialisation and Interaction Process*. London: Routledge and Kegan Paul.

Pfeffer, N. (1993) *The Stork and the Syringe: A History of Medicine and Infertility in England in the 20th Century*. Cambridge: Polity.

Pollard, S. (1965) *The Genesis of Modern Management. A Study of the Industrial Revolution in Great Britain*. London: Arnold.

Reinfelder, M. (1980) Introduction: breaking the spell of technicism, in P. Slater (ed.) *Outlines of a Critique of Technology*, pp. 9–37. London: Ink Links.

Reiser, S. J. (1978) *Medicine and the Reign of Technology*. Cambridge: Cambridge University Press.

Richards, J. L. (1979) The reception of a mathematical theory: Non-Euclidean geometry in England, 1868–1883, in B. Barnes and S. Shapin (eds) *Natural Order: Historical Studies of Scientific Culture*, pp. 143–63. Beverly Hills, CA: Sage.

Robinson, J. P. and Godley, G. (1997) *Time for Life: The Surprising Ways Americans Use their Time*. University Park, PA: Pennsylvania State Press.

Rose, H. (1994) *Love, Power and Knowledge: Towards a Feminist Transformation of the Sciences*. Bloomington and Indianapolis, IN: Indiana University Press.

Rosen, S. (ed.) (1973) *Testing the Theory of the Military-Industrial Complex*. Lexington, MA: D. C. Heath.

Rosenberg, N. (1976) *Perspectives on Technology*. Cambridge: Cambridge University Press.

Rosenberg, N. (1982) *Inside the Black Box: Technology and Economics*. Cambridge: Cambridge University Press.

Rüdig, W. (1983) Capitalism and nuclear power: a reassessment. *Capital and Class*, 20: 117–56.

Sahal, D. (1981a) Alternative concepts of technology. *Research Policy*, 10: 3–24.

Sahal, D. (1981b) *Patterns of Technological Innovation*. Reading, MA: Addison-Wesley.

Sandberg, L. (1969) American rings and English mules: the role of economic rationality. *Quarterly Journal of Economics*, 83: 25–43.

Sapolsky, H. M. (1972) *The Polaris System Development: Bureaucratic and Programmatic Success in Government*. Cambridge, MA: Harvard University Press.

Sarkesian, S. C. (ed.) (1972) *The Military-Industrial Complex. A Reassessment*. Beverly Hills, CA and London: Sage.

Saul, S. B. (ed.) (1970) *Technological Change: The United States and Britain in the Nineteenth Century*. London: Methuen.

Schon, D. (1963) *Displacement of Concepts*. London: Tavistock.

Schon, D. (1982) The fear of innovation, in B. Barnes and D. Edge (eds) *Science in Context: Readings in the Sociology of Science*, pp. 290–302. Milton Keynes: Open University Press.

Schrum, W. *et al.* (eds) (1995) *Science, Technology, and Society in the Third World: An Annotated Bibilography*. Metuchen, NJ: Scarecrow Press.

Schumacher, E. F. (1973) *Small is Beautiful: A Study of Economics as if People Mattered*. London: Abacus.

Schumpeter, J. (1934) *The Theory of Economic Development: An Inquiry into Profits, Capital, Credit, Interest, and the Business Cycle*. Cambridge, MA: Harvard University Press.

Schumpeter, J. (1939) *Business Cycles: A Theoretical, Historical, and Statistical Analysis of the Capitalist Process*. New York: McGraw-Hill.

Schumpeter, J. (1943) *Capitalism, Socialism and Democracy*. London: Allen & Unwin.

Schumpeter, J. (1951) *Imperialism and Social Classes*. Oxford: Blackwell.

Shapin, S. (1972) The Pottery Philosophical Society, 1819–1835: an examination of the cultural uses of provincial science. *Science Studies*, 2: 311–36.

Shapin, S. (1982) History of science and its sociological reconstructions. *History of Science*, 20: 157–211.

Silverman, D. (1970) *The Theory of Organisations*. London: Heinemann.

Simpson, J. (1983) *The Independent Nuclear State: The United States, Britain and the Military Atom*. London and Basingstoke: Macmillan.

Smith, D. K. and Alexander, R. C. (1988) *Fumbling the Future: How Xerox Invented, then Ignored, the First Personal Computer*. New York: William Morrow & Company, Inc.

Smith, M. R. and Marx, L. (eds) (1994) *Does Technology Drive History? The Dilemma of Technological Determinism*. Cambridge, MA: MIT Press.

Snow, R. (1994) Each to her own: investigating women's response to contraception, in G. Sen and R. C. Snow (eds) *Power and Decision: The Social Control of Reproduction*. Boston, MA: Harvard University Press.

Staudenmaier, J. M. (1980) *Design and Ambience: Historians and Technology, 1958–1977*. Ph.D. thesis, University of Pennsylvania.

Staudenmaier, J. M. (1985) *Technology's Storytellers: Reweaving the Human Fabric*. Cambridge, MA: MIT Press.

Stern, P. C. and Fineberg, H. (1996) *Understanding Risk: Informing Decisions in a Democratic Society*. Washington, DC.: National Academy Press.

Stoneman, P. (ed.) (1995) *Handbook of the Economics of Innovation and Technological Change*. Oxford: Blackwell.

Strasser, S. (1982) *Never Done: A History of American Housework*. New York: Pantheon.

Summerfield, P. (1977) Women workers in the Second World War. *Capital and Class*, 1: 27–42.

Thompson, E. P. (1967) Time, work-discipline and industrial capitalism. *Past and Present*, 38: 56–97.

Thompson, E. P. (1982) Notes on exterminism, the last stage of civilization, in E. P. Thompson, *Exterminism and Cold War*, pp. 1–33. London: Verso.

Tobey, R. C. (1996) *Technology as Freedom: The New Deal and the Electrical Modernization of the American Home*. Berkeley, CA: University of California Press.

Trescott, M. M. (ed.) (1979) *Dynamos and Virgins Revisited: Women and Technological Change in History*. Metuchen, NJ and London: Scarecrow Press.

Uselding, P. (1977) Studies of technology in economic history. *Research in Economic History*, supplement 1: 159–219.

Usher, A. P. (1954) *A History of Mechanical Inventions*. (Revised edn, first published 1929). Cambridge, MA: Harvard University Press.

Vanek, J. (1974) Time spent on housework. *Scientific American*, 231 (5): 116–20.

Vincenti, W. G. (1990) *What Engineers Know and How They Know It: Analytical Studies from Aeronautical History*. Baltimore, MD: Johns Hopkins University Press.

Wainwright, M. (1998) North-east to bank 2,000 new jobs. *Guardian*, 27 February.

Wajcman, J. (1991a) *Feminism Confronts Technology*. University Park PA: Pennsylvania State University Press.

Wajcman, J. (1991b) Patriarchy, technology and conceptions of skill. *Work and Occupations*, 18 (1): 29–45.

Webster, J. (1990) *Office Automation: The Labour Process and Women's Work in Britain*. Hemel Hempstead: Harvester Wheatsheaf.

Webster, J. (1996) *Shaping Women's Work: Gender, Employment and Information Technology*. London: Longman.

Westrum, R. (1991) *Technologies and Society: The Shaping of People and Things*. Belmont, CA: Wadsworth.

White, L. Jr. (1978) *Medieval Technology and Social Change* (first published 1962). New York: Oxford University Press.

Williams, R. (1997) Globalisation and contingency: tensions and contradictions in the mutual shaping of technology and work organization, in I. McLoughlin and D. Mason (eds) *Innovation, Organizational Change and Technology*. London: International Thompson Business Press.

Winner, L. (1977) *Autonomous Technology: Technics-out-of-Control as a Theme in Political Thought*. Cambridge, MA, and London: MIT Press.

Winner, L. (1993) Upon opening the black box and finding it empty: social constructivism and the philosophy of technology. *Science, Technology, & Human Values*, 18: 362–78.

Wood, S. (ed.) (1982) *The Degradation of Work? Skill, Deskilling and the Labour Process*. London: Hutchinson.

Woolgar, S. (1991) The turn to technology in social studies of science. *Science, Technology, & Human Values*, 16: 20–50.

Worrall, J. (1982) The pressure of light: the strange case of the vacillating 'Crucial Experiment'. *Studies in the History and Philosophy of Science*, 13: 133–71.

Zuckerman, S. (1980) Science advisers and scientific adviser. *Proceedings of the American Philosophical Society*, 124: 241–55.

▶ Index